DYNAMIC ANALYSIS OF ROBOT MANIPULATORS

A Cartesian Tensor Approach

THE KLUWER INTERNATIONAL SERIES IN ENGINEERING AND COMPUTER SCIENCE

ROBOTICS: VISION, MANIPULATION AND SENSORS

Consulting Editor: **Takeo Kanade**

ROBOTIC GRASPING AND FINE MANIPULATION, M. Cutkosky
ISBN: 0-89838-200-9

SHADOWS AND SILHOUETTES IN COMPUTER VISION, S. Shafer
ISBN: 0-89838-167-3

PERCEPTUAL ORGANIZATION AND VISUAL RECOGNITION, D. Lowe
ISBN: 0-89838-172-X

ROBOT DYNAMICS ALGORITHMS, F. Featherstone
ISBN: 0-89838-230-0

THREE- DIMENSIONAL MACHINE VISION, T. Kanade (editor)
ISBN: 0-89838-188-6

KINEMATIC* MODELING, IDENTIFICATION AND CONTROL OF *ROBOT MANIPULATORS, H.W. Stone
ISBN: 0-89838-237-8

OBJECT RECOGNITION USING VISION AND TOUCH, P. Allen
ISBN: 0-89838-245-9

INTEGRATION, COORDINATION AND CONTROL OF MULTI-SENSOR ROBOT SYSTEMS, H.F. Durrant-Whyte
ISBN: 0-89838-247-5

MOTION UNDERSTANDING: Robot and Human Vision, W.N. Martin and J. K. Aggrawal (editors)
ISBN: 0-89838-258-0

BAYESIAN MODELING OF UNCERTAINTY IN LOW-LEVEL VISION, R. Szeliski
ISBN 0-7923-9039-3

VISION AND NAVIGATION: THE CMU NAVLAB, C. Thorpe (editor)
ISBN 0-7923-9068-7

***TASK-DIRECTED SENSOR FUSION AND PLANNING: A Computational* Approach,** G. D. Hager
ISBN: 0-7923-9108-X

COMPUTER ANALYSIS OF VISUAL TEXTURES, F. Tomita and S. Tsuji
ISBN: 0-7923-9114-4

DATA FUSION FOR SENSORY INFORMATION PROCESSING *SYSTEMS,* J. Clark and A. Yuille
ISBN: 0-7923-9120-9

PARALLEL ARCHITECTURES AND PARALLEL ALGORITHMS FOR INTEGRATED VISION SYSTEMS, A.N. Choudhary, J. H. Patel
ISBN: 0-7923-9078-4

ROBOT MOTION PLANNING, J. Latombe
ISBN: 0-7923-9129-2

DYNAMIC ANALYSIS OF ROBOT MANIPULATORS
A Cartesian Tensor Approach

C.A. Balafoutis and R.V. Patel
Concordia University
Montreal, Canada

Springer-Science+Business Media, LLC

Library of Congress Cataloging-in-Publication Data

Balafoutis, C. A. (Constantinos A.)
Dynamic analysis of robot manipulators : a Cartesian tensor approach / C.A. Balafoutis and R.V. Patel.
p. cm. — (The Kluwer international series in engineering and computer science ; SECS 131)
Includes bibliographical references and index.
ISBN *978-0-7923-9145-6* ISBN 978-1-4615-3952-0 (eBook)
DOI 10.1007/978-1-4615-3952-0
1. Manipulator (Mechanism) 2. Robotics. I. Patel, Rajnikant V.
II. Title III. Series.
TJ211.B335 1991
629.8'92—dc20 90-27818
CIP

Printed on acid-free paper.

Contents

List of Figures

List of Tables

Preface

The purpose of this monograph is to present computationally efficient algorithms for solving basic problems in robot manipulator dynamics. In particular, the following problems of rigid-link open-chain manipulator dynamics are considered : *i*) computation of inverse dynamics, *ii*) computation of forward dynamics, and *iii*) generation of linearized dynamic models. Computationally efficient solutions of these problems are prerequisites for real-time robot applications and simulations.

Cartesian tensor analysis is the mathematical foundation on which the above mentioned computational algorithms are based. In particular, it is shown in this monograph that by exploiting the relationships between second order Cartesian tensors and their vector invariants, a number of new tensor-vector identities can be obtained. These identities enrich the theory of Cartesian tensors and allow us to manipulate complex Cartesian tensor equations effectively. Moreover, based on these identities the classical vector description for the Newton-Euler equations of rigid body motion are rewritten in an equivalent tensor formulation which is shown to have computational advantages over the classical vector formulation. Thus, based on Cartesian tensor analysis, a conceptually simple, easy to implement and computationally efficient tensor methodology is presented in this monograph for studying classical rigid body dynamics.

Application of this tensor methodology to the dynamic analysis of rigid-link open-chain robot manipulators is simple and leads to an efficient formulation of the dynamic equations of motion. Moreover, the use of generalized and augmented links enables us to devise modeling schemes which are very much suited for the dynamic analysis of the aforementioned class of robot manipulators, since they allow us to compute off-line as many as possible of the configuration independent dynamic parameters of a manipulator. The resulting algorithms are recursive in nature. This is in line with other computationally efficient methods appearing in the literature, the most recent of which have been listed in the references at the end of the chapters.

In this monograph it is assumed that the reader is familiar with basic vector analysis and introductory (undergraduate) statics and dynamics. Knowledge of Cartesian tensor analysis, robot or rigid-body dynamics would be helpful, but is not a prerequisite for following the material in this monograph.

Montreal, Canada

CONSTANTINOS A. BALAFOUTIS
RAJNIKANT V. PATEL

Chapter 1

Introduction

The science of robotics began less than thirty years ago, when the first computer-controlled manipulator was demonstrated by Unimation Inc. Since that time, scientists and engineers have designed hundreds of different manipulators and the study of robotics has become a highly complex and interdisciplinary field which encompasses a number of topics taken from other "classical" fields such as: mathematics, mechanical and electrical engineering, computer science, etc. Today, with the advances made over the last decade, robots have come to symbolize high-level automation in almost every aspect of human activity. Applications of robots can be found almost everywhere: from hazardous environments such as in space and oceans, to more pleasant home environments [1-7]. However, by far the majority of applications of robots to date has been in the automotive manufacturing and metalworking industries [8,9]. A few typical applications of the so called *industrial robots* include *spray painting, welding, material handling, machine loading, assembly.*

Exactly what constitutes an industrial robot is still debatable not only from the view point of social science experts, but also from that of robotics specialists. For example, the Robot Institute of America defines [9] an industrial robot as:

> *"A reprogrammable multifunctional manipulator designed to move materials, parts, tools or specialized devices through variable programmed motions for the performance of a variety of tasks".*

On the other hand, the Japan Industrial Robot Industry Association uses a broader definition of an industrial robot:

> *"An all-purpose machine equipped with a memory device and a terminal, and capable of rotation and of replacing human labor by automatic performance of movements."*

This debate on the definition of an industrial robot simply indicates the continuous evolution which the field of robotics is undergoing year after year.

Independent of any specific definition, robotics specialists agree that a *robot manipulator*, which is the most important form of industrial robots, consists of the following physical components: a *mechanical system*, *sensors* and a *controller*. In the mechanical system the basic components are the *arm*, the *end-effector* and the *actuating mechanisms*. The arm usually consists of six rigid-links connected together in an open kinematic chain by revolute or prismatic joints, and allows the robot to position the end-effector in different locations in the workspace. The end-effector (gripper, welding torch, electro-magnet, etc.) provides the means of manipulating objects or performing various other tasks in the workspace. The actuating mechanisms consist of power source(s), actuators (electric, hydraulic, pneumatic) and drive mechanisms (chains, gears, etc.). The sensors (visual, acoustic, force) measure and determine the state (positions, orientations, velocities) of the manipulator links and the end-effector. Furthermore, sensors measure and determine forces and moments exerted by the manipulated object on the manipulator. Finally, the controller is the device which supervises and regulates the programmed motion.

From a mathematical point of view, the study of robot manipulators includes topics such as: *modeling and design; robot arm kinematics, dynamics and control; trajectory planning; sensors; robot vision; robot control languages;* etc. Each of these topics can be studied on its own in great depth, as part of the education of a robotics specialist, or as an application area in different aspects of engineering. However, although each one of these topics is very important in robotics applications, a deep study of manipulator kinematics and dynamics is the cornerstone of successful utilization of today's robots and those which are going to be used in the future.

Robot manipulator kinematics deal with the geometry and the time-dependent manipulator motion without consideration of forces and/or moments that cause the motion. In other words, they deal with the spatial configuration of the manipulator in physical space. In particular, kinematic analysis of robot manipulators is concerned with *configuration* and *motion* analysis. Configuration analysis deals with possible mathematical descriptions of the manipulator's spatial configuration as a function of time, and motion analysis deals with the first and second time derivatives of these configuration functions. The dynamics of a robot manipulator deal with the relation between actuator torques or forces and the manipulator's motion, considering its mass and inertial properties. These relations define the *dynamic equations of motion* of a robot manipulator which are fundamental to any robotic application. In particular, in the dynamic analysis of robot manipulators, we deal with the following basic problems.

1.1 Basic Problems in Robot Manipulator Dynamics

As is well known, the dynamical performance of an n degrees-of-freedom system of rigid bodies can generally be described by n second order, usually coupled, nonlinear differential equations which can be represented by a second order n-dimensional coupled nonlinear vector differential equation. These differential equations are known as the *dynamic equations of motion* of the system and denote its *dynamic model.*

In a dynamic model of a system there are two main aspects with which one is concerned: *motion* and *forces*. The motion of a system is called its *trajectory* and consists of a sequence of desired *positions, velocities,* and *accelerations* of some point or points in the system. Forces are usually characterized as *internal* (or *constraint*) forces and *external* (or applied) forces. The external forces are the ones which cause motion.

In robotics, a dynamic robot model usually describes relationships between robot motion and forces causing that motion, so that given one of these quantities, we can determine the other. There are, therefore, the following problems to be considered.

i) **Forward Dynamics:** The *Forward* or *direct dynamics* problem (FDP) is one where the forces which act on a robot are given and we wish to solve for the resulting motion. In its simplest form, the forward dynamics problem can be expressed symbolically as a vector differential equation of the form

$$\ddot{\mathbf{q}} = \mathbf{h}(\mathbf{q}, \dot{\mathbf{q}}, \boldsymbol{\tau}, \text{manipulator parameters}) \tag{1.1.1}$$

where, $\mathbf{q}$ is the vector of generalized coordinates (joint variables), $\dot{\mathbf{q}}$ and $\ddot{\mathbf{q}}$ are its derivatives with respect to time, $\boldsymbol{\tau}$ is the (input) generalized force vector, i.e., the vector of joint torques and/or joint forces and the "manipulator parameters" are all those parameters which characterize the particular geometry and dynamics of a robot manipulator.

The importance of forward dynamics in robotics stems mainly from its use in simulation [10]. Simulation of robot motion is a way of testing control strategies or manipulator designs prior to the expensive task of working with the actual manipulator. In general, as we shall see later, equation (1.1.1) is not a simple equation for which an analytic solution can be provided easily. For a general robot manipulator, equation (1.1.1) is very complex since it is highly nonlinear with strong coupling between the joint variables. Hence, the solution of (1.1.1) for $\mathbf{q}$ requires complex procedures for evaluating $\mathbf{h}$ and for performing numerical integration. Fortunately, a solution for equation (1.1.1) is rarely required in practical applications. More often, we are interested in the following "inverse" problem.

ii) **Inverse Dynamics:** The *inverse dynamics* problem (IDP) is one in which we need to determine the generalized forces that will produce a specified motion trajectory. The inverse dynamics problem can be described mathematically by an equation of the form

$$\boldsymbol{\tau} = \mathbf{f}\,(\mathbf{q}\,,\dot{\mathbf{q}}\,,\ddot{\mathbf{q}}\,,\text{manipulator parameters}\,) \tag{1.1.2}$$

where, as in (1.1.1), the manipulator parameters describe the particular robot manipulator, $\boldsymbol{\tau}$ is the vector of the unknown generalized forces and $(\mathbf{q}\,,\dot{\mathbf{q}}\,,\ddot{\mathbf{q}})$ is the given manipulator trajectory.

Inverse dynamics is very important in practical robot applications because it enables us to determine the profile of the generalized forces necessary to achieve a desired robot trajectory. Efficient computation of the inverse dynamics becomes particularly important when $\boldsymbol{\tau}$ has to be evaluated online. This can arise in several practical situations, e.g., when the robot payload varies, or when the desired trajectory has to be modified online (e.g. for collision avoidance). Also, inverse dynamics plays an important role in many advanced robot control strategies where the inverse dynamics are used in the feedforward or feedback paths and may need to be computed online [11]. Moreover, to ensure convergence of the control scheme, the inverse dynamics computations may have to be performed very frequently. Consequently, the formulation and evaluation of these equations of motion affect the servo rate of the robot controller and partially determine the feasibility of implementing many control schemes online.

The forward and inverse dynamics problems are two problems which constitute what is usually known as *robot manipulator dynamics* [12,13]. However, since both problems are described by highly nonlinear and dynamically coupled equations, it can be of great assistance in many robotic applications if we have available the *linearized dynamic equations* of a robot manipulator. Thus, besides forward and inverse dynamics, we may also include the following problem in robot manipulator dynamics.

iii) **Linearized Robot Dynamics:** As is well known, the linearized dynamics of a nonlinear system can be described by the following first order vector differential equation (in state-space form)

$$\delta\dot{\mathbf{x}} = \mathbf{A}(t)\delta\mathbf{x} + \mathbf{B}(t)\delta\mathbf{u} \tag{1.1.3}$$

where, the matrices $\mathbf{A}(t)$ and $\mathbf{B}(t)$ are functions of time and $\delta\mathbf{u}$ and $\delta\mathbf{x}$ denote small perturbations in the input $\mathbf{u}$ and state $\mathbf{x}$, respectively, about some nominal (given) trajectory. Equation (1.1.3) describes the perturbed motion (for sufficiently small perturbations) of a dynamical system and is usually derived from the actual nonlinear dynamic equation (1.1.2) by using a Taylor series expansion about a nominal trajectory [14-16]. The Taylor series expansion is applicable to nonlinear robot dynamics because, as can be easily shown, the nonlinearities in robot dynamics are analytic functions of their arguments. Therefore, the derivation of (1.1.3) from a nonlinear dynamic robot model, at least in principle, does not present any problems. However, applying the Taylor series expansion to a nonlinear system which has the complexity of a general robot manipulator dynamics is a challenging problem, especially if one attempts to derive efficient *computational* algorithms for determining the coefficient matrices of the linearized model.

Linearized robot dynamics may be used in manipulator control [17]. This is best illustrated by the following example. In ideal situations, equation (1.1.2) provides the generalized forces which will drive a manipulator along a desired trajectory. However, in practice, because of perturbations resulting from modeling errors, unpredicted working conditions or payload variations, this cannot be achieved without the application of some control strategies, which are designed to compensate against these perturbations. Currently, there are many well established control strategies in the linear systems area. However, direct application of these linear control strategies to robotics is not possible, since, as we have already mentioned, the dynamic models defined by equation (1.1.2) are dynamically coupled and highly nonlinear. Therefore, one way in which these linear control schemes can be used is by obtaining linearized robot dynamic models derived from equation (1.1.2). Another application of linearized robot dynamic models is in carrying out parameter sensitivity analysis of robot manipulator dynamics for the purpose of efficient manipulator design [14].

1.2 General Remarks on Robot Manipulator Dynamics

In principle, solving forward or inverse dynamics for rigid-link robot manipulators presents no difficulty. A robot manipulator is just a system of rigid bodies, and the equations of motion of such systems have been known for a long time. The real problem in robot dynamics is a practical one, namely, that of finding formulations for the equations of motion that lead to efficient computational algorithms. To derive these equations, we can use well established procedures from classical mechanics [18,19] such as those based on the equations of *Newton* and *Euler, Euler* and *Lagrange, Kane*, etc. The choice of a particular procedure determines the nature of the analysis and the amount of effort needed to state the equations of motion in the form of a computational algorithm. For example, in the Newton-Euler approach, the derivation of the equations of motion is based on direct application of Newton's and Euler's laws, while in the Lagrangian approach, the equations of motion are derived from two scalar quantities, namely, the *kinetic* and *potential energy*. Also, in the Newton-Euler approach, physical coordinate systems (usually Cartesian) are employed to express the equations of motion. Some of the coordinates may not be independent but may be related to others by kinematic constraints which are employed simultaneously with the equations of motion. In contrast, the Lagrangian approach usually employs linearly independent generalized coordinates. Therefore, the analysis and consequently the effort needed to derive the equations varies. However, irrespective of the approach, the equations of motion for rigid-link open-chain robot manipulators can be stated in the following forms:

In a *closed-form* formulation, the equations of motion are usually described by the equation

$$\boldsymbol{\tau} = \mathbf{D}(\mathbf{q})\ddot{\mathbf{q}} + \mathbf{C}(\mathbf{q}, \dot{\mathbf{q}}) + \mathbf{G}(\mathbf{q}) \tag{1.2.1}$$

where $\boldsymbol{\tau}$ is the vector of generalized forces, $(\mathbf{q}, \dot{\mathbf{q}}, \ddot{\mathbf{q}})$ denotes the joint trajectories, $\mathbf{D}(\mathbf{q})$ is the generalized inertia tensor of the manipulator, and $\mathbf{C}(\mathbf{q}, \dot{\mathbf{q}})$ and $\mathbf{G}(\mathbf{q})$ are the Coriolis and centrifugal, and gravitational vectors respectively. A closed-form representation, such as (1.2.1), can be used directly for solving the IDP, or it can be adapted easily for the FDP by solving for $\ddot{\mathbf{q}}$. This is probably the most attractive feature of closed-form formulations for the equations of motion of a robot manipulator. But since these formulations are computationally inefficient it is preferable to use more efficient recursive formulations for the equations of motion in practical real-time robot applications.

In a *recursive formulation*, the equations of motion of a robot manipulator are expressed implicitly in terms of recurrence relations between quantities describing various properties of the robotic system [13]. Recursive formulations do not have the compact representation of the closed-form one but they too solve the IDP directly and, what is more important, they can be implemented in a very efficient manner. However, it is not possible to solve the FDP with the same recursive equations without major modifications. But this is not a drawback because, even with major modifications, we can solve the FDP efficiently. From the foregoing, it is not surprising that most of the research effort for solving manipulator dynamics has been directed at deriving efficient recursive formulations.

From the work that has been done to date on developing algorithms for computing manipulator dynamics, it appears that there is a general misconception that the computational efficiency of the algorithms depends on the formulations used for their derivation. Thus for example, it has been believed for some time that the algorithms derived from the Newton-Euler formulation are computationally more efficient than those derived using the Lagrangian formulation. It seems that this confusion results from a lack of deeper understanding of the mathematical representations used to describe the equations of motion. For example, in the Newton-Euler approach, the time variation in the orientation is generally described by using the *angular velocity vector*, and in the Lagrangian approach it is described by using the time derivative of a *rotation tensor*. But it can be shown [20], that the Lagrangian formulation will yield a similar algorithm to that obtained using the Newton-Euler formulation, if an equivalent representation of angular velocity is employed. Obviously, this result should be expected because the Lagrange equations can be derived from the Newton-Euler equations based on arguments of virtual work.

Therefore, in computing efficient robot manipulator dynamics, the issue is not which procedure from classical mechanics to use in the analysis. With proper analysis we can derive [21] exactly the same computational algorithms for solving manipulator dynamics. The real issue, in terms of computational efficiency, is which mathematical representation to use for expressing various physical quantities when the nature of the quantities allows us to use more than one representation. Obviously, a particular representation dictates a certain mathematical analysis which leads to descriptions of the basic dynamic equations whose structure corresponds to that particular analysis. Then, since the implementation of an algorithm depends on such structure, it follows that the computational efficiency of a particular algorithm will depend on the mathematical representation used to describe these physical quantities. Therefore, in searching for efficient computational algorithms to solve problems in robot manipulator dynamics, we have to search for a mathematical representation of the basic physical quantities of motion which will allow us to describe rigid body motion more efficiently.

1.3 Objectives and Motivation

Among the problems of robot manipulator dynamics, the IDP is the more important one. An efficient solution of this problem is a prerequisite for real-time robot applications, which in turn is necessary for flexible automation in a dynamically changing environment. Therefore, the main objectives of this monograph are the analysis of the computational cost of solving the inverse dynamics problem and the development of algorithms with significantly reduced computational complexity.

In the last decade, a large number of algorithms has been proposed for solving the inverse dynamics problem. The emphasis in most of these algorithms is placed on reducing their computational complexity by using analytical organization procedures and customization [22,23]. However, particular analytical organization procedures and customization are generally used in implementing the set of equations of an algorithm and not for deriving them. Moreover, in many cases, analytic procedures and customization are restricted to robot manipulators with a specific geometry. In this monograph, the emphasis is placed on improving the computational efficiency of the algorithms through a more efficient formulation of the dynamic equations of motion and not through better implementation of existing formulations. Thus, our intention is to devise a methodology for analysis and formulation of the dynamic equations of rigid body motion which has to be conceptually simple, easy to implement, and computationally efficient. To this end, we shall apply this methodology for solving in a computationally efficient manner the problems of inverse and forward dynamics of rigid-link, open-chain robot manipulators. Also, the methodology will be used for the derivation of linearized robot dynamic models in a computationally efficient manner.

As we mentioned above, the representation of the physical quantities which are involved in the formulation of the equations of motion of a rigid body system determines the kind of mathematical analysis that will be used in deriving these equations. Therefore, a better understanding of the mathematical representations used to describe basic physical quantities is essential for the analysis of the equations of motion.

For example, in the classical Newtonian formulation of rigid body dynamics (which has been applied successfully in deriving computational algorithms for solving inverse dynamics [24]), vectors are normally used to represent most of the physical quantities and, therefore, vector analysis is used for deriving the equations of rigid body motion. In particular, vector analysis is imposed on classical Newtonian dynamics from the consideration that *angular rates* (i.e., linearly independent rates of change of rigid body orientation) constitute the components of a vector quantity, the *angular velocity vector*. This consideration also assigns a vector character to other physical quantities which are defined in terms of the angular velocity vector such as: *angular acceleration, angular momentum, external torque*, etc. However, as is well known [20-21,25-27], angular velocity can also be described by a second order skew-symmetric Cartesian tensor, the *angular velocity tensor*. Obviously then, the tensor representation of angular velocity calls for Cartesian tensor analysis to be applied in rigid body dynamics. Application of Cartesian tensor analysis (within the framework of the Newtonian approach) in rigid body dynamics requires that all the other physical quantities which are defined in terms of the angular velocity be treated as Cartesian tensors instead of vectors. Thus, we have to examine if a Cartesian tensor representation of basic physical quantities such as: angular acceleration, angular momentum, external torque, etc., simplifies the equations of rigid body motion and leads to more efficient computational algorithms.

The use of tensor analysis is obviously known in rigid body dynamics. However, most of the time the analysis is performed in the *configuration space* of the rigid body system which is generally a *Riemanian space*, i.e., a non-Euclidean *manifold* [28,29]. In this monograph, we shall use tensors to analyze the motion of a system of rigid bodies, but the analysis will be carried out in Euclidean space instead of on nonlinear manifolds, i.e., we shall use *Cartesian tensor analysis* [26,27]. To do this, we shall need to review basic results from Cartesian tensor analysis and, in particular, we shall need to understand the relations between three dimensional vectors and second order skew-symmetric Cartesian tensors.

Finally, almost all existing algorithms which solve forward or inverse manipulator dynamics have a structure which requires that all quantities involved in these algorithms be computed online (except the configuration independent geometric and dynamic parameters of the individual links). This obviously is a consequence of the underlying modeling schemes which have been used to derive the algorithms, because the structure of an algorithm depends on the underlying scheme. However, from a computational point of view it is desirable to devise algorithms which allow us to compute *off-line* as many quantities as possible and at the same time, to keep the *online* computations as simple as possible. Therefore, in order to derive computationally efficient algorithms for solving the inverse and forward manipulator dynamics problems, we have to examine if it is possible to devise a modeling scheme for the robot manipulators which allows us to compute off-line as many as possible of the configuration independent kinematic or dynamic parameters of the robot manipulator.

1.4 Preview

This monograph presents a new methodology for the analysis and formulation of computationally efficient algorithms for solving basic problems of robot manipulator dynamics. The layout of the monograph is as follows: Chapter 2 contains a brief review of some basic manipulator terminology and concepts. Chapter 3 is concerned with Cartesian tensor analysis based on which the new methodology is devised, and Chapter 4 demonstrates how this theory can be applied to rigid body motion. New algorithms for solving the problems of inverse and forward dynamics of rigid-link open-chain robot manipulators are proposed in Chapters 5 and 6 respectively, while Chapter 7 deals with linearized dynamic robot models. The main contents of each chapter are as follows:

Chapter 2: This chapter introduces the notation to be used throughout the monograph. The configuration analysis of rigid bodies is briefly reviewed. Also, this chapter presents some relevant robot manipulator terminology as well as a configuration analysis of rigid-link open-chain robot manipulators.

Chapter 3: This chapter introduces relevant definitions, some basic algebraic Cartesian tensor operations and outlines the structural symmetries of second order Cartesian tensors. Also, based on tensor-vector invariants, 1–1 operators between vectors and second order skew-symmetric Cartesian tensors are defined and important propositions are stated which establish some basic tensor identities.

Chapter 4: Based on Cartesian tensor analysis, kinematic and dynamic aspects of rigid body motion are considered in this chapter. In particular, the angular velocity and acceleration tensors are shown to be very powerful tools for describing the motion of a rigid body since, by using these two tensors, a tensor representation for the angular momentum and external torque surfaces naturally and leads to a Cartesian tensor description for the Newtonian formulation of rigid body motion. This Cartesian tensor formulation of rigid body motion has the same simplicity as the classical vector formulation but (as shown in the chapter) can be implemented far more efficiently.

Chapter 5: A brief survey of existing methods for solving the inverse dynamics problem for rigid-link open-chain manipulators is followed by some observations and remarks on various issues concerning the computational efficiency of some "classical" algorithms for solving the problem. It is then shown that by using the Cartesian tensor description of the Newtonian formulation of rigid body motion and utilizing two different modeling schemes, computationally efficient algorithms for solving the IDP can be devised. The computational complexity of these algorithms is shown to be significantly less than that of other algorithms which are based on the classical vector formulation of rigid body motion. Also, it is demonstrated that these algorithms can be cast in a form where their computational efficiency is actually independent of the particular procedure of classical mechanics which has been used for their derivation.

Chapter 6: The Cartesian tensor analysis and the modeling scheme which have been proven successful in solving the IDP efficiently are used in this chapter to facilitate the solution of the FDP for rigid-link open-chain manipulators. After a brief review of the composite rigid body method, we introduce a new algorithm for computing the generalized inertia tensor of rigid-link open-chain robot manipulators. By combining this algorithm with efficient algorithms for solving the IDP, we improve significantly the computational efficiency of solving the problem of forward dynamics.

Chapter 7: This chapter is concerned with the linearization of the dynamic equations of motion for rigid-link open-chain manipulators. Using a Taylor series expansion, we derive the associated linearized dynamic models of the nonlinear models presented in Chapter 5. It is then shown that the coefficient sensitivity matrices of these linearized models can be computed efficiently using appropriate Cartesian tensor formulations. Also, in this chapter Cartesian space descriptions of the equations of motion for rigid-link open-chain manipulators are reviewed and a method for deriving their associated Cartesian space linearized dynamic models is proposed.

1.5 References

[1] A. Cohen, and J. D. Erickson, "Future Uses of Machine Intelligence and Robotics for the Space Station and Implications for the U.S. Economy", *IEEE J. Robotics and Automation*, RA-1, No. 3, pp. 117-123, 1985.

[2] A. K. Bejczy, and Z. Szakaly, "Universal Computer Control System (UCCS) for Space Telerobots", in *Proc. IEEE Int. Conf. on Robotics and Automation*, pp. 318-125, Raleigh, NC, March 31-April 3, 1987.

[3] J. H. Smith, J. Estus, C. Heneghan, and C. Nainan, "The Space Station Freedom Evolution-Phase: Crew-EVA Demand for Robotic Substitution by Task Primitive", in *Proc. IEEE Int. Conf. on Robotics and Automation*, pp. 1478-1485, Scottsdale, AR, May 14-19, 1989.

[4] K. Edahiro, "Development of Underwater Robot Cleaner for Marine Live Growth in Power Station", in *Proc. '83 ICAR Int. Conf. on Advanced Robotics*, pp. 99-106, Tokyo, Japan, Sept. 1983.

[5] R. C. Mann, W. R. Hamel, and C. R. Weisbin, "The Development of an Intelligent Nuclear Maintenance Robot", in *Proc. IEEE Int. Conf. on Robotics and Automation*, pp. 621-623, Philadelphia, PA, April 24-29, 1988.

[6] K. G. Engelhardt, "Applications of Robots to Health and Human Services", in *Proc. Robots 9: Current Issues, Future Concerns*, pp. 14-48 to 14-65, Detroit, Michigan, June, 1985.

[7] G. N. Saridis, "Robotic Control to Help the Disabled" in *Recent Advances in Robotics*, C. Ben and S. Hackwood, *Eds.*, John Wiley, New York, 1985.

[8] V. Shimon, *Eds.*, *Handbook of Industrial Robotics*, John Wiley, New York, 1985.

[9] R. K. Miller, *Industrial Robot Handbook,* Fairmont Press, Indian Trail, NY, 1987.

[10] M. W. Walker, and D. E. Orin, "Efficient Dynamic Computer Simulation of Robotic Mechanisms", *ASME J. Dynamic Systems, Measurement and Control,* Vol. 104, pp. 205-211, 1982.

[11] J.Y.S. Luh, M.W. Walker, and R. P. Paul, "Resolved Acceleration Control for Mechanical Manipulators", *ASME J. Dyn., Sys., Meas. and Contr.,* Vol. 102. pp. 69-76, 1980.

[12] M. Brady *et al.,* (Eds.), *Robot Motion: Planning and Control,* MIT Press, Cambridge, MA, 1982.

[13] R. Featherstone, *Robot Dynamics Algorithms,* Kluwer Academic Publishers, Boston MA, 1987.

[14] C. P. Neuman, and J. J. Murray, "Linearization and Sensitivity Functions of Dynamic Robot Models", *IEEE Trans. Systems, Man, and Cybernetics,* Vol. SMC-14, no. 6, pp. 805-818, 1984.

[17] C. A. Balafoutis, P. Misra, and R. V. Patel, "Recursive Evaluation of Linearized Dynamic Robot Models", *IEEE J. Robotics and Automation,* RA-2, pp. 146-155, 1986.

[16] C. A. Balafoutis, and R. V. Patel, "Linearized Robot Models in Joint and Cartesian Spaces", *CSME Transactions,* Vol. 13, No. 4, pp. 103-112, 1989; also presented at the *9th-Symposium on Engineering Applications of Mechanics,* pp. 587-594, London, Ontario, May 27-31, 1988.

[17] P. Misra, R. V. Patel, and C. A. Balafoutis, "Robust Control of Linearized Dynamic Robot Models", in *Robot Manipulators: Modeling, Control and Education,* M. Jamshidi, J. Y. S. Luh, and M. Shahinpur, *Eds.,* North-Holland Publishing Co., Inc., New York, 1986.

[18] H. Goldstein, *Classical Mechanics 2nd ed.,* Addison-Wesley, Reading, MA, 1980.

[19] T. R. Kane, P. W. Likins, and D. A. Levinson, *Spacecraft Dynamics*, McGraw-Hill, New York, 1983.

[20] W. M. Silver, "On the Equivalence of Lagrangian and Newton-Euler Dynamics for Manipulators", *Int. J. Robotics Research*, Vol. 1, No. 2, pp. 60-70, 1982.

[21] C. A. Balafoutis, R. V. Patel, and J. Angeles, "A Comparative Study of Lagrange, Newton-Euler and Kane's Formulation for Robot Manipulator Dynamics" in *Robotics and Manufacturing: Recent Trends in Research, Education, and Applications*, M. Jamshidi, J. Y. S. Luh, H. Seraji, and G. P. Starr, *Eds.*, ASME Press, New York, 1988.

[22] J. W. Burdick, "An Algorithm for Generation of Efficient Manipulator Dynamic Equations", in *Proc. 1986 IEEE Int. Conf. Robotics and Automation*, pp. 212-218, San Francisco, CA, Apr. 1986.

[23] J. J. Murray, and C. P. Neuman, "Organizing Customized Robot Dynamics Algorithms for Efficient Numerical Evaluation", *IEEE Trans. on Systems, Man, and Cybernetics*, Vol. SMC-18, No. 1, pp. 115-125, 1988.

[24] J. Y. S. Luh, M. W. Walker, and R. P. Paul, "On-Line Computational Scheme for Mechanical Manipulators", *ASME J. Dynamic Systems, Measurement and Control*, Vol. 102, pp. 69-79, 1980.

[25] O. Bottema, and B. Roth, *Theoretical Kinematics*, North-Holland Publishing Co., Amsterdam, 1978.

[26] H. Jeffreys, *Cartesian Tensors*, Cambridge University press, Cambridge, 1961.

[27] A. M. Goodbody, *Cartesian Tensors: With Applications to Mechanics, Fluid Mechanics and Elasticity*, Ellis Horwood, England, 1982.

[28] L. Brillouin, *Tensors in Mechanics and Elasticity*, Academic Press, New York, 1964.

[29] I. S. Sokolnikoff, *Tensor Analysis: Theory and Applications to Geometry and Mechanics of Continua,* John Wiley & sons, New York, 1965.

Chapter 2

Notation, Terminology and Background Material

This chapter introduces the notation, presents some basic concepts from *rigid body kinematics*, defines relevant robot terminology, and deals with the *configuration* analysis of *rigid-link, open-chain robot manipulators.* The chapter has two main sections: Section 2.2 contains results from rigid body kinematics. In particular, the *configuration* of a rigid body in the real world or physical space is defined and its *finite displacement* in this space is reviewed. Section 2.3 is concerned with the geometric description of rigid-link open-chain robot manipulators and defines the *joint* and *Cartesian* space descriptions for their configurations.

2.1 Notation

Throughout the text, boldface lower case roman letters are used to denote position vectors. Subscripts indicate, in order, the tail and the head of a position vector, and a superscript indicates the coordinate system with respect to which the position vector is expressed. Upper case boldface roman letters are used to denote second order tensors or vectors of forces and moments. From the context it will be clear if a tensor or a vector is considered. A second order skew-symmetric Cartesian tensor associated with a vector will be denoted with a tilde (˜) above the boldface lower or upper case roman letter denoting this vector. Subscripts denote a point on a link with respect to which the tensors (or the force and moment vectors) are defined, and superscripts denote the coordinate system with respect to which the tensors (or the force and moment vectors) are expressed. The superscript for tensors or vectors expressed in the base frame (inertial frame) is omitted. The *coordinate matrix*, associated with a tensor or a vector, will be denoted by the corresponding lower or upper case italic letter or by including the quantity in question in square brackets ([ ]).

2.2 Rigid Bodies and their Finite Displacement

The main objective in robotics is to manipulate objects in a static or dynamically changing environment, and one of the basic requirements for achieving this goal is to describe effectively, i.e., simply and accurately, objects such as *points* and *rigid bodies* relative to some coordinate system. In this section, we deal with the configuration analysis of points and rigid bodies. In particular, we review some of the possible approaches for describing the location and displacement of points and rigid bodies in physical space.

2.2.1 The Configuration of Points and Rigid Bodies in Physical Space

A description of the static or instantaneous location of an object in a space relative to a reference coordinate system is referred to as the *configuration* of the object [1-4]. The configuration is usually expressed as a function of a number of linearly independent variables which are known as *generalized coordinates.* The number of linearly independent generalized coordinates defines how many *degrees-of-freedom* the object has relative to the reference coordinate system.

In general, the complexity of the function, which defines the configuration of an object in terms of the generalized coordinates, depends on the reference coordinate system. The reference coordinate system characterizes the space where the object belongs and can be *linear* or *curvilinear.* Usually, the configuration of an object, in a curvilinear reference coordinate system is very complex, as opposed to a linear or Cartesian reference coordinate system where its configuration usually has a simple form. In robotics, it is of prime concern to describe the configuration of objects in the real world and, to our advantage, the real world or physical space can be modeled as a *three dimensional Euclidean space.* This allows us to consider a Cartesian coordinate system as a reference system relative to which, as we shall demonstrate later, the configuration of points or rigid bodies assumes relatively simple forms.

The configuration of a point in a general space is completely specified by its point-coordinates relative to a reference coordinate system. For points in three dimensional physical space, their point-coordinates are functions of three generalized coordinates. Moreover, in physical space which is a Euclidean space, we can identify points with vectors, since in this space, transformations of point-coordinates are identical to transformations of vector components [5]. This identification allows us to specify the configuration (i.e., the location) of a point in physical space by using the components of a vector. This vector is referred to as the *position vector* of the point under consideration. Therefore, the configuration of a point in physical space can be described, in an orthonormal Cartesian coordinate system, by a three dimensional vector, its position vector.

A description for the configuration of a rigid body in physical space is more involved, compared to that of a point. As is well known [1-4,10-14], a rigid body in physical space possesses six degrees-of-freedom. This implies that its configuration will be described in a reference coordinate system in terms of six generalized coordinates. Now, if we consider the configuration of a rigid body to be described by a vector (as we did in the case of a point), this vector must have six independent components. It is obvious then, that this six dimensional vector does not belong to physical space. It belongs to a six dimensional space which is not Euclidean and is known as the *configuration space.* Therefore, by assuming a vector description for the configuration of a rigid body we have to use a non-Euclidean space and hence curvilinear reference coordinate systems. This approach of describing the configuration of a rigid body leads us, in general, to very complex functional expressions.

Conventionally, we overcome these difficulties by grouping the six generalized coordinates, which describe the configuration of a rigid body into two sets. One set contains the generalized coordinates which describe the *orientation* of the rigid body. The other set contains the remaining three generalized coordinates and they describe the configuration (position) of a point on the rigid body.

To implement this scheme, we first associate a *frame* with the rigid body. A frame is a representation for a coordinate system so that the representation includes the possibility that the coordinate system may be displaced (translated) and/or rotated with respect to another coordinate system. In other words a frame contains a coordinate system whose orientation defines the "frame orientation" and is known as the *frame coordinate system*, and a position vector, which defines the *origin* of the frame coordinate system. The frame coordinate system is assumed to have a fixed relationship with the rigid body. Therefore, the frame coordinate system is sometimes referred to as the *body coordinate system.* This allows us to identify the orientation of the rigid body with that of the frame. Moreover, to simplify the description, we consider the frame coordinate system to be an *orthonormal Cartesian system.* Therefore, it is obvious that the configuration of the rigid body will be completely specified, relative to a reference coordinate system, if we specify the orientation of an orthonormal Cartesian coordinate system (the frame coordinate system) and its position vector.

As we noted above, position vectors can be described easily. Therefore, to complete the description for the configuration of a rigid body, we need to describe the orientation of a frame coordinate system relative to a reference one. This can be easily accomplished by considering an intermediate coordinate system which has the same orientation as the reference system, but whose origin is the same as that of the frame coordinate system. Then, we need only to describe the orientation of the frame coordinate system relative to the intermediate one, i.e., we need to describe the relative orientation of two Cartesian coordinate systems with a common origin.

There are many ways of specifying the orientation of a Cartesian coordinate system relative to another one with a common origin. As is well known [1-4], the orientation of two Cartesian coordinate systems with a common origin is described by a linear transformation. Here, since the coordinate systems are orthonormal, the linear transformation will be an orthogonal one. Moreover, from physical considerations (-coordinate systems represent orientations of rigid bodies), the linear transformation is *proper* (i.e., it has determinant equal to one) and so it represents a rotation. Therefore, this results in the problem of how to describe a *rotation.*

One of the most common methods to be found in the literature [1-4,6-7,10-14] which describes a rotation, is that of using a second order rotation tensor which relative to a basis has a *real orthonormal* 3×3 matrix representation. The entries of this matrix are the *direction cosines* which relate the axes of the two coordinate systems -the reference and the rotated one. A set of nine direction cosines completely specify the rotation between any two

Cartesian systems. Of course, the set of the nine direction cosines does not form a set of independent generalized coordinates, since as is well known [1-4], they satisfy six orthogonality relationships. However, the use of direction cosines to describe the orientation of one Cartesian coordinate system with respect to another has a number of important advantages. The most obvious is that they permit the use of Cartesian coordinate systems in describing the orientation of a rigid body, thereby avoiding the need for a curvilinear coordinate system for describing the configuration of a rigid body.

As mentioned above, the nine direction cosines of a 3×3 real orthonormal matrix have only three degrees-of-freedom which may be specified in terms of three linearly independent parameters. However, it is a well known fact [6], that there is no 1-1 global representation for a rotation matrix in terms of three independent variables (generalized coordinates). Nevertheless, in many practical applications and for a restricted domain it is possible to find three linearly independent variables, which can serve as generalized coordinates to describe a rotation. These generalized coordinates can be chosen in a number of ways. A common approach is to choose a particular sequence of rotation angles (α, β, γ) about the axes of an orthonormal coordinate system. The *roll, pitch and yaw* or the **z−y−x** *Euler angles* are examples of this approach. Finally, an alternative way of describing rotations is to use more than three variables, which obviously will not be linearly independent, and the description will not be 1−1. For example, we can use *quaternions, spinors, Pauli spin matrices, special unity* 2×2 and 3×3 matrices [1-4,6-9,11-14], *geometric* or *frame invariants* [10], etc.

When we use generalized coordinates to describe the configuration of a rigid body, for notational convenience, we sometimes consider the three "generalized" angles α, β, γ which describe a rotation (over a restricted domain) as the components of an "orientation vector". This "orientation vector" is then combined with the position vector of a point to produce a six dimensional "vector" χ (with coordinate matrix $\chi = [\, x\ y\ z\ \alpha\ \beta\ \gamma\,]^T$) which describes the configuration of a rigid body. We usually refer to χ as the *Cartesian space configuration vector* for the rigid body. The set of all Cartesian configuration vectors then defines the *Cartesian (configuration) space* for the rigid body.

Remark 2.1: As mentioned above, the "orientation vector" is created merely for notational convenience. It is not a valid representation for a rotation. Mathematically, a rotation is a *second order tensor* which is not a skew-symmetric tensor. Therefore, it cannot be represented by a vector, which is a first order tensor (see Chapter III). Another, probably simpler way to see that the "generalized" angles ($\alpha\ \beta\ ,\gamma$) do not form the components of a vector is the following. The composition (multiplication) of finite rotations is known to be associative but not commutative. Now, the only vector operations which produce a vector are addition and vector cross product. But vector addition is commutative and the vector cross product is not associative. Therefore, it is obvious that vectors do not represent finite rotations, because neither vector addition nor vector cross product is compatible with the composition of rotations. Hence, the "orientation vector" does not exist. Therefore, the Cartesian configuration vectors are not "real" vectors. This implies that the terminology "Cartesian space" or "Cartesian vector" is used in robotics in a broader sense than that used in linear algebra.

In the following section, we shall analyze briefly the relationships between the configurations of the same rigid body in two different locations in physical space.

2.2.2 On Finite Displacement of a Rigid Body

As is well known, the difference between position vectors for the same point on a rigid body at two different locations of this body in physical space is referred to as the displacement of that point in space. In this monograph, we shall define displacement in a broader sense to also include the difference between two orientations of the same rigid body.

A finite *displacement* in physical space is expressed mathematically as a transformation of the 3-D Euclidean space $\mathbf{E}^3$ into itself, with the property that it preserves the Euclidean distance. To elaborate, if $\mathbf{W} : \mathbf{E}^3 \rightarrow \mathbf{E}^3$ is a transformation and

$$\mathbf{p}' = \mathbf{W}\mathbf{p} \tag{2.2.1}$$

denotes the action of $\mathbf{W}$ on a point $\mathbf{p}$ in $\mathbf{E}^3$, then $\mathbf{W}$ defines a displacement if $(\mathbf{p}' - \mathbf{r}')^2 = (\mathbf{p} - \mathbf{r})^2$ for all $\mathbf{p}$ and $\mathbf{r}$ in $\mathbf{E}^3$. Clearly, $\mathbf{W}$ has been defined as a point-transformation, but since in Euclidean space points are identified by their position vectors, we can view $\mathbf{W}$ as a vector transformation of $\mathbf{E}^3$ into itself.

It can be shown [3] that displacements form a group. It is a subgroup of the group of transformations and it consists of those transformations which leave the distance of any two points *invariant.* It can also be shown [3] that displacements are angle-preserving transformations. In particular, right angles correspond to right angles. This implies that not only is the distance between two points invariant but also that the distance between a point and a linear subspace, and the distance between two parallel subspaces are invariant under displacement. Finally, if for a certain displacement $\mathbf{W}$, a point $\mathbf{p}$ coincides with its image $\mathbf{p}'$, this point is called a *fixed* point of $\mathbf{W}$.

The concept of displacement is fundamental to rigid body kinematics. It provides the mathematical apparatus for the study of rigid body motion. To see this, we note that a rigid body is defined as a system of mass points subject to the holonomic constraints that the distances between all pairs of points remain constant throughout any motion. Now, since displacement is a distance and angle preserving transformation, it is obvious that rigid body motion can be described mathematically as a displacement between two distinct configurations of the rigid body.

A general displacement or motion of a rigid body is best analyzed by considering the following two special displacements.

Translation: The transformation $\mathbf{T_d}$, which is defined by the equation

$$\begin{aligned} \mathbf{p}' &= \mathbf{T_d p} \\ &\triangleq \mathbf{p} + \mathbf{d} \end{aligned} \tag{2.2.2}$$

where $\mathbf{d}$ is a fixed vector, is obviously a displacement. It is the simplest displacement and is called a *translation.* The vector $\mathbf{d}$ in (2.2.2) is called the *vector of translation.* As we can see from (2.2.2), if $\mathbf{p}$ and $\mathbf{r}$ are two points, then the vector $\mathbf{w}' = \mathbf{p}' - \mathbf{r}'$ not only has the same length as $\mathbf{w} = \mathbf{p} - \mathbf{r}$ but is also parallel to it. Also, it is obvious from (2.2.2) that translations form a commutative group, which is a subgroup of the group of displacements. Moreover, from equation (2.2.2) it can also be seen that a translation is not a linear transformation (it does not map the origin of the space into itself) and has no fixed points.

The second special displacement is the familiar rotation which can be defined as follows.

Rotation: A displacement $\mathbf{A}$ for which a point $\mathbf{o}$ is a fixed point is called a *rotation* about $\mathbf{o}$.

It is well known [3], that rotations about a point $\mathbf{o}$ constitute a subgroup of the displacement group. The rotation group is not commutative. Also, in contrast with translations, rotations are linear transformations.

A rotation, in general, can be interpreted in two ways. First, we consider a rotation $\mathbf{A}$ as an *operator* which acts on a vector $\mathbf{p}$ and produces another vector $\mathbf{p}'$. This is an *active* point of view. In this approach the space is described in an invariant coordinate system, relative to which all vectors are rotated. So, $\mathbf{p}$ and $\mathbf{p}'$ are two different vectors, expressed in the same coordinate system. In the second approach, we consider the same rotation $\mathbf{A}$ as a transformation which acts on a reference coordinate system $\{ \mathbf{e} \}$ and produces a new reference coordinate system $\{ \mathbf{e}' \}$. This is a *passive* point of view. In this approach, the actual vector remains invariant and only the reference coordinate system is rotated. An invariant vector is represented by $\mathbf{p}$ relative to the old coordinate system, and by $\mathbf{p}'$ relative to the new one. Both interpretations of a rotation define the same action on a vector and both interpretations are described mathematically using the same algebra. We express the action of a rotation $\mathbf{A}$ by writing

$$\mathbf{p}' = \mathbf{A}\,\mathbf{p} \tag{2.2.3}$$

Rotations can be defined on any n-dimensional Euclidean space $\mathbf{E}^n$. However, it should be notice that the dimension n of the Euclidean space where a rotation is defined is fundamental to the analysis of rotations. Thus for n even there are, in general, no fixed points different from $\mathbf{o}$. For n odd ($n = 2m + 1, m = 0, 1, \cdots$), we always have at least one line of fixed points, the *axis of the rotation.* Therefore, a consequence of the general theory of rotations in odd dimensional Euclidean spaces is the following Theorem.

Theorem 2.1 (Euler): If a rigid body undergoes a displacement, leaving fixed one of its points $\mathbf{o}$, then a set of points of the body lying on a line that passes through $\mathbf{o}$ remains fixed as well.

A corollary to Euler's theorem, sometimes called *Chasles' theorem,* states the following result.

Theorem 2.2 (Chasles): The most general displacement of a rigid body is a translation plus a rotation.

Proofs for these theorems can be found in any book on classical mechanics, e.g. see [1-4,10]. An important consequence of Chasles' Theorem is that a general displacement of a rigid body can be written as

$$\mathbf{p}' = \mathbf{A}\,\mathbf{p} + \mathbf{d} \tag{2.2.4}$$

where $\mathbf{A}$ is a pure rotation and $\mathbf{d}$ is the translation vector of a pure translation. Equation (2.2.4) is very important in rigid body kinematics. It expresses a general displacement explicitly in terms of a rotation $\mathbf{A}$ about a point **o** and a translation of **o** by **d**. However, a compact representation for a general displacement, as that in equation (2.2.1), is more appealing, especially when one deals with a series of displacements, as is often the case in robotics. To achieve a compact representation for a general displacement, in terms of a rotation and a translation, we proceed as follows.

Let **W** be a general displacement, **o** an arbitrary point and **o**′ its image under **W**. Now, if $\mathbf{T_d}$ is a translation with a vector **d** which transfers **o** into **o**′, then $\mathbf{T_d^{-1}W}$ is clearly a rotation **A** about **o**. Here, $\mathbf{T_d^{-1}}$ denotes the inverse translation of $\mathbf{T_d}$. Now, if we write $\mathbf{T_d^{-1}W} = \mathbf{A}$ it follows that

$$\mathbf{W} = \mathbf{T_d A} \tag{2.2.5}$$

i.e., a general displacement can be written as the product of a rotation and a translation. We usually refer to a transformation which may not only change the orientation but also the origin of a coordinate system as a *homogeneous transformation.* From the foregoing, equation (2.2.5) defines a homogeneous transformation. Obviously, the homogeneous transformation **W**, as a product of a nonlinear and a linear transformation, is a nonlinear transformation.

Homogeneous transformations are best analyzed in terms of *homogeneous coordinates* [11-14]. The coordinates of a point, line or plane are called homogeneous if the entity they determine is not altered when the coordinates are multiplied by the same scalar. As is well known, a three dimensional vector **p** in a coordinate system has a (3×1) column matrix representation. In a homogeneous coordinate system a three dimensional vector has a (4×1) column matrix representation. The last entry of the column contains a scaling factor which can be chosen to be equal to 1. With the scaling factor equal to 1, the homogeneous coordinate matrix of **p** is given by [12]

$$\begin{aligned} p_h &= [\, p_x \; p_y \; p_z \;\; 1 \,]^T \\ &= [\, p^T \mid 1 \,]^T \end{aligned} \tag{2.2.6}$$

where $p = [\, p_x \, p_y \, p_z \,]^T$ is the coordinate matrix of **p** in three dimensional Euclidean space.

A translation homogeneous transformation $\mathbf{T_d}$ with a vector of translation **d** has a (4×4) matrix representation in homogeneous coordinates given by [12]

$$T_d = \begin{bmatrix} 1 & 0 & 0 & | & d_x \\ 0 & 1 & 0 & | & d_y \\ 0 & 0 & 1 & | & d_z \\ - & - & - & - & - \\ 0 & 0 & 0 & | & 1 \end{bmatrix} \tag{2.2.7}$$

where $d = [\, d_x \, d_y \, d_z \,]^T$ is the coordinate matrix of **d** in three dimensional Euclidean space.

Similarly, a rotation homogeneous transformation **A** has a (4x4) matrix representation in homogeneous coordinates given by

$$A = \begin{bmatrix} A & | & 0 \\ \hline 0^T & | & 1 \end{bmatrix} \tag{2.2.8}$$

where, A is the usual (3×3) matrix representation (via direction cosines) of a rotation, and 0 is the three dimensional zero vector. Note that we use the same notation for the (3x3) matrix and the homogeneous matrix representations of a rotation. We shall rely on the context to distinguish between the two representations.

Now, as we can see by using equation (2.2.7) and (2.2.8), the homogeneous transformation **W** defined by (2.2.5) has a (4x4) homogeneous matrix representation given by

$$W = \left[\begin{array}{c|c} A & d \\ \hline 0^T & 1 \end{array}\right] \tag{2.2.9}$$

Equation (2.2.9) allows us to express the general displacement of a vector (i.e., equation (2.2.1)) in homogeneous coordinates as follows.

$$\left[\begin{array}{c} p' \\ \hline 1 \end{array}\right] = \left[\begin{array}{c|c} A & d \\ \hline 0^T & 1 \end{array}\right] \left[\begin{array}{c} p \\ \hline 1 \end{array}\right] \tag{2.2.10}$$

Equation (2.2.10) is equivalent to the two equations

$$p' = Ap + d \tag{2.2.11a}$$

$$1 = 1 \tag{2.2.11b}$$

where obviously (2.2.11a) gives the matrix representation in the three dimensional space of equation (2.2.4).

From the foregoing, we have two ways of analyzing a general rigid body displacement transformation. Either equation (2.2.4) or equation (2.2.1) can be used.

As we shall see in later chapters, equation (2.2.4), with a matrix representation given by equation (2.2.11a), is suitable for kinematic analysis of rigid body motion when computational issues are of main concern. Equation (2.2.1), with a matrix representation given by (2.2.10), leads to compact representations, but with significantly higher computational complexity. The two representations are, of course, equivalent and equations (2.2.10) and (2.2.11) provide the bridge between them.

Remark 2.2: Besides describing general displacements, homogeneous transformations are often used in robotics [12] to represent coordinate frames, i.e., relative configurations of rigid bodies. Thus, for example, equation (2.2.9) can be used to define the homogeneous coordinate matrix representation for the configuration of a rigid body relative to another reference coordinate system.

Based on these preliminaries, we can now introduce the physical system on whose dynamic analysis this monograph is focused.

2.3 Robot Manipulators

As we mentioned in Chapter 1, robot manipulators or robot arms are the most important form of robotic systems in use today. The dynamic analysis of such robots is therefore of practical importance. A general description for the physical components of a robot manipulator has been given in Chapter 1. In this section, we provide in more details, a "geometric" description for the arm of a robot manipulator and introduce some relevant terminology.

2.3.1 Description of Robot Manipulators

A robot manipulator is essentially a mechanical device that can be programmed to automatically manipulate objects in physical space (the real world). The arm or articulate portion of a robot is usually constructed as a series of coupled bodies, known as *links*, which together constitute what is called a *kinematic chain.* If every link is connected to at least two other links, the kinematic chain is said to be *closed*, and such a mechanism is called a *linkage.* If, however, some of the links are connected to only one other link, then the kinematic chain is said to be *open* or *serial-type* and such a mechanism is called a *manipulator.* Therefore, depending on their articulate portion, we can have robots with closed or open kinematic chains. However, since the kinematic and dynamic analyses of closed kinematic chains is more involved [15-19], most industrial robots have open kinematic chains i.e., a manipulator, with some form of *end-effector* attached to the final link. Depending on the intended applications, the end-effector can be a *gripper*, a *welding torch* or other device. We usually refer to this class of industrial robots as *robot manipulators.*

Although, in reality, all mechanical devices are flexible to a degree, the links of present day industrial robot manipulators are made of quite heavy and rigid material and are usually modeled as rigid bodies. This provides a realistic approximation which allows us to simplify considerably their kinematic and dynamic analyses. However, in recent years, some research has been directed towards modeling and analysis of robot manipulators with flexible links [20-23]. Flexible link manipulators are made of light weight material and may be useful for space applications but have not yet become popular in industry.

A *kinematic pair* is the coupling of two adjacent links. In current industrial manipulators the most frequently encountered kinematic pairs (and the simplest ones) are the *revolute pair,* which allows only relative *rotational motion* about a single axis (the *joint*), and the *prismatic pair,* which allows only relative *translational motion* along a single axis (the joint). For these kinematic pairs, since motion is allowed in a single direction only, one parameter (variable) is sufficient for specifying the relative motion between two adjacent links. This implies that revolute or prismatic pairs are characterized by one *degree of freedom.* The corresponding variable which measures the linear or rotational relative motion of a kinematic pair is referred to as the *generalized coordinate* of that kinematic pair or joint.

As we saw in Section 2.2, there are six degrees-of-freedom associated with the configuration of a rigid object. Therefore, if the links of a manipulator are connected by only revolute and/or prismatic joints (as is the case with most current industrial manipulators), then there must be at least six such links (and hence joints) if the manipulator is to be capable of arbitrarily positioning objects in a three dimensional space. Otherwise stated, any manipulator must have at least six degrees-of-freedom (links and/or joints) in order for it to achieve arbitrary real world configurations. There are, however, many manipulators that have fewer than six degrees-of-freedom because they are designed to perform tasks which do not require such freedom. Also, there are robot manipulators which have been designed to have more than six degrees-of-freedom. Such manipulators are called *redundant* arm manipulators [24-28]. These manipulators are particularly useful in environments where collision avoidance [24,25] is important. However, in this monograph we shall be concerned only with the kinematic and dynamic analyses of non-redundant robot manipulators with rigid links.

To be able to identify the links and the joints of a robot manipulator, we number the links (and implicitly the joints) from zero to n successively - zero being the first link, which is known as the *base*, and is fixed, and n the last one which corresponds to the free end. With this scheme, the joint which connects the $(i-1)$-th and the i-th links is referred to as the i-th joint. The end-effector (if it exists) is usually considered as the $(n+1)$-th link which is rigidly attached to the n-th link with no joint between them.

Also, to be able to specify the configuration, of each link relative to an *inertial* or any other coordinate system, we associate with each link a frame which we denote by $\{ \mathbf{e}_i \} \equiv \{\mathbf{x}_i \; \mathbf{y}_i \; \mathbf{z}_i \; \mathbf{o}_i \}$ $i = 0,1,2 \ldots n$, and refer to as the i-th link frame. The i-th frame is composed of the frame coordinate system (which we denote by $\{ \mathbf{e}_i \} \equiv \{\mathbf{x}_i \; \mathbf{y}_i \; \mathbf{z}_i \}$ and refer to as the i-th *link coordinate system*) and the position vector of the origin of this coordinate system relative to the inertial or any other coordinate system. The i-th link coordinate system is rigidly attached to the i-th link, and so the i-th frame defines the configuration of the i-link relative to the inertial or any other reference system. The frame which is associated with link 0 is often referred to as the *base frame*. When the base frame is considered as an inertial reference frame (which is often the case) it serves as a universal frame relative to which everything we discuss can be referred. On the end-effector, we attach a frame to which we assign the number $n+1$ and which we call the *tool frame*. Note that the tool frame has a constant configuration relative to the n-th frame.

The configuration of the tool frame, relative to an inertial frame, is usually considered as the configuration of the robot manipulator. This is justified, since the end-effector is that part of a robot which is designed to make contact with the environment for the purpose of executing some task. However, in the actual kinematic and dynamic analyses of a robot manipulator, we consider the configuration of its last link as the configuration of the robot manipulator. This is acceptable for two reasons: First, the end-effector is always attached rigidly to the last link, and thus has a constant configuration relative to that of the last link; and second, we want the analysis to be general and not specific to a particular end-effector. Moreover, to make the analysis independent of a particular environment where the robot may be used, we choose the inertial or reference coordinate system to be attached to the base of the robot. Thus, we usually choose the basis frame $\{ \mathbf{e}_0 \}$ to be the inertial or universal reference frame. Therefore, in this monograph, unless indicated otherwise, we shall take the configuration of a robot manipulator to mean the configuration of its last link relative to the base coordinate frame $\{ \mathbf{e}_0 \}$. Also, when a quantity such as a position or a velocity vector is defined relative to the inertial reference frame, the term *absolute* will be used. When the reference frame is another link frame, we shall use the term *relative*.

2.3.2 Geometric Description of a Link

To obtain the proper kinematic and dynamic equations for a robot manipulator, it is important to know the exact geometric characteristic of each link. This will enable us to define the absolute or relative configuration of any link in the articulated portion of a robot manipulator.

A geometric description of a link is mainly concerned with what relationship exists between two neighboring joint axes of a manipulator. Mathematically, joint axes are defined by lines in three dimensional space. Thus the joint axis i is defined to be the line or vector direction in space about which the i-th link rotates or translates relative to the $(i-1)$-th link. Note that this definition for the joint axes implies that link i $(i \neq n)$ connects two joint axes, the i-th and the $(i+1)$-th. In particular, the joint axis i about or along which link i moves is called the *proximal joint*, and the joint axis $(i+1)$ which connects link i with link $(i+1)$ is the *distal joint* associated with the i-th link.

The relative location (configuration) of two axes in three dimensional space is defined by specifying the following two quantities:

Link length: For any two axes in three dimensional space, there exists a well-defined distance between them. This distance is measured along a line which is mutually perpendicular to both axes. This distance always exists and is unique except when the two axes are parallel, in which case there are many mutually perpendicular lines of equal length.

As we have mentioned, the i-th link is associated with the i-th and the $(i+1)$-th joints. Therefore, the mutually perpendicular line between the i-th and the $(i+1)$-th joints allows us to define what is called the *length* of the i-th link.

Link twist: Between two axes in three dimensional space, we can always define an angle. There are two cases to examine. In the first case, the axes are assumed to be parallel, and we consider that a zero angle exists between them. In the second case, we assume that the axes are not parallel and we define an angle between them as follows: We consider a plane normal to the mutually perpendicular line which exists between the two non-parallel axes. Now, by considering the projection of these two axes on the normal plane, we obtain two non-parallel lines on the plane. At the point of their intersection we can choose one of the four angles to be the *twist angle* of the two axes. Based on this angle, we shall define the twist angle of a link in the next section.

2.3.3 Description of Link Connections and the Configuration of a Robot Manipulator

The primary purpose of this section is to describe the "configuration" transformation, which defines the relative displacement of two neighboring links of a robot manipulator, as well as to derive a matrix representation for it.

A special description for the configuration transformation of two neighboring members of a spatial kinematic chain has been established over the years and is known as the Denavit-Hartenberg (D-H) description or convention. The D-H convention has proven to be very practical in robotics because it allows for a systematic description of a spatial kinematic chain, in particular when it is of an open-loop structure. The D-H convention was first introduced by Denavit and Hartenberg [29] in 1955 for the purpose of analyzing spatial linkages, and was specialized to open-loop spatial kinematic chains by Kahn [30] in 1969. The usual D-H convention, as originally designed for kinematic analysis, has some disadvantages and can lead to ambiguities when it is used in robots with links having more than two joints [18]. The scheme which we describe here is well suited for open loop spatial kinematic chains and can be easily adapted for spatial kinematic chains with tree or closed-loop structures [18].

The D-H convention associates a frame rigidly with every link (or joint). In particular, the i-th frame is associated with the i-th link and its coordinate system defines the orientation of the i-th link (frame) relative to the $(i-1)$-th link (frame). Also its position vector defines the origin displacement of the i-th coordinate system relative to the $(i-1)$-th coordinate system. In the D-H convention, to assign the i-th coordinate frame on the i-th link, we use the following two basic assumptions:

i) The $\mathbf{z}_i$ basis unit vector of the i-th frame coordinate system is always parallel to either the proximal or the distal axis of the i-th link.

ii) The $\mathbf{x}_i$ basis unit vector of the i-th frame coordinate system is always parallel to the mutual perpendicular between the i-th and the $(i+1)$-th joint axes

Based on these assumptions we define the i-th link coordinate system. We assume that the $\mathbf{z}_i$ basis vector is parallel to the proximal joint of the link, i.e., it is parallel to the i-th joint axis. The origin o_i of the i-th frame coordinate system is located at the intersection of the i-th joint axis and the mutual perpendicular between the i-th and the $(i+1)$-th joint axes. When this point is not unique (in the case of parallel joint axes) we choose the one which minimizes the relative distance between the origin of the $(i-1)$ and i-th coordinate systems. The $\mathbf{x}_i$ basis unit vector is on the mutual perpendicular to the axis $\mathbf{z}_i$ and $\mathbf{z}_{i+1}$ directed from the former to the latter. The $\mathbf{y}_i$ basis unit vector of the i-th frame coordinate system is chosen as the unique perpendicular to both $\mathbf{z}_i$, and $\mathbf{x}_i$ at the point o_i, which defines a *right-hand* oriented coordinate system.

In the D-H convention four parameters are needed to specify completely the relative configuration of two neighboring frames. As is shown in Figure 2.1, the D-H parameters for the i-th frame are defined as follows:

α_i : The angle about $\mathbf{x}_{i-1}$, between $\mathbf{z}_{i-1}$ and $\mathbf{z}_i$.

a_i : The distance along $\mathbf{x}_{i-1}$, between $\mathbf{z}_{i-1}$ and $\mathbf{z}_i$.

d_i : The distance along $\mathbf{z}_i$, between $\mathbf{x}_{i-1}$ and $\mathbf{x}_i$.

θ_i : The angle about $\mathbf{z}_i$, between $\mathbf{x}_{i-1}$ and $\mathbf{x}_i$.

Remark 2.3: The angle α_i and the distance a_i, are associated with the $(i-1)$-th link, and can be used to define the twist angle and the length, respectively, of this link. Since these two quantities are associated with the $(i-1)$-th link, some authors [13] use the notation α_{i-1} and a_{i-1} instead of α_i and a_i. We prefer to use α_i and a_i, because, as we shall see later, it leads to a uniform notation for the matrix representation of the configuration transformation between two neighboring links.

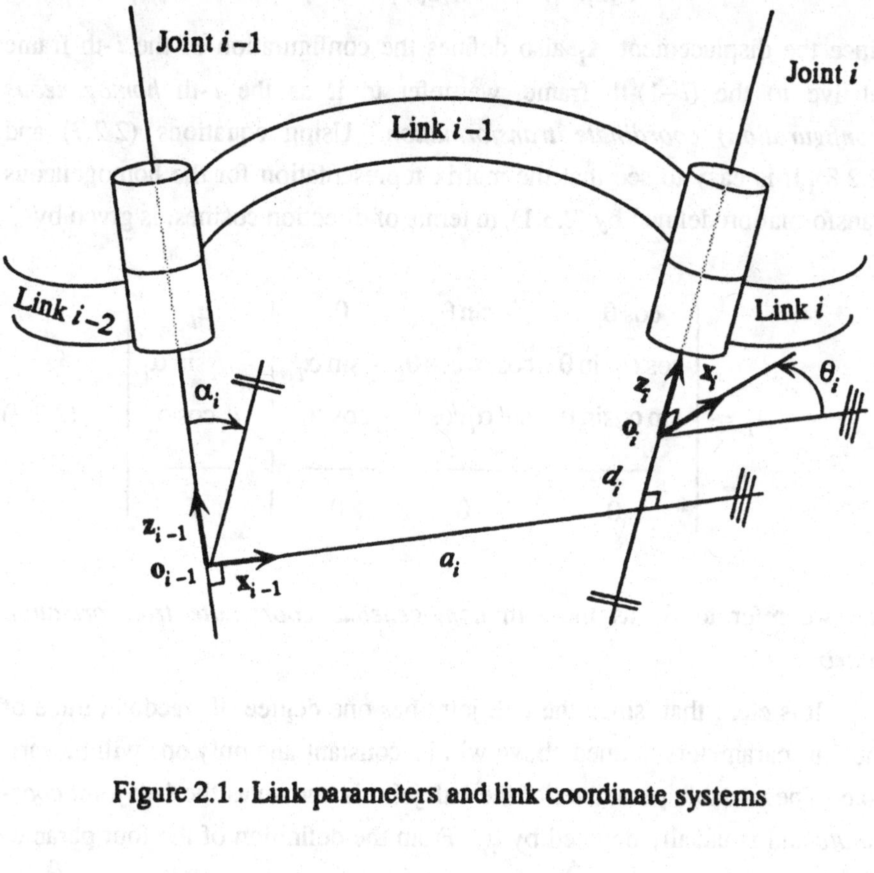

Figure 2.1 : Link parameters and link coordinate systems

Each of the parameters defined above, can be used to define a pure translation or a pure rotation displacement. We shall call these displacements *elementary displacements*, since they occur along or about a coordinate axis, and we denote them by *Trans*(axis , var) and *Rot*(axis , var) respectively, where "axis" is the coordinate axis of the displacement and "var" denotes the variable of the displacement.

Now, using elementary displacement operations, we can describe [12-14] the displacement of the i-th link relative to the $(i-1)$-th link, which we denote by $\mathbf{A}_i \equiv {}^{i-1}\mathbf{A}_i$, as follows

$$\mathbf{A}_i = Rot(\mathbf{x}_{i-1},\alpha_i)Trans(\mathbf{x}_{i-1},a_i)Rot(\mathbf{z}_i,\theta_i)Trans(\mathbf{z}_i,d_i) \tag{2.3.1}$$

Since the displacement $\mathbf{A}_i$ also defines the configuration of the i-th frame relative to the $(i-1)$-th frame, we refer to it as the i-th *homogeneous (configuration) coordinate transformation*. Using equations (2.2.7) and (2.2.8), it is easy to see that the matrix representation for the homogeneous transformation defined by (2.3.1), in terms of direction cosines, is given by

$$A_i = \left[\begin{array}{ccc|c} \cos\theta_i & -\sin\theta_i & 0 & a_i \\ \cos\alpha_i \sin\theta_i & \cos\alpha_i \cos\theta_i & -\sin\alpha_i & -d_i \sin\alpha_i \\ \sin\alpha_i \sin\theta_i & \sin\alpha_i \cos\theta_i & \cos\alpha_i & d_i \cos\alpha_i \\ \hline 0 & 0 & 0 & 1 \end{array}\right] \tag{2.3.2}$$

and we refer to A_i as the i-th *homogeneous coordinate transformation matrix*.

It is clear that, since the i-th joint has one degree of freedom, three of the four parameters defined above will be constant and only one will be variable. The variable parameter of the i-th joint is known as the i-th *joint coordinate* and is usually denoted by q_i. From the definition of the four parameters, it is obvious, that $q_i \triangleq \theta_i$, when the i-th joint is revolute and $q_i \triangleq d_i$, when the i-th joint is prismatic. Therefore, with the appropriate definition for the joint variable q_i, the transformation $\mathbf{A}_i \triangleq \mathbf{A}_i(q_i)$, defined by (2.3.1), describes completely the one degree-of-freedom displacement (motion) of the i-th link relative to the $(i-1)$-th link.

As we mentioned in Section 2.2, a general displacement can be expressed as a pure rotation and a pure translation. Thus, by comparing equations (2.2.9) and (2.3.2), we can see that the rotation $\mathbf{A}_i$ of the i-th link coordinate system relative to the $(i-1)$-th, has a (3×3) matrix representation, in terms of direction cosines, which is given by

$$A_i = \begin{bmatrix} \cos\theta_i & -\sin\theta_i & 0 \\ \cos\alpha_i \sin\theta_i & \cos\alpha_i \cos\theta_i & -\sin\alpha_i \\ \sin\alpha_i \sin\theta_i & \sin\alpha_i \cos\theta_i & \cos\alpha_i \end{bmatrix} \tag{2.3.3}$$

Similarly, the origin displacement or position vector of the i-th coordinate system relative to the $(i-1)$-th coordinate system, has a (3×1) matrix representation which is given by

$$^{i-1}s_{i-1,i} = \begin{bmatrix} a_i \\ -d_i \sin\alpha_i \\ d_i \cos\alpha_i \end{bmatrix} \tag{2.3.4}$$

Therefore, the homogeneous transformation $\mathbf{A}_i$, or the rotation $\mathbf{A}_i$ together with the translation $\mathbf{T}_{\mathbf{s}_{i-1,i}}$, completely describes the configuration of the i-th link relative to the $(i-1)$-th link.

Now, we can use the relative configuration between two neighboring links to define the absolute configuration of any link of a robot manipulator. We do that here in terms of homogeneous transformations since this leads to a compact description.

It is well known [12] that, in general, if we postmultiply a homogeneous transformation representing a frame by a second homogeneous transformation describing a rotation and/or a translation, we make that rotation and/or translation with respect to the frame which is described by the first transformation. Thus, the homogeneous transformation

$$^{i-1}\mathbf{W}_{i+1} = \mathbf{A}_i \mathbf{A}_{i+1} \tag{2.3.5}$$

describes the configuration of the $(i+1)$-th frame (link) relative to the $(i-1)$-th frame. Therefore, the absolute configuration of the i-th link is described by the transformation,

$$
\begin{aligned}
{}^0\mathbf{W}_i \equiv \mathbf{W}_i &= \mathbf{A}_1\mathbf{A}_2 \dots \mathbf{A}_i \\
&= \mathbf{W}_{i-1}\mathbf{A}_i \qquad\qquad i = 1,2, \dots, n \qquad\qquad (2.3.6)
\end{aligned}
$$

and obviously the absolute configuration of the manipulator (i.e., the configuration of the n-th link) is given by $\mathbf{W}_n$.

As can be seen from equation (2.3.6), the homogeneous transformation $\mathbf{W}_n$ is a function of the joint coordinates q_i, $i = 1, 2, \dots, n$. The joint coordinates q_i are linearly independent, and this allows us to view them as the components of an n dimensional vector $\mathbf{q}$, relative to some curvilinear coordinate system which describes the *joint space* of the robot manipulator. The vector $\mathbf{q}$ is referred to as the *joint space vector* and since $\mathbf{W}_n \triangleq \mathbf{W}_n(\mathbf{q})$, $\mathbf{W}_n$ provides a *joint space description* for the configuration of a robot manipulator.

The joint space description for the configuration of a robot manipulator provided by $\mathbf{W}_n$ can be viewed as an "internal" description, since implicitly it contains the configuration of all the individual links of the manipulator. Now, from section 2.2, we recall that the configuration of a rigid body (and therefore that of the last link or the end-effector of a robot manipulator) can be described in Cartesian space in terms of a "Cartesian vector" $\boldsymbol{\chi}$† . The Cartesian space description for the configuration of a robot manipulator can be viewed as an "external" description, since it does not take into account the configuration of the individual links in the manipulator chain.

---

† Note that sometimes in robotics, more general spaces (e.g. the *operational space* [31]) are used to describe the configuration of a robot manipulator.

Both, the joint and Cartesian space descriptions for the configuration of a robot manipulator are fundamental in robotics. The Cartesian space description is useful, mainly to human operators, since it allows for an easy description of the motion of the end-effector between two different locations in space. However, the motion of a robot manipulator is realized in joint space and therefore a joint space description is necessary for the robot controller. Hence, some of the basic kinematic problems in robotics (eg. forward and inverse kinematics) deal with transformations between joint space and Cartesian space descriptions for the configuration of a robot manipulator.

Besides configuration analysis, "motion" kinematic analysis is also needed for the dynamic analysis of any mechanical system. However, before we deal with motion kinematics of robot manipulators we need first to develop a methodology which will allow us to study kinematics and dynamics in a simple and efficient manner. Since this methodology will be based on Cartesian tensors, the next chapter is devoted to Cartesian tensor analysis.

2.4 References

[1] H. Goldstein, *Classical Mechanics* , 2*nd* ed. Reading, MA:, Addison Wesley, 1980.

[2] J. L. Synge, "Classical Dynamics", in *Encyclopedia of Physics*, S. Flugge Ed., Vol. III, Springer-Verlag, Berlin-Gottingen-Heidelberg, 1960.

[3] O. Bottema and B. Roth, *Theoretical Kinematics*, North-Holland Publishing Co., Amsterdam, 1978.

[4] J. Angeles, *Spatial Kinematics Chains : Analysis, Synthesis, Optimization*, Springer-Verlag, New York, 1982.

[5] D. Lovelock and H. Rund, *Tensors, Differential Forms and Variational Principles*, John Wiley & Sons, New York, 1965.

[6] J. Stuelpnagel, "On the Parametrization of the Three-Dimensional Rotation Group" , *SIAM REVIEW*, Vol. 6, No. 4, pp. 422-430, October 1964.

[7] J. Rooney, "A Survey of Representations of Spatial Rotations About a Fixed Point", *Environment and Planning B*, Vol. 4, pp. 185-210, 1977.

[8] D. Hestenes, "Vectors, Spinors, and Complex Numbers in Classical and Quantum Physics", *J. Math. Phys.* Vol. 39, pp 1013-1027, 1971.

[9] R. A. Wehage, "Quaternions and Euler Parameters - A Brief Exposition", in *Computer Aided Analysis and Optimization of Mechanical System Dynamics*, E. J. Haug, Ed., Springer-Verlag, New York, 1984.

[10] J. Angeles, *Rational Kinematics*, Springer-Verlag, New York, 1988.

[11] L. Brand, *Vector and Tensor Analysis*, John Wiley & Sons, New York, 1947.

[12] R. P. Paul, *Robot Manipulator : Mathematics, Programming and Control*, MIT Press, Cambridge, MA, 1981.

[13] J.J. Craig, *Introduction to Robotics : Mechanics & Control*, Addison-Wesley, Reading, MA: 1986.

[14] W. A. Wolovich, *Robotics : Basic Analysis and Design*, Holt, Rinehart and Winston, New York, 1987.

[15] B. Paul, "Analytical Dynamics of Mechanisms: A Computer Oriented Overview", *Mechanism and Machine Theory*, Vol. 10, pp. 481-507, 1975.

[16] J.Y.S. Luh and Y.F. Zheng, "Computation of Input Generalized Forces for Robots with Closed Kinematic Chain Mechanisms", *IEEE J. Robotics and Automation*, RA-1, No 2, pp 95-103 , 1985.

[17] T.R. Kane and H. Faessler, "Dynamics of Robots and Manipulators Involving Closed Loops", in *Theory and Practice of Robots and Manipulators* , A. Morecki, G. Bianchi and K. Kedzior, Eds., MIT Press, Cambridge, MA, 1985.

[18] W. Khalil and J. F. Kleinfinger, "A New Geometric Notation for Open and Closed-Loop Robots", in *Proc. IEEE Int. Conf. on Robotics and Automation*, pp. 1174-1179, San Francisco, CA, 1986.

[19] Y. Nakamura and M. Ghodoussi, "A Computational Scheme of Closed Link Robot Dynamics Derived by D'Alembert's Principle", in *Proc. IEEE Int. Conf. on Robotics and Automation*, pp. 1353-1360, Philadelphia PA, 1988.

[20] W.J. Book, "Recursive Lagrangian Dynamics of Flexible Manipulators", *The Int. J. Robotics Research*, vol. 3, no. 3, pp. 87-101, 1984.

[21] M. Geradin, G. Robert and C. Bernardin, "Dynamic Modeling of Manipulators with Flexible Members", in *Advanced Software in Robotics* , A. Danthine and M. Geradin, Eds., Elsevier Science Pub. Co., New York, 1984.

[22] G. B. Yang and M. Donath, "Dynamic Model of a One-Link Robot Manipulator with Both Structural and Joint Flexibility", in *Proc. IEEE Int. Conf. on Robotics and Automation*, pp. 476-481, Philadelphia PA, 1988.

[23] W. J. Book and S. Cetinkunt, "Performance Limitations of Joint Variable-Feedback Controllers Due to Manipulator Structural Flexibility", *IEEE Trans. on Robotics and Automation*, Vol. 6, No. 2, pp. 219-231, 1990.

[24] A. Maciejewski and C. A. Klein, "Obstacle Avoidance for Kinematically Redundant Manipulators in Dynamically Varying Environments", *The Int. J. of Robotics Research*, Vol. 4, No. 3, pp 109-117, 1985.

[25] J. Baillieul, ''Avoiding Obstacles and Resolving Kinematic Redundancy'', in *Proc. IEEE Int. Conf. on Robotics and Automation*, pp. 1698-1704, San Francisco, CA, 1986.

[26] C. A. Klein and A. I. Chirco, ''Dynamic Simulation of a Kinematically Redundant Manipulator System'', *Journal of Robotic Systems*, Vol. 4, No. 1, pp. 5-23, 1987.

[27] P. Hsu, J. Hauser, and S. Sastry, ''Dynamic Control of Redundant Manipulators'', In *Proc. IEEE Int. Conf. on Robotics and Automation*, pp. 183-187, Philadelphia PA, 1988.

[28] H. Seraji, ''Configuration Control of Redundant Manipulators; Theory and Implementation'', *IEEE Trans. on Robotics and Automation*, Vol. 5, No. 4, pp. 472-490, 1989.

[29] J. Denavit and R. S. Hartenberg, ''A Kinematic Notation for Lower-Pair Mechanisms Based on Matrices", *ASME J. of Appl. Mechanics*, Vol. 23, pp. 215-221, 1955.

[30] M. E. Kahn, ''The Near-Minimum-Time Control of Open-Loop Articulated Kinematic Chains", Stanford Artificial Intelligence Project, *Memo. AIM-106*, Dec. 1969.

[31] O. Khatib, ''A Unified Approach for Motion and Force Control of Robot Manipulators : The Operational Space Formulation'', *IEEE J. Robotics and Automation*, Vol. RA-3, No. 1, pp. 43-53, 1987.

Chapter 3

Cartesian Tensor Analysis

3.1 Introduction

As we mentioned in Chapter 1, our intention is to describe the dynamic equations of rigid body motion by using *Cartesian tensors.* Cartesian tensor analysis, being more general than vector analysis, is powerful and, if properly used, can result in a tensor formulation for the equations of general motion of a dynamic system. As we shall show in Chapter 5, such a formulation will enable us to derive computationally efficient algorithms for the dynamic equations of motion of rigid-link open-chain robot manipulators. In this chapter, we provide an introduction to the theory of Cartesian tensors. Moreover, based on 1-1 operators between three dimensional vectors and second order skew-symmetric Cartesian tensors, this theory is extended here by establishing a number of tensor-vector identities. These identities, as we shall see in the following chapters, will allow for easy algebraic manipulations of the equations of motion of a complex dynamic system such as those of a robotic system.

Historically, the ideas and symbolism of tensor calculus originated in differential geometry, and were invented by the Italian mathematicians Ricci and Levi-Civita [1]. Gradual introduction and assimilation of these ideas and symbols were greatly accelerated by their use by Einstein in his general theory of relativity; and today tensor analysis forms a well established field which provides the only appropriate language for studying differential geometry and related topics such as the theory of general relativity [2-5].

But if tensor calculus is a necessity for studying differential geometry, for applications in continuum and classical mechanics [6-26] it is a great convenience, because it enables one to express geometrical or physical relationships of tensor entities in a concise manner which does not depend on the introduction of a coordinate system. Moreover, even in cases where we have to introduce coordinate systems, because measurements are required or for other reasons, tensor equations are *formally* the same in *all* admissible coordinate systems. This fact will help us later when tensor equations have to be written in the various coordinate systems used to derive dynamic models for robot manipulators.

In the general theory of tensor analysis, the space or environment where a tensor is defined is a *manifold* [3] which is characterized in terms of curvilinear coordinate systems. But, since the environment or physical space for physical systems is a Euclidean point space which can be associated with a three dimensional Euclidean vector space (the *translation* space [8]), we restrict our attention to the study of tensors in Euclidean vector spaces. In a Euclidean vector space, orthogonal Cartesian coordinate systems are sufficient for tensor analysis. Tensors analyzed in orthogonal Cartesian coordinate systems are referred to as *Cartesian tensors* [7-10, 20-25]. Therefore, since in our analysis we shall use *orthonormal* Cartesian coordinate systems, when we write tensors we shall mean (second order) Cartesian tensors. The use of orthonormal Cartesian coordinate systems will simplify our analysis, since the distinction between *covariant* and *contravariant* components, which is necessary in *curvilinear* coordinate systems, disappears in orthonormal Cartesian coordinates. Moreover, since the Euclidean vector space is a *flat space*, terms arising from *curvature* are zero in the theory of Cartesian tensors.

The outline of this chapter is as follows: In Section 3.2 second order tensors are defined and basic algebraic tensor operations are introduced. Section 3.3 outlines the structural symmetries of second order tensors based on *Cartesian* and *Spectral decompositions*, and an important 1-1 operator between three dimensional vectors and second order skew-symmetric Cartesian tensors is defined. Finally, based on the said operator, we state and prove a number of propositions which establish some basic tensor-vector identities.

3.2 Second Order Cartesian Tensors

In this section, we define Cartesian tensors of order 2. We also introduce basic tensor algebraic operations which will be used later in algebraic manipulations. Here, we restrict our attention to Cartesian tensor analysis in three dimensional (3-D) Euclidean vector spaces. However, this theory can be extented to Euclidean vector spaces of higher dimensions in a straightforward manner.

3.2.1 On The Definition of the Second Order Cartesian Tensors.

Tensor analysis, in general, is concerned with mathematical or physical entities which although of different nature have common characteristics and properties. These common properties allow us to classify them into common classes and to refer to them with such names as *scalars, vectors* or in general *tensors* of certain order.

There are two basic approaches which one can take to define a Cartesian tensor, or the class where a particular physical entity belongs.

In the first approach, which is usually used to define tensors in general (i.e., not necessary Cartesian), we analyze a physical quantity from an "external" point of view and define its tensor character based on their observed properties. These properties are determined by considering how a physical quantity is related to some environment and how this relationship changes under controlled changes of this environment. In this approach, we may proceed as follows.

We can introduce first a frame of reference, i.e., a coordinate system, which describes the environment and expresses (by considering a set of ordered numbers or functions, known as components) the quantity at hand relative to that coordinate system. Then we make a change (by considering a map) in the coordinate system and analyze how the components of this quantity are related relative to the old and new coordinate system. If the changes in the components follow a definite law, we define the quantity under consideration to be a tensor (of certain order). This is a *passive* approach, since the tensor quantities are actually independent of the coordinate systems used to describe the environment and represent them. Under a change of coordinate systems it is their components that change, not they themselves. Based on what we have said, we can give the following definition:

Definition 3.1: Suppose that the abstract object $\mathbf{T}$ when it is associated with an orthonormal Cartesian coordinate system $\{\mathbf{e}\}$, of a 3-D Euclidean space, can be represented by the set of components T_{ij}, where the subscripts i, j are ordered and can take the values 1,2,3. Let T'_{ij} be the corresponding set of components representing $\mathbf{T}$ when it is associated with another orthonormal Cartesian coordinate system $\{\mathbf{e}'\}$, which is related to the coordinate system $\{\mathbf{e}\}$ by an orthogonal transformation $\mathbf{A}$, i.e., the unit vectors of the coordinate system $\{\mathbf{e}'\}$ satisfy the equations

$$[\mathbf{e}_l']_j = A_{ji}[\mathbf{e}_l]_i \qquad l = 1,2,3 \qquad (3.2.1)$$

where A_{ji} are the entries (direction cosines) of the coordinate matrix A which represents the transformation **A** relative to the two coordinate systems $\{ \mathbf{e}' \}$ and $\{ \mathbf{e} \}$. Then if the equation

$$T'_{rs} = A_{ri} A_{sj} T_{ij} \tag{3.2.2}$$

is valid, we say that **T** is a second order Cartesian tensor.

Remark 3.1: Sometimes in tensor analysis the nomenclatures "order" or "rank" are used to denote the exponent (r) to which the dimension (n) of the space, on which the tensor is defined, must be raised to give the number of components (coefficients) of a tensor. To put it another way, the order of a tensor denotes how many copies of the original Euclidean space we need to consider for producing the environment (a *tensor product space*) where the tensor is defined. In the case of the second order tensor **T** defined on a 3-dimensional Euclidean space, we consider two copies of the original Euclidean space. Therefore, for any tensor in this space we must have $3^2 = 9$ components and this is the case as (3.2.2) indicates. We shall continue to use the word *order* with this meaning. With the word *rank*, we shall associate an *intrinsic* property of a tensor, which we shall define later.

In the second approach, which is commonly used to define Cartesian tensors, we define the tensor character of a physical quantity by considering its *action* on some environment, i.e., we treat, in this approach, the physical quantity as an *operator*. If the operator has certain properties, then we are able to say that the quantity under consideration is a tensor. From this point of view we can give the following definition:

Definition 3.2: We say that the *linear vector transformation* **T** is a second order tensor if we can compute the action of **T** on any vector **r** and denote that action of **T** on **r** by writing $\mathbf{T}(\mathbf{r})$ or $\mathbf{T} \cdot \mathbf{r}$ or simple $\mathbf{T}\mathbf{r}$.

Definition 3.2 is not complete as stated, since we have not mentioned explicitly the domain and range of the linear vector transformation. The domain and range of a second order tensor are obvious from the context. For example, in the equation $\mathbf{L}_c = \mathbf{I}_c\boldsymbol{\omega}$ the domain of the inertia tensor ($\mathbf{I}_c$) is the space of angular velocities ($\boldsymbol{\omega}$) and its range is the space of angular momentum ($\mathbf{L}_c$).

Remark 3.2: Although the terms "linear transformation" and "tensor" in Definition 3.2 refer to mathematical functions of the same kind they are not completely synonymous, because they have different connotations in applications. The term "tensor" is always used when describing certain physical quantities. Thus, we often say a rotation tensor or a rotation transformation, but we never call the *inertia tensor*, $\mathbf{I}_c$, "inertia linear transformation", although it defines the linear transformation $\mathbf{L}_c = \mathbf{I}_c\boldsymbol{\omega}$.

As an example, of how we can use Definition 3.2 to define a tensor, let us consider the following vector equation,

$$\mathbf{u} \otimes \mathbf{v}(\mathbf{r}) = \mathbf{u}(\mathbf{v} \cdot \mathbf{r}). \tag{3.2.3}$$

It is clear that $\mathbf{u} \otimes \mathbf{v}(\mathbf{r})$ is linear in $\mathbf{r}$ since

$$\mathbf{u}[\mathbf{v} \cdot (\lambda_1 \mathbf{r}_1 + \lambda_2 \mathbf{r}_2)] = \lambda_1 \mathbf{u}(\mathbf{v} \cdot \mathbf{r}_1) + \lambda_2 \mathbf{u}(\mathbf{v} \cdot \mathbf{r}_2) \tag{3.2.4}$$

Therefore, the quantity $\mathbf{T} = \mathbf{u} \otimes \mathbf{v}$ is said to be a second order tensor. Since the tensor $\mathbf{u} \otimes \mathbf{v}$ is defined by using a "kind of product" between the two vectors $\mathbf{u}$ and $\mathbf{v}$ it is called the *tensor product* of $\mathbf{u}$ and $\mathbf{v}$. Many authors prefer to write equation (3.2.3) as

$$(\mathbf{u}\mathbf{v}) \cdot \mathbf{r} = \mathbf{u}(\mathbf{v} \cdot \mathbf{r}). \tag{3.2.5}$$

In this case the tensor $\mathbf{u}\mathbf{v}$ is called a *dyad* of $\mathbf{u}$ and $\mathbf{v}$. According to Gibbs [17], who developed the theory of dyadics (a dyadic is a sum of dyads), a dyad or indeterminate product is a purely symbolic quantity which requires a determinate physical meaning only when used as a linear operator. Therefore, by Definition (3.2), dyads are second order tensors. In [17], Gibbs has also shown that in a 3-D Euclidean space any linear vector transformation can be written as the sum of at most three dyads of which either the first or the second vector, but not both, may be arbitrarily chosen provided they are linearly independent. This implies that to each tensor we can associate k ($k \leq 3$) *linearly independent vectors* (or *directions*). Or, to use the approach of Hestenes' [31], to each second order tensor in a 3-D Euclidean space we can associate a direction of k dimensions ($k \leq 3$). Based on this, we can give the following definition for the term "rank".

Definition 3.3: We define the *rank* of a tensor to be the number of linearly independent directions which a tensor possesses.

The definition of rank as given here, coincides with what we call rank in linear algebra. In particular, the rank of a second order tensor (or a linear transformation) **T** coincides with the rank of the matrix T which represents **T** in some coordinate system. This uniformity in the meaning of the term "rank" is not possible if we identify rank with order as is often done in textbooks on tensor analysis.

Remark 3.3: Although in the general theory of tensor analysis, scalars and vectors are treated as tensors of order *zero* and *one* respectively, we continue to refer to them as scalars and vectors and reserve the word "tensor" for tensors of second and higher order.

3.2.2 The Linear Space Structure for Second Order Cartesian Tensors

In order to use tensors in an efficient manner, we have to define algebraic structures on them by introducing basic algebraic operations. Thus, based on Definition 3.2, we can see that the following tensors and algebraic operations are well defined.

The *zero* and *unity (identity)* tensors are denoted and defined, respectively, by

$$\mathbf{0v} = \mathbf{0} \quad , \quad \forall\, \mathbf{v} \tag{3.2.6}$$

$$\mathbf{1v} = \mathbf{v} \quad , \quad \forall\, \mathbf{v} \tag{3.2.7}$$

The algebraic operations of *addition* and *scalar multiplication* can be defined as usual. Thus the addition of two second order tensors $\mathbf{T}$ and $\mathbf{S}$ is defined by

$$(\mathbf{T}+\mathbf{S})\mathbf{v} = \mathbf{T}\mathbf{v} + \mathbf{S}\mathbf{v} \quad , \forall \mathbf{v} \tag{3.2.8}$$

and the multiplication of $\mathbf{T}$ by a scalar λ is defined by

$$(\lambda\mathbf{T})\mathbf{v} = \lambda(\mathbf{T}\mathbf{v}) \quad , \forall \lambda \tag{3.2.9}$$

Also, we say that two second order tensors $\mathbf{T}$ and $\mathbf{S}$ are equal if

$$\mathbf{T} = \mathbf{S} \quad \Leftrightarrow \quad \mathbf{T}\mathbf{v} = \mathbf{S}\mathbf{v} \qquad \forall \mathbf{v} \tag{3.2.10}$$

or equivalently

$$\mathbf{T} = \mathbf{S} \quad \Leftrightarrow \quad \mathbf{v}\cdot\mathbf{T}\mathbf{u} = \mathbf{v}\cdot\mathbf{S}\mathbf{u} \qquad \forall \mathbf{u},\mathbf{v} \tag{3.2.11}$$

It is now easy to see that the set of all tensors of second order, together with the two algebraic operations of addition and scalar multiplication, constitutes a linear space over a scalar field, which is assumed here to be the field of real numbers. To find a basis for the linear space of second order tensors, defined on a 3-D Euclidean space and the components of a tensor $\mathbf{T}$ relative to it, let us consider an orthonormal basis $\{\mathbf{e}\} = \{\mathbf{e}_1, \mathbf{e}_2, \mathbf{e}_3\}$ of a 3-D Euclidean space. Then for any vector $\mathbf{r}$ we can write

$$\begin{aligned} \mathbf{T}\mathbf{r} &= \mathbf{T}(r_1\mathbf{e}_1 + r_2\mathbf{e}_2 + r_3\mathbf{e}_3) \\ &= r_1\mathbf{T}\mathbf{e}_1 + r_2\mathbf{T}\mathbf{e}_2 + r_3\mathbf{T}\mathbf{e}_3. \end{aligned} \tag{3.2.12}$$

But $\mathbf{T}\mathbf{e}_1$, $\mathbf{T}\mathbf{e}_2$ and $\mathbf{T}\mathbf{e}_3$ are vectors and therefore may be expressed in terms of the Cartesian components as follows:

$$\mathbf{T}\mathbf{e}_1 = T_{11}\mathbf{e}_1 + T_{21}\mathbf{e}_2 + T_{31}\mathbf{e}_3 \tag{3.2.13a}$$

$$\mathbf{T}\mathbf{e}_2 = T_{12}\mathbf{e}_1 + T_{22}\mathbf{e}_2 + T_{32}\mathbf{e}_3 \tag{3.2.13b}$$

$$\mathbf{T}\mathbf{e}_3 = T_{13}\mathbf{e}_1 + T_{23}\mathbf{e}_2 + T_{33}\mathbf{e}_3 \tag{3.2.13c}$$

where the coefficients $T_{11}, T_{21}, \cdots, T_{33}$ can be computed by using the equation

$$T_{ij} = \mathbf{e}_i \cdot \mathbf{T}\mathbf{e}_j \tag{3.2.14}$$

Now, using (3.2.13) in (3.2.12) and the equations

$$\mathbf{e}_i r_i = \mathbf{e}_i (\mathbf{e}_i \cdot \mathbf{r}) = \mathbf{e}_i \otimes \mathbf{e}_i (\mathbf{r}),$$

for $i = 1, 2, 3$, we get after a few manipulations

$$\mathbf{T}\mathbf{r} = (T_{11}\mathbf{e}_1 \otimes \mathbf{e}_1 + T_{12}\mathbf{e}_1 \otimes \mathbf{e}_2 + \cdots + T_{33}\mathbf{e}_3 \otimes \mathbf{e}_3)\mathbf{r}. \tag{3.2.15}$$

Therefore, since the vector $\mathbf{r}$ is arbitrary, we have from (3.2.15) that

$$\begin{aligned} \mathbf{T} = {} & T_{11}\mathbf{e}_1 \otimes \mathbf{e}_1 + T_{12}\mathbf{e}_1 \otimes \mathbf{e}_2 + T_{13}\mathbf{e}_1 \otimes \mathbf{e}_3 \\ & + T_{21}\mathbf{e}_2 \otimes \mathbf{e}_1 + T_{22}\mathbf{e}_2 \otimes \mathbf{e}_2 + T_{23}\mathbf{e}_2 \otimes \mathbf{e}_3 \\ & + T_{31}\mathbf{e}_3 \otimes \mathbf{e}_1 + T_{32}\mathbf{e}_3 \otimes \mathbf{e}_2 + T_{33}\mathbf{e}_3 \otimes \mathbf{e}_3. \end{aligned} \tag{3.2.16}$$

or, if we use Einstein's† notation for the summation, we can write

$$\mathbf{T} = T_{ij}\mathbf{e}_i \otimes \mathbf{e}_j, \qquad i, j = 1, 2, 3. \tag{3.2.17}$$

Now, it can be shown [15], that this representation of $\mathbf{T}$ is unique and that the set of second order tensors $\{\mathbf{e}_1 \otimes \mathbf{e}_1, \mathbf{e}_1 \otimes \mathbf{e}_2, \cdots, \mathbf{e}_3 \otimes \mathbf{e}_3\}$ forms a basis for the aforementioned tensor linear space. From the foregoing, this linear space has dimension 3^2. We shall denote this linear space by $E_3 \otimes E_3 \equiv E_3^{(2)}$ and we may call it the second tensorial power of E_3. Relative to the basis $\{\mathbf{e}_1 \otimes \mathbf{e}_1, \mathbf{e}_1 \otimes \mathbf{e}_2, \cdots, \mathbf{e}_3 \otimes \mathbf{e}_3\}$ the components of $\mathbf{T} \in E_3^{(2)}$ are the coefficients $T_{11}, \cdots, T_{33}$ defined by (3.2.14). These coefficients can be put in the matrix form

$$T = \begin{bmatrix} T_{11} & T_{12} & T_{13} \\ T_{21} & T_{22} & T_{23} \\ T_{31} & T_{32} & T_{33} \end{bmatrix} \tag{3.2.18}$$

† In Einstein's notation any expression in which two or more indices $i, j \cdots$ are each

and we call T the coordinate matrix of $\mathbf{T}$ relative to the basis defined by the set $\{\mathbf{e}_1 \otimes \mathbf{e}_1, \mathbf{e}_1 \otimes \mathbf{e}_2, \cdots, \mathbf{e}_3 \otimes \mathbf{e}_3\}$. Now, it is easy to see that there is a one-to-one correspondence between the orthonormal basis $\{\mathbf{e}\} = \{\mathbf{e}_1, \mathbf{e}_2, \mathbf{e}_3\}$ and the basis of the tensor linear space $E_3^{(2)}$ which is defined by the set $\{\mathbf{e}_1 \otimes \mathbf{e}_1, \mathbf{e}_1 \otimes \mathbf{e}_2, \cdots, \mathbf{e}_3 \otimes \mathbf{e}_3\}$. Therefore, by slightly abusing the notation, we shall refer to (3.2.18) as the coordinate matrix of the tensor $\mathbf{T}$ relative to the Cartesian basis $\{\mathbf{e}\} = \{\mathbf{e}_1, \mathbf{e}_2, \mathbf{e}_3\}$. With this in mind, we shall say that a second order tensor is defined in a 3-D Euclidean vector space E_3 instead of the 9-D linear space $E_3^{(2)}$.

3.2.3 More Algebraic Operations.

Besides the addition (or subtraction) and scalar multiplication defined above, there are two other algebraic operations which one can define on tensors, namely, the tensor product and contraction. These operations are defined [12,16-21] as follows:

Tensor product: Let $\mathbf{T}$ and $\mathbf{S}$ be two second order tensors whose components, referred to a coordinate system $\{\mathbf{e}\}$, are t_{ij} and s_{kl}. Then it can be shown that the 3^4 scalars

$$u_{ijkl} = t_{ij} s_{kl} \tag{3.2.19}$$

form the components of a tensor $\mathbf{U}$, say, of order 4. We call $\mathbf{U}$ the tensor product of $\mathbf{T}$ and $\mathbf{S}$ and we write $\mathbf{U} = \mathbf{T} \otimes \mathbf{S}$.

Contraction: Given a tensor of order $r \geq 2$, we may select a pair of indices and replace them by two identical indices. This action by virtue of Einstein's convention implies summation over the possible values of the identical

---

repeated is to be interpreted as the sum of all the values which it can take, as $i, j, \cdots$ take the values 1, 2, 3.

indices. This process is known as contraction and the quantities obtained by contraction constitute the components of a tensor of order $r-2$.

Note: In the special case where $r=2$, the contraction operator is synonymous with the familiar *trace* operator since, as we can see, the contraction of the two indices produces a tensor of order zero, i.e., a scalar. Note, also, that the algebraic operations of tensor product and contraction can be performed on any tensor not necessarily Cartesian.

Some other important algebraic operations which may be termed as *contracted multiplications*, and which are applicable to second order Cartesian tensors, are defined [22] as follows.

The left and right dot products: Let $\mathbf{T}$ be a second order tensor and $\mathbf{v}$ be a vector, with coordinate matrices T and v respectively relative to the same coordinate system. Then the equation

$$u_i = T_{ij} v_j \tag{3.2.20}$$

which is computed by considering first the tensor product of $\mathbf{T}$ and $\mathbf{v}$ and then contracting the second index for the components of $\mathbf{T}$ and the index for the components of $\mathbf{v}$, defines the *right dot product* or *post-multiplication* of $\mathbf{T}$ and $\mathbf{v}$. Similarly using the equation:

$$w_i = v_j T_{ji} \tag{3.2.21}$$

we can define the vector $\mathbf{w}$. Equation (3.2.21) defines the *left dot product* or *pre-multiplication* of $\mathbf{v}$ and $\mathbf{T}$. In tensor form, equations (3.2.20) and (3.2.21) are written as

$$\mathbf{u} = \mathbf{T} \cdot \mathbf{v} \equiv \mathbf{T}\mathbf{v} \tag{3.2.22}$$

$$\mathbf{w} = \mathbf{v} \cdot \mathbf{T} \equiv \mathbf{v}\mathbf{T} \tag{3.2.23}$$

respectively. Note that we can use the transpose operation to reorder the factors in the dot products defined above. Thus, for example, we can write

$$\mathbf{w} = \mathbf{v} \cdot \mathbf{T} = \mathbf{T}^T \cdot \mathbf{v} \tag{3.2.24}$$

In terms of their coordinate matrices, equations (3.2.22) and (3.2.23) are written as

$$u = Tv \tag{3.2.25}$$

and

$$w = v^T T \tag{3.2.26}$$

and they define the familiar post- and pre- multiplication in matrix theory.

Finally, we can define two products between two second order tensors **T** and **S** as follows:

The dot product: Let T and S be the coordinate matrices, relative to the same coordinate system, of second order tensors **T** and **S**, respectively. Then the equation

$$U_{ij} = T_{il} S_{lj} \tag{3.2.27}$$

defines the components of the second order tensor **U** which we call the *dot product* or *multiplication* of **T** and **S**, and we write

$$\mathbf{U} = \mathbf{T} \cdot \mathbf{S} \equiv \mathbf{T}\mathbf{S} \tag{3.2.28}$$

In terms of their coordinate matrices equation, (3.2.28) is written as

$$U = TS \tag{3.2.29}$$

and agrees with the usual matrix multiplication.

The double dot or inner product: The double dot product of two tensors **T** and **S** is given by

$$\mathbf{T} : \mathbf{S} = tr(\mathbf{T} \cdot \mathbf{S}) \tag{3.2.30}$$

where tr denotes the trace operator and produces a scalar.

Remark 3.4: Strictly speaking some operations introduced above are defined by using coordinate matrix representations of second order tensors and vectors. However, once these operations have been established, the actual coordinate matrices used to represent the tensors or vectors are of no consequence and we can speak of these operations as being defined on the tensors themselves without any ambiguity.

The double dot product is a generalization of the familiar vector dot product, and this allows us to define a norm for a tensor and the "angle" between two tensors. In particular, we can define the norm of a tensor **T** as follows

$$\|\mathbf{T}\|_F \triangleq \sqrt{\mathbf{T} : \mathbf{T}^T} \tag{3.2.31}$$

and we may call this norm the *Frobenius norm* since, when we use a coordinate matrix T of a tensor **T** to evaluate equation (3.2.31), we end up with the familiar Frobenius matrix norm [30]

$$\|T\|_F \triangleq \sqrt{tr(TT^T)}. \tag{3.2.32}$$

With this definition for the norm of a tensor the set of all second order Cartesian tensors, regarded as a linear space of dimension 3^2, becomes an inner-product space. Now, using the double dot product and the Frobenius norm, we can define the cosine of the "angle" θ between two non-zero tensors **T** and **S** as follows

$$\cos(\theta) = \frac{\mathbf{T} : \mathbf{S}^T}{\|\mathbf{T}\|_F \|\mathbf{S}\|_F} \qquad 0 \le \theta \le \pi \tag{3.2.33}$$

Equation (3.2.33) is a generalization of the following familiar definition for the cosine of the angle θ between two non-zero vectors **t** and **s**

$$\cos(\theta) = \frac{\mathbf{t} \cdot \mathbf{s}}{\|\mathbf{t}\| \|\mathbf{s}\|} \qquad 0 \le \theta \le \pi \tag{3.2.34}$$

Actually, based on dual (or axial) vectors (see Section 3.3) and using Proposition 3.6, it can be shown that equation (3.2.33) is essentially identical to equation (3.2.34) when it is applied to skew-symmetric tensors. Based on this definition for the angle between two tensors, we shall say that two tensors **T** and **S** are *orthogonal* to each other if and only if their double dot product is zero.

As we have mentioned above, the double dot product (and not the dot product) between two Cartesian tensors is the generalization of the familiar dot product between two vectors. The dot product between two second order tensors is not commutative as is the case with the familiar dot product between vectors, but it is associative and distributive over addition. This allows us to view the linear space of second order Cartesian tensors, when supplied with this dot product, as a *linear algebra*. (This also follows from Definition 3.2 which identifies second order tensors with linear transformations. With this identification the tensor dot product is equivalent to the composition of linear transformations). This algebra is isomorphic to the matrix algebra. Thus we can use the well-established matrix algebra to carry out calculations for various operations with tensors. Since the matrix elements are scalars (real numbers), matrix algebra has the advantage of reducing all such calculations to addition and multiplication of real numbers. However, it has the disadvantage of requiring that a basis be introduced (which defines an isomorphism between tensors and matrices) and which may be quite irrelevant to the problem at hand and this often obscures the physical or geometrical meaning of the tensor involved. Moreover, this matrix representation may lead us to perform irrelevant and unnecessary calculations.

In the next sections, we analyze some basic properties of second order Cartesian tensors which will allow us to perform algebraic manipulations with tensors without resorting to their matrix representations.

3.3 Properties of Second Order Cartesian Tensors

In this section we analyze the structural symmetries of second order Cartesian tensors by considering their *Cartesian* and *Spectral decompositions* and define their *scalar* and *vector invariants*. Also, we introduce very important dual correspondences between second order skew symmetric (or pseudo) tensors and vectors defined on a 3-D Euclidean space.

3.3.1 Isotropic Cartesian Tensors.

As we mentioned in section 3.2.1, tensors themselves are independent of coordinate systems; but the numerical values for the components of a tensor, in general, depend on coordinate systems. Therefore, if we use an orthogonal transformation **A** to change the basis from $\{\mathbf{e}\}$ to $\{\mathbf{e}'\}$, i.e., if

$$\mathbf{e}_j' = \mathbf{A}\mathbf{e}_i \tag{3.3.1}$$

then the coordinate matrices T and T' of a tensor $\mathbf{T}$, relative to the old and new Cartesian bases respectively, are different. By Definition 3.1, they are related and this relationship is given by

$$T' = ATA^T. \tag{3.3.2}$$

A tensor which has the *same* coordinate matrix in *all* Cartesian coordinate systems, or that is invariant under orthogonal transformations is called an *isotropic* tensor. An example of an isotropic tensor is the Kronecker or *delta* tensor $\boldsymbol{\delta}$, with components

$$\delta_{ij} = \begin{cases} 1 & \text{if } i = j \\ 0 & \text{if } i \neq j \end{cases} \tag{3.3.3}$$

relative to any orthonormal Cartesian coordinate system. Actually it can be shown [23] that the Kronecker tensor is the only isotropic tensor of order 2 (apart from a scalar multiple). Since the Kronecker tensor $\boldsymbol{\delta}$ is equivalent to the unit tensor, in many cases we shall use the symbol $\mathbf{1}$ to denote it.

Although we are concerned with second order tensors, we mention here a 3rd-order tensor which is also isotropic and which we shall use on some occasions. This is the *Levi-Civita* or *alternating* tensor $\boldsymbol{\varepsilon}$. The components of $\boldsymbol{\varepsilon}$ in any Cartesian basis are defined by

$$\varepsilon_{ijk} = \begin{cases} +1 & \text{if } (i,j,k) \text{ is an even permutation of } (1,2,3) \\ -1 & \text{if } (i,j,k) \text{ is an odd permutation of } (1,2,3) \\ 0 & \text{if } (i,j,k) \text{ has any other set of values} \end{cases} \tag{3.3.4}$$

With this definition the non-zero components of ε are the following,

$$\varepsilon_{123} = \varepsilon_{231} = \varepsilon_{312} = 1$$

$$\varepsilon_{321} = \varepsilon_{213} = \varepsilon_{132} = -1$$

Moreover, as we can see from the above discussion, the Levi-Civita tensor is skew-symmetric (or antisymmetric) with respect to any two of its indices.

A number of important relationships between the two isotropic tensors ε and δ can be stated. However, here we shall consider only the following.

$$\varepsilon_{ijk}\varepsilon_{rsk} = \delta_{ir}\delta_{js} - \delta_{is}\delta_{jr} \tag{3.3.5}$$

The truth of (3.3.5) may be established as follows [21].

If $i = j$ or $r = s$, the right-hand side of (3.3.5) is zero and the left-hand side also vanishes by the definition of the Levi-Civita tensor. Consider the case when $i \neq j$ and $r \neq s$. Without loss of generality we may choose $i = 1$ and $j = 2$. Using the definition of the Levi-Civita tensor, the left-hand side of (3.3.5) then becomes

$$\varepsilon_{121}\varepsilon_{rs1} + \varepsilon_{122}\varepsilon_{rs2} + \varepsilon_{123}\varepsilon_{rs3} = \varepsilon_{rs3}$$

The right-hand side of (3.3.5) becomes

$$\delta_{1r}\delta_{2s} - \delta_{1s}\delta_{2r} = k.$$

where, k is a scalar. As $r \neq s$, there are just the following possibilities to consider:

$r = 3$ in which case $k = 0$ for all s ;

$s = 3$ in which case $k = 0$ for all r ;

$r = 3, s = 2$, giving $k = 1$;

$r = 2, s = 1$, giving $k = -1$;

Hence $k = \varepsilon_{rs3}$, and (3.3.5) is proved.

Furthermore, using equation (3.3.5) we can also show that

$$\varepsilon_{ijk}\varepsilon_{rjk} = 2\delta_{ir} \qquad (3.3.6)$$

3.3.2 Cartesian and Spectral Decomposition of Second Order Tensors.

An arbitrary Cartesian tensor **T** defined on an n-dimensional Euclidean space may always be decomposed into the sum of a symmetric and a skew-symmetric tensor, as follows

$$\mathbf{T} = \mathbf{T}_+ + \mathbf{T}_- \qquad (3.3.7)$$

where $\mathbf{T}_+$ is symmetric and $\mathbf{T}_-$ is skew-symmetric. This equation is usually referred to as the *Cartesian decomposition* of a second order tensor. The symmetric and the skew-symmetric parts of equation (3.3.7) are defined as follows:

$$\mathbf{T}_+ = \frac{1}{2}[\mathbf{T} + \mathbf{T}^T], \quad \mathbf{T}_- = \frac{1}{2}[\mathbf{T} - \mathbf{T}^T]. \qquad (3.3.8)$$

As we saw earlier, second order Cartesian tensors, defined on a 3-D Euclidean space, constitute a linear space of dimension 3^2. It is easy to see now that the set of all symmetric Cartesian tensors form a $\frac{1}{2}3(3+1) = 6$ dimensional subspace and the set of all skew-symmetric Cartesian tensors form a $\frac{1}{2}3(3-1) = 3$ dimensional subspace of the linear space of second order Cartesian tensors. Also, it is easy to see that bases for these two sub-spaces can be formed by considering the following tensor and *wedge* products of the vectors of a basis $\{ \mathbf{e}_1, \mathbf{e}_2, \mathbf{e}_3 \}$.

$$\mathbf{e}_i \otimes \mathbf{e}_j + \mathbf{e}_j \otimes \mathbf{e}_i , \qquad i \le j$$

and

$$\mathbf{e}_i \wedge \mathbf{e}_j \qquad i < j$$

respectively. Note that the wedge or *exterior* (or *outer*) product of two vectors $\mathbf{u}$ and $\mathbf{v}$, which is defined by the equation

$$\mathbf{u} \wedge \mathbf{v} = \mathbf{u} \otimes \mathbf{v} - \mathbf{v} \otimes \mathbf{u} \tag{3.3.9}$$

produces a skew-symmetric tensor.

From the foregoing, we can view the tensor vector space as a *direct sum* of two subspaces, namely, the subspace of symmetric and skew-symmetric tensors. Moreover, Proposition 3.7 (see Section 3.4) reveals that these two subspaces are also orthogonal, relative to the double dot product, and so one is the *orthogonal complement* of the other.

An important class of tensors is that of *projection* tensors. A tensor $\mathbf{T}$ is called a projection if it is *idempotent*, i.e., if $\mathbf{T}^2 = \mathbf{T}$. Moreover, a projection is called a *perpendicular* (orthogonal) projection if it is also symmetric, i.e., if $\mathbf{T}^T = \mathbf{T}$. A projection tensor is the simplest nontrivial example of a second order tensor which we can construct from the tensor product of two vectors. In particular, given two vectors $\mathbf{e}$ and $\mathbf{v}$, where $\mathbf{e}$ is a unit direction vector, the orthogonal projection of $\mathbf{v}$ on $\mathbf{e}$ is denoted and defined by

$$\mathbf{P}_\mathbf{e} \mathbf{v} \equiv \mathbf{e} \otimes \mathbf{e} \cdot \mathbf{v} = \mathbf{e}(\mathbf{e} \cdot \mathbf{v}). \tag{3.3.10}$$

The tensor

$$\mathbf{P}_\mathbf{e} = \mathbf{e} \otimes \mathbf{e} \tag{3.3.11}$$

which is defined as the tensor product of the unit vector $\mathbf{e}$, is the orthogonal projection tensor along the direction of $\mathbf{e}$. Moreover, as we can see (from the theory of projections or by direct verification), the tensor

$$\mathbf{P}^{\mathbf{e}} = \mathbf{1} - \mathbf{P}_{\mathbf{e}} \tag{3.3.12}$$

is also an orthogonal projection tensor which projects any vector **v** onto a plane perpendicular to **e**. From the foregoing, it is obvious that a vector **v** can be decomposed into the following orthogonal components with respect to **e**

$$\mathbf{v}_{\parallel} = \mathbf{P}_{\mathbf{e}} \cdot \mathbf{v} \tag{3.3.13}$$

$$\mathbf{v}_{\perp} = \mathbf{P}^{\mathbf{e}} \cdot \mathbf{v} \tag{3.3.14}$$

Now, given an orthonormal Cartesian coordinate system with basis vectors $\mathbf{e}_i$, we can define the orthogonal projection tensors $\mathbf{P}_i \equiv \mathbf{P}_{\mathbf{e}_i}$, $i = 1, 2, 3$. It is easy to see that these projection tensors, besides being symmetric and idempotent, also satisfy the following properties.

$$(a)\ \textit{orthogonality}:\ \mathbf{P}_i : \mathbf{P}_j = 0 \quad \text{if } i \neq j \tag{3.3.15}$$

and

$$(b)\ \textit{completeness}:\ \mathbf{P}_1 + \mathbf{P}_2 + \mathbf{P}_3 = \mathbf{1}. \tag{3.3.16}$$

Based on orthogonal projections, we can define an important decomposition for symmetric tensors. As is well known [7,8], any symmetric tensor **T** has real eigenvalues (λ_i) with a complete set of orthonormal eigenvectors. Therefore, if $\mathbf{P}_k$ is the orthogonal projection tensor along the unit direction of the k-th eigenvector, we can write for **T** the following *canonical form* or *spectral decomposition* [7]

$$\mathbf{T} = \lambda_1 \mathbf{P}_1 + \lambda_2 \mathbf{P}_2 + \lambda_3 \mathbf{P}_3 \equiv \lambda_i \mathbf{P}_i \tag{3.3.17}$$

The spectral decomposition of a symmetric tensor, will allow us to give simple proofs for some basic propositions in Section 3.4.

3.3.3 Tensor Invariants

Scalar Invariants: As we saw in section 3.3.1, except for isotropic tensors, the individual components of a tensor are not invariant. They depend on the coordinate systems. There are, however, a number of scalar invariants associated with every second order tensor, i.e., scalars which depend on the tensor itself, and not on the matrix representing it or its individual components. These numbers are known as *scalar invariants*.

Functional expressions for the scalar invariants of a tensor can be written in different forms. Thus, for example, for a general second order tensor $\mathbf{T}$, which has matrix representation T relative to some basis, we can define [21] the following scalar invariants:

$$i)\quad I_1 = t_{ii} \tag{3.3.18}$$

$$ii)\quad I_2 = t_{22}t_{33} - t_{23}t_{32} + t_{33}t_{11} - t_{13}t_{31} + t_{11}t_{22} - t_{12}t_{21} \tag{3.3.19}$$

$$iii)\quad I_3 = t_{11}t_{22}t_{33} + t_{12}t_{23}t_{31} + t_{13}t_{21}t_{32} - t_{11}t_{23}t_{32} - t_{22}t_{13}t_{31} - t_{33}t_{12}t_{21} \tag{3.3.20}$$

$$iv)\quad I_4 = t_{ik}t_{ik} \tag{3.3.21}$$

where, obviously, I_1 is the trace of $\mathbf{T}$, I_2 is the sum of its principal minors, and I_3 is its determinant.

In the case of symmetric tensors, I_4 is not independent of I_1, I_2, I_3. The three independent scalar invariants of a symmetric tensor are also known as *principal invariants* and in terms of the eigenvalues (λ_i) of the symmetric tensor are given by

$$I_1 = \lambda_1 + \lambda_2 + \lambda_3 = tr(\mathrm{T}) \tag{3.3.22}$$

$$I_2 = \lambda_1\lambda_2 + \lambda_2\lambda_3 + \lambda_3\lambda_1 \tag{3.3.23}$$

$$I_3 = \lambda_1\lambda_2\lambda_3 \tag{3.3.24}$$

Remark 3.5: For a symmetric tensor $\mathbf{T}$, we can define the scalar invariants without resorting to the components of a matrix representation of $\mathbf{T}$. It can be shown [21,27] that the scalars $F_1 = \mathrm{tr}(\mathbf{T})$, $F_2 = \mathrm{tr}(\mathbf{T}^2)$, $F_3 = \mathrm{tr}(\mathbf{T}^3)$ are invariant. Obviously, these scalar invariants are not independent of the principal invariants. It can be verified that

$$I_1 = F_1,\ I_2 = \frac{1}{2}(F_1^2 - F_2) \quad \text{and} \quad I_3 = \frac{1}{6}(2F_3 + F_1^3 - 3F_1F_2)$$

Vector Invariants: Besides scalar invariants, we can associate to any second order tensor defined on a 3-D Euclidean space, a vector which belongs to the 3-D Euclidean space and is defined as follows.

Let t_{kj} be the components of a tensor $\mathbf{T}$ relative to a Cartesian orthonormal basis $\{\mathbf{e}\}$. By considering the tensor product of the Levi-Civita tensor $\boldsymbol{\varepsilon}$ and the tensor $\mathbf{T}$ and contracting twice over two common indices we have

$$t_i = \varepsilon_{ijk} t_{kj}, \tag{3.3.25}$$

which, by using equation (3.3.4) for $i = 1, 2, 3$ gives

$$t_1 = t_{32} - t_{23} \tag{3.3.26a}$$

$$t_2 = t_{13} - t_{31} \tag{3.3.26b}$$

$$t_3 = t_{21} - t_{12}. \tag{3.3.26c}$$

Equation (3.3.26) implies that $\mathbf{t} = 0$, if $\mathbf{T}$ is symmetric, and $\mathbf{t}$ has components which are numerically twice those of $\mathbf{T}$ if $\mathbf{T}$ is skew-symmetric. Therefore, the vector

$$t_i = \frac{1}{2}\varepsilon_{ijk} t_{kj} \tag{3.3.27}$$

is uniquely defined when the tensor $\mathbf{T}$ is given. We denote this vector by writing

$$\mathbf{t} = vect(\mathbf{T}) = vect(\mathbf{T}_{\_}) \tag{3.3.28}$$

where, $vect(\cdot)$ denotes the ***tensor valued vector operator*** which is defined by equation (3.3.27). The vector $\mathbf{t}$ is referred to as the *vector (geometric) invariant* of $\mathbf{T}$ [27].

Dual Vectors and Dual tensors: As we can see from equation (3.3.26), the *kernel* (or *null space*) of the vect operator is the subspace of the symmetric tensors, and therefore is a non-empty set. Hence, the vect operator in not a 1–1 operator. Therefore, if we are seeking a 1–1 correspondence between tensors and their vector invariants, we have to consider the restriction of the vect operator onto the subspace of the skew-symmetric tensors. To do this, we shall restrict our attention to the subspace of skew-symmetric tensors.

As we saw above, to each skew-symmetric tensor there corresponds a vector - its vector invariant. Conversely, as we shall show, to any vector $\mathbf{t}$ in a 3-D Euclidean space, there corresponds a skew-symmetric tensor, which we shall denote by $\tilde{\mathbf{t}}$. Moreover, the skew-symmetric tensor $\tilde{\mathbf{t}}$ has the important property that its vector invariant is the vector $\mathbf{t}$, from which the tensor $\tilde{\mathbf{t}}$ has been generated. To see this, given a vector $\mathbf{t}$, we define the tensor $\tilde{\mathbf{t}}$ by considering the tensor product of the Levi-Civita tensor $\boldsymbol{\varepsilon}$ with the vector $\mathbf{t}$ and contract the first index of $\boldsymbol{\varepsilon}$ with that of $\mathbf{t}$, i.e., we consider the equation

$$t_{kj} = \varepsilon_{ijk} t_i \tag{3.3.29}$$

The tensor $\tilde{\mathbf{t}}$ is skew-symmetric, since the Levi-Civita tensor is antisymmetric with respect to the indices j and k. Moreover, if we multiply equation (3.3.29) by ε_{rjk} and use equation (3.3.6) we get

$$\varepsilon_{rjk} t_{kj} = 2\delta_{ri} t_i ,$$

which, by using the definition of the Kroneker tensor δ_{ri}, can also be written as

$$t_r = \frac{1}{2} \varepsilon_{rjk} t_{kj} .$$

This equation is equivalent to equation (3.3.27) and therefore, $\mathbf{t}$ is indeed the vector invariant of the skew-symmetric tensor $\tilde{\mathbf{t}}$.

From the foregoing, by considering the restriction of the vect operator onto the subspace of skew-symmetric tensors, we can see that equation (3.3.27) (together with equation (3.3.29)) defines a 1–1 correspondence between skew-symmetric tensors and vectors. To express this 1–1 correspondence between skew-symmetric tensors and vectors, we introduce the following tensor-valued tensor operator,

$$dual(\cdot) \triangleq vect(\cdot)\Big|_{\left\{\text{second order skew-symmetric tensors}\right\}} \tag{3.3.30}$$

which is a 1–1 operator, as can be seen from equation (3.3.26).

The dual operator can also be defined (simultaneously with its inverse) in a component-wise manner, as follows:

Definition 3.4: The *dual operator* is a 1–1 *tensor-valued tensor operator* which has the following property. When this operator is evaluated at a tensor of order one (i.e., a vector) we get a skew-symmetric tensor of order two, and when it is evaluated at a second order skew-symmetric tensor, we get a tensor of order one. We define the action of the dual operator on a vector or a skew-symmetric tensor, component-wise, using the following 1–1 correspondence

$$\begin{bmatrix} u_1 \\ u_2 \\ u_3 \end{bmatrix} \Leftrightarrow \begin{bmatrix} 0 & -u_3 & u_2 \\ u_3 & 0 & -u_1 \\ -u_2 & u_1 & 0 \end{bmatrix}. \tag{3.3.31}$$

Symbolically, we denote the action of the dual operator on a vector $\mathbf{u}$ by writing

$$\tilde{\mathbf{u}} \triangleq dual(\mathbf{u}), \tag{3.3.32}$$

Similarly

$$\mathbf{u} \stackrel{\Delta}{=} dual(\tilde{\mathbf{u}}) \tag{3.3.33}$$

denotes the action of the dual operator on a skew-symmetric tensor $\tilde{\mathbf{u}}$.

It is important to note here that, for simplicity, we write $dual(\cdot)$ to denote both, the vector-valued and the tensor-valued dual operators. We shall rely on the argument to distinguish between the two cases (i.e., the "direct" or the "inverse" operator), and we shall refer to the tensor $\tilde{\mathbf{u}}$, defined by (3.3.32), as the *dual tensor* of the vector $\mathbf{u}$. Similarly, we shall refer to the vector $\mathbf{u}$, defined by equation (3.3.33), as the *dual vector*† of the tensor $\tilde{\mathbf{u}}$.

It can be verified that the dual vector $\mathbf{u}$ of the skew-symmetric tensor $\tilde{\mathbf{u}}$ is simply its vector invariant as defined by equation (3.3.27). Therefore, Definition 3.4 and equation (3.3.30), both introduce the same tensor-valued tensor operator.

In associating a skew-symmetric tensor of order 2 with a vector, a sign convention or a relative orientation clearly arises. Therefore, the 1–1 correspondence which has been established above is not the only possible 1–1 correspondence between vectors and skew-symmetric tensors in a 3-D Euclidean space. For example, if instead of equation (3.3.31), we consider the following correspondence

$$\begin{bmatrix} u_1 \\ u_2 \\ u_3 \end{bmatrix} \Leftrightarrow \begin{bmatrix} 0 & u_3 & -u_2 \\ -u_3 & 0 & u_1 \\ u_2 & -u_1 & 0 \end{bmatrix}, \tag{3.3.34}$$

---

† The terminology "dual vector" or "dual tensor" is not standard in the literature. For example, to define the same idea, the term dual vector has been used in [9] and the older terminology "axial vector" has been used in [7,8,10]. Here this terminology has been introduced to express the particular 1–1 relationship which exists between vectors in E_3 and skew-symmetric tensors in $E_3^{(2)}$.

which assigns the opposite relative orientation to the components of the skew-symmetric tensor, we can define, as in Definition 3.4, another operator which also establishes a 1–1 correspondence between vectors and skew-symmetric tensors in 3-D Euclidean space.

The tensor-valued tensor operator which describes this new 1–1 correspondence between vectors and skew-symmetric tensors will again be referred to as a dual operator. But, to express it symbolically, we can use a different notation from that used for the dual operator given by Definition 3.4. In particular, to denote the action of the dual operator, which is introduced here by the correspondence (3.3.34), on a vector or a skew-symmetric tensor, we shall write

$$\bar{\mathbf{u}} \triangleq (\mathbf{u})dual, \tag{3.3.35}$$

and

$$\mathbf{u} \triangleq (\bar{\mathbf{u}})dual, \tag{3.3.36}$$

respectively. In the following, we shall rely on the notation or the context to make clear which correspondence, i.e., (3.3.31) or (3.3.34), has to be used for the dual operator. However, in either case, we can say that the corresponding skew-symmetric tensor is *oriented* relative to its dual vector, or simply, that the skew-symmetric tensor is a *relatively oriented* tensor.

Remark 3.6: As is well known [16,22], the components of a skew-symmetric tensor change sign when the "handedness" of the coordinate system is changed, from right-handed to left-handed, say. This is also true for quantities whose orientation or sense is established by convention, such as the familiar *axial vectors*. From the foregoing, it is easy to see that a dual vector is actually an axial vector and this implies that the axial vector of classical vector analysis is nothing more than a second order skew-symmetric tensor disguised (based on a dual correspondence) as an axial vector. Therefore, physical quantities which in classical mechanics are described by axial vectors can be described equally well by using second order skew-symmetric Cartesian tensors.

3.4 Cartesian Tensor Algebraic Identities

In this section we shall prove a number of propositions which define important tensor equations. These equations will allow us to manipulate second order tensors very efficiently as abstract objects, without the need to resort to coordinate bases. Moreover, if needed, the transition from these tensor equations to the corresponding coordinate matrix equations is effortless. This is because the basic tensor algebraic operations, as defined in section 3.2.3, are formally the same as the basic algebraic operations in matrix theory.

In the following, unless mentioned otherwise, by a dual operator, we mean the dual operator which is defined by the correspondence (3.3.31). Based on this dual operator, we shall state a number of propositions which define important tensor-vector identities. Obviously, similar propositions can be stated also in terms of the dual operator which is defined by the correspondence (3.3.34). To prove most of the propositions, which are presented in this section, we shall use the following well known equation [7-26]

$$\mathbf{u} \times \mathbf{v} = \tilde{\mathbf{u}} \cdot \mathbf{v} \tag{3.4.1}$$

which expresses the vector cross product of two vectors $\mathbf{u}$ and $\mathbf{v}$ as a right dot product between the dual tensor $\tilde{\mathbf{u}}$ and the vector $\mathbf{v}$.

Proposition 3.1: The dual operator is linear, i.e., it satisfies the following relations:

$$dual(k\mathbf{u}) = k\tilde{\mathbf{u}} \tag{3.4.2}$$

$$dual(\mathbf{u} + \mathbf{v}) = \tilde{\mathbf{u}} + \tilde{\mathbf{v}} \tag{3.4.3}$$

where, k is a scalar and $\mathbf{u}$, $\mathbf{v}$ are vectors.

Proof: The result follows from the correspondence (3.3.31). □

Proposition 3.2: The vector $\mathbf{u}$ provides a basis for the null space of its dual tensor $\tilde{\mathbf{u}}$, i.e.,

$$\tilde{\mathbf{u}} \cdot \mathbf{u} = 0 \tag{3.4.4}$$

$$\tilde{\mathbf{u}} \cdot \mathbf{v} = 0 \Leftrightarrow \mathbf{u} \parallel \mathbf{v} \tag{3.4.5}$$

Proof: Equation (3.4.4) follows from the fact that $\mathbf{u} \times \mathbf{u} = 0$. Equation (3.4.5) follows from equation (3.4.4) which implies that the one dimensional null space of $\tilde{\mathbf{u}}$ is spanned by the vector $\mathbf{u}$. □

The following proposition defines some equations which allow us to reorder the factors of a left or right dot product between dual tensors and vectors. This is often desirable when algebraic manipulations are needed for simplifying other complex tensor equations.

Proposition 3.3: The right or left dot product of a dual tensor and a vector satisfies the following equations

$$\tilde{\mathbf{u}} \cdot \mathbf{v} = -\tilde{\mathbf{v}} \cdot \mathbf{u} \tag{3.4.6}$$

$$\mathbf{v} \cdot \tilde{\mathbf{u}} = -\tilde{\mathbf{u}} \cdot \mathbf{v} \tag{3.4.7}$$

$$\mathbf{v} \cdot \tilde{\mathbf{u}} = \tilde{\mathbf{v}} \cdot \mathbf{u} \tag{3.4.8}$$

Proof: Equation (3.4.6) follows from equation (3.4.1) and the anticommutativity of the vector cross product. Equation (3.4.7) follows from equation (3.2.24) and the fact that a dual tensor is skew-symmetric. Finally, equation (3.4.8) follows from equations (3.4.6) and (3.4.7). □

Proposition 3.4: The dot product of $\mathbf{a} \times \mathbf{u}$ and $\mathbf{v} \times \mathbf{b}$ can be computed by using the following identity:

$$(\mathbf{a} \times \mathbf{u}) \cdot (\mathbf{v} \times \mathbf{b}) = \mathbf{a} \cdot \tilde{\mathbf{u}} \tilde{\mathbf{v}} \cdot \mathbf{b} \tag{3.4.9}$$

Proof : Equation (3.4.9) follows from equations (3.4.1) and (3.4.8). □

Proposition 3.5: The dot product of two skew-symmetric tensors can be written in terms of the tensor and dot product of their dual vectors, as follows:

$$\tilde{\mathbf{u}}\tilde{\mathbf{v}} = \mathbf{v}\otimes\mathbf{u} - \mathbf{u}\cdot\mathbf{v}\,\mathbf{1}. \tag{3.4.10}$$

Proof: Using equation (3.4.1) we can write the double vector cross product $\mathbf{u}\times(\mathbf{v}\times\mathbf{r})$ as

$$\mathbf{u}\times(\mathbf{v}\times\mathbf{r}) = \tilde{\mathbf{u}}\cdot(\tilde{\mathbf{v}}\cdot\mathbf{r}) = \tilde{\mathbf{u}}\tilde{\mathbf{v}}\cdot\mathbf{r} \tag{3.4.11}$$

Also, the same double vector cross product can be written [17,18] as

$$\begin{aligned}\mathbf{u}\times(\mathbf{v}\times\mathbf{r}) &= \mathbf{v}(\mathbf{u}\cdot\mathbf{r}) - (\mathbf{u}\cdot\mathbf{v})\mathbf{r}\\ &= (\mathbf{v}\otimes\mathbf{u} - \mathbf{u}\cdot\mathbf{v}\,\mathbf{1})\cdot\mathbf{r}\end{aligned} \tag{3.4.12}$$

Therefore, equating (3.4.11) with (3.4.12) we have

$$\tilde{\mathbf{u}}\tilde{\mathbf{v}}\cdot\mathbf{r} = (\mathbf{v}\otimes\mathbf{u} - \mathbf{u}\cdot\mathbf{v}\,\mathbf{1})\cdot\mathbf{r} \tag{3.4.13}$$

from where the identity (3.4.10) follows, since (3.4.13) is true for every vector $\mathbf{r}$. □

Another useful form of equation (3.4.10) is the following

$$\mathbf{v}\otimes\mathbf{u} = \tilde{\mathbf{u}}\tilde{\mathbf{v}} + (\mathbf{v}\cdot\mathbf{u})\mathbf{1} \tag{3.4.14}$$

which expresses the tensor product of two vectors in terms of their dot product and the dot product of their dual tensors.

Proposition 3.6: The double dot product of two skew-symmetric tensors $\tilde{\mathbf{u}}$ and $\tilde{\mathbf{v}}$ is related to the dot product of their dual vectors $\mathbf{u}$ and $\mathbf{v}$ by the equation

$$\tilde{\mathbf{u}}:\tilde{\mathbf{v}} = -2\mathbf{u}\cdot\mathbf{v} \tag{3.4.15}$$

Proof: From the definition of the double dot product we have

$$\tilde{\mathbf{u}}:\tilde{\mathbf{v}} = tr[\tilde{\mathbf{u}}\tilde{\mathbf{v}}].$$

Therefore, using equation (3.4.10) we can write,

$$\begin{aligned}\tilde{\mathbf{u}}:\tilde{\mathbf{v}} &= tr[\mathbf{v}\otimes\mathbf{u} - \mathbf{u}\cdot\mathbf{v}\,\mathbf{1}]\\ &= tr[\mathbf{v}\otimes\mathbf{u}] - \mathbf{u}\cdot\mathbf{v}\,tr[\mathbf{1}]\\ &= \mathbf{u}\cdot\mathbf{v} - 3\mathbf{u}\cdot\mathbf{v}\\ &= -2\mathbf{u}\cdot\mathbf{v}\end{aligned}$$

□

Another usuful identity, which follows from the proof of Proposition (3.6), is the following

$$\mathbf{u}\cdot\mathbf{v} = -\frac{1}{2}tr[\tilde{\mathbf{u}}\tilde{\mathbf{v}}] \tag{3.4.16}$$

Proposition (3.6) implies that two skew-symmetric tensors are orthogonal (with respect to the double dot or inner product) if and only if their dual vectors are orthogonal.

Proposition 3.7: Symmetric and a skew-symmetric second order Cartesian tensors are always orthogonal. In other words, the double dot product of a symmetric tensor $\mathbf{I}$ and a skew-symmetric tensor $\mathbf{S}$ is always zero, i.e.,

$$\mathbf{I}:\mathbf{S} = tr[\mathbf{I}\mathbf{S}] = 0 \tag{3.4.17}$$

Proof: It is well known [28] that $tr[\mathbf{A}\mathbf{B}] = tr[\mathbf{B}\mathbf{A}]$ and $tr[\mathbf{A}] = tr[\mathbf{A}^T]$ for any tensor $\mathbf{A}$ and $\mathbf{B}$. Then, for $\mathbf{I}$ symmetric and $\mathbf{S}$ skew-symmetric, we have

$$\begin{aligned}tr[\mathbf{I}\mathbf{S}] &= tr[(\mathbf{I}\mathbf{S})^T]\\ &= tr[\mathbf{S}^T\mathbf{I}^T]\\ &= -\,tr[\mathbf{S}\mathbf{I}]\\ &= -\,tr[\mathbf{I}\mathbf{S}]\end{aligned}$$

which implies that $tr[\mathbf{I}\mathbf{S}] = 0$, and thus equation (3.4.17) is valid. □

Proposition 3.8: The dual tensors $\tilde{\mathbf{u}}$ and $\tilde{\mathbf{v}}$ satisfy the following equations.

$$\tilde{\mathbf{v}}\tilde{\mathbf{u}}\tilde{\mathbf{u}} + \tilde{\mathbf{u}}\tilde{\mathbf{u}}\tilde{\mathbf{v}} = -(\mathbf{u}\cdot\mathbf{u})\tilde{\mathbf{v}} - (\mathbf{v}\cdot\mathbf{u})\tilde{\mathbf{u}} \tag{3.4.18}$$

$$\tilde{\mathbf{u}}\tilde{\mathbf{v}}\tilde{\mathbf{u}} = -(\mathbf{v}\cdot\mathbf{u})\tilde{\mathbf{u}} \tag{3.4.19a}$$

$$= \tilde{\mathbf{v}}\tilde{\mathbf{u}}\tilde{\mathbf{u}} + \tilde{\mathbf{u}}\tilde{\mathbf{u}}\tilde{\mathbf{v}} - \frac{1}{2}tr[\tilde{\mathbf{u}}\tilde{\mathbf{u}}]\tilde{\mathbf{v}} \tag{3.4.19b}$$

$$\tilde{\mathbf{v}}\tilde{\mathbf{v}}\tilde{\mathbf{u}}\tilde{\mathbf{u}} - \tilde{\mathbf{u}}\tilde{\mathbf{u}}\tilde{\mathbf{v}}\tilde{\mathbf{v}} = (\mathbf{u}\cdot\mathbf{v})[\tilde{\mathbf{u}}\ ,\tilde{\mathbf{v}}] \tag{3.4.20}$$

where $[\tilde{\mathbf{u}}\ ,\tilde{\mathbf{v}}]$ is the *Lie bracket* or *commutator* of two skew-symmetric tensors and is defined [29] as follows

$$[\tilde{\mathbf{u}}\ ,\tilde{\mathbf{v}}] = \tilde{\mathbf{u}}\tilde{\mathbf{v}} - \tilde{\mathbf{v}}\tilde{\mathbf{u}} \tag{3.4.21}$$

Proof: Using equations (3.4.6) and (3.4.10), the left-hand side of equation (3.4.18) can be simplified as follows:

$$\begin{aligned}\tilde{\mathbf{v}}\tilde{\mathbf{u}}\tilde{\mathbf{u}} + \tilde{\mathbf{u}}\tilde{\mathbf{u}}\tilde{\mathbf{v}} &= \tilde{\mathbf{v}}(\mathbf{u}\otimes\mathbf{u} - \mathbf{u}\cdot\mathbf{u}\,\mathbf{1}) + \tilde{\mathbf{u}}(\mathbf{v}\otimes\mathbf{u} - \mathbf{v}\cdot\mathbf{u}\,\mathbf{1})\\ &= (\tilde{\mathbf{v}}\mathbf{u})\otimes\mathbf{u} - (\mathbf{u}\cdot\mathbf{u})\tilde{\mathbf{v}} + (\tilde{\mathbf{u}}\mathbf{v})\otimes\mathbf{u} - (\mathbf{v}\cdot\mathbf{u})\tilde{\mathbf{u}}\\ &= -(\tilde{\mathbf{u}}\mathbf{v})\otimes\mathbf{u} - (\mathbf{u}\cdot\mathbf{u})\tilde{\mathbf{v}} + (\tilde{\mathbf{u}}\mathbf{v})\otimes\mathbf{u} - (\mathbf{v}\cdot\mathbf{u})\tilde{\mathbf{u}}\\ &= -(\mathbf{u}\cdot\mathbf{u})\tilde{\mathbf{v}} - (\mathbf{v}\cdot\mathbf{u})\tilde{\mathbf{u}}\end{aligned}$$

and this gives equation (3.4.18). Equation (3.4.19a) results from the following manipulations

$$\begin{aligned}\tilde{\mathbf{u}}\tilde{\mathbf{v}}\tilde{\mathbf{u}} &= \tilde{\mathbf{u}}(\mathbf{u}\otimes\mathbf{v} - (\mathbf{v}\cdot\mathbf{u})\mathbf{1})\\ &= (\tilde{\mathbf{u}}\mathbf{u})\otimes\tilde{\mathbf{v}} - (\mathbf{v}\cdot\mathbf{u})\tilde{\mathbf{u}}\\ &= -(\mathbf{v}\cdot\mathbf{u})\tilde{\mathbf{u}}\,.\end{aligned}$$

Equation (3.4.19b) follows from equations (3.4.16), (3.4.18) and (3.4.19a). Finally, to prove equation (3.4.20), we pre- and post-multiply equation (3.4.18) by $\tilde{\mathbf{v}}$ and subtract the resulting equations. Then, after cancelling some terms we derive equation (3.4.20). □

When the vectors $\mathbf{u}$ and $\mathbf{v}$ are perpendicular, since $\mathbf{v}\cdot\mathbf{u} = 0$, equations (3.4.18), (3.4.19a) and (3.4.20) are simplified to

$$\tilde{\mathbf{v}}\tilde{\mathbf{u}}\tilde{\mathbf{u}} + \tilde{\mathbf{u}}\tilde{\mathbf{u}}\tilde{\mathbf{v}} = -(\mathbf{u}\cdot\mathbf{u})\tilde{\mathbf{v}} \tag{3.4.22}$$

$$\tilde{\mathbf{u}}\,\tilde{\mathbf{v}}\,\tilde{\mathbf{u}} = 0 \tag{3.4.23}$$

and

$$\tilde{\mathbf{v}}\,\tilde{\mathbf{v}}\,\tilde{\mathbf{u}}\,\tilde{\mathbf{u}} - \tilde{\mathbf{u}}\,\tilde{\mathbf{u}}\,\tilde{\mathbf{v}}\,\tilde{\mathbf{v}} = 0 \tag{3.4.24}$$

respectively. Also, when the vectors $\mathbf{u}$ and $\mathbf{v}$ are parallel (i.e., $\mathbf{u} = k\mathbf{v}$, where k is a scalar) equations (3.4.18) and (3.4.19a) become

$$\tilde{\mathbf{u}}\,\tilde{\mathbf{u}}\,\tilde{\mathbf{u}} = -(\mathbf{u}\cdot\mathbf{u})\tilde{\mathbf{u}} \tag{3.4.25}$$

and equation (3.4.20) equals zero. Therefore, when the vectors $\mathbf{u}$ and $\mathbf{v}$ are either parallel or orthogonal than the symmetric tensors $\tilde{\mathbf{u}}\,\tilde{\mathbf{u}}$ and $\tilde{\mathbf{v}}\,\tilde{\mathbf{v}}$ commute.

Proposition 3.9: The dual tensor $dual(\tilde{\mathbf{u}}\,\mathbf{v})$ can be written in one of the following equivalent forms:

$$dual(\tilde{\mathbf{u}}\cdot\mathbf{v}) = \mathbf{v}\otimes\mathbf{u} - \mathbf{u}\otimes\mathbf{v} \tag{3.4.26a}$$

$$= \tilde{\mathbf{u}}\,\tilde{\mathbf{v}} - \tilde{\mathbf{v}}\,\tilde{\mathbf{u}} \tag{3.4.26b}$$

$$= [\tilde{\mathbf{u}}\ ,\tilde{\mathbf{v}}] \tag{3.4.26c}$$

$$= \mathbf{v}\ \wedge \mathbf{u} \tag{3.4.26d}$$

where $[\tilde{\mathbf{u}}\ ,\tilde{\mathbf{v}}]$ is the Lie bracket of $\tilde{\mathbf{u}}$ and $\tilde{\mathbf{v}}$ and $\mathbf{v}\ \wedge \mathbf{u}$ is the exterior or outer product of $\mathbf{v}$ and $\mathbf{u}$.

Proof: To prove (3.4.26a), we shall use the double vector cross product $(\mathbf{u}\times\mathbf{v})\times\mathbf{r}$. From vector analysis it is known [17] that

$$(\mathbf{u}\times\mathbf{v})\times\mathbf{r} = \mathbf{v}(\mathbf{u}\cdot\mathbf{r}) - \mathbf{u}(\mathbf{v}\cdot\mathbf{r})$$

$$= \mathbf{v}\otimes\mathbf{u}\cdot\mathbf{r} - \mathbf{u}\otimes\mathbf{v}\cdot\mathbf{r} \qquad \text{(by (3.2.3))}$$

$$= (\mathbf{v}\otimes\mathbf{u} - \mathbf{u}\otimes\mathbf{v})\cdot\mathbf{r} \tag{3.4.27}$$

Moreover, by using equation (3.4.1), the term $(\mathbf{u}\times\mathbf{v})\times\mathbf{r}$ can also be written as

$$(\mathbf{u}\times\mathbf{v})\times\mathbf{r} = dual(\tilde{\mathbf{u}}\cdot\mathbf{v})\cdot\mathbf{r} \tag{3.4.28}$$

Then, since (3.4.27) and (3.4.28) are true for every vector $\mathbf{r}$ we can state that

$$dual(\tilde{\mathbf{u}} \cdot \mathbf{v}) = \mathbf{v} \otimes \mathbf{u} - \mathbf{u} \otimes \mathbf{v}$$

which implies that equation (3.4.26a) is true. Equation (3.4.26b) follows from equations (3.4.26a) by using equation (3.4.14). Equation (3.4.26c) follows from equation (3.4.26b) and the definition of the Lie bracket, i.e., equation (3.4.21). Finally, equation (3.4.26d) follows from equation (3.4.26a) and the definition of the outer product of two vectors i.e., equation (3.3.9). □

Proposition (3.9) is important because it allows us to derive relationships between the cross product of two vectors and the Lie bracket of two skew-symmetric tensors, as well as between the vector cross product and the outer and the tensor product of two vectors. Thus, for example, from equation (3.4.26c), it is obvious that

$$\mathbf{u} \times \mathbf{v} = dual[\tilde{\mathbf{u}}\ ,\tilde{\mathbf{v}}] \tag{3.4.29}$$

and from equation (3.4.26d) it follows that

$$\mathbf{u} \times \mathbf{v} = dual(\mathbf{v} \wedge \mathbf{u}) \tag{3.4.30}$$

Also, since for any second order Cartesian tensor $\mathbf{A}$ we have

$$vect(\mathbf{A}^T) = -vect(\mathbf{A}), \tag{3.4.31}$$

it is easy to see that from equations (3.4.26a) and (3.4.26b) we have

$$\mathbf{u} \times \mathbf{v} = -2vect(\mathbf{u} \otimes \mathbf{v}) \tag{3.4.32}$$

and

$$\mathbf{u} \times \mathbf{v} = 2vect(\tilde{\mathbf{u}}\,\tilde{\mathbf{v}}) \tag{3.4.33}$$

respectively.

Proposition 3.10: The dual tensor $dual(\tilde{\mathbf{w}}\,\tilde{\mathbf{u}}\,\mathbf{v})$ satisfies the following equations.

$$dual(\tilde{\mathbf{w}}\,\tilde{\mathbf{u}}\,\mathbf{v}) = \tilde{\mathbf{u}}(\mathbf{w} \cdot \mathbf{v}) - \tilde{\mathbf{v}}(\mathbf{w} \cdot \mathbf{u}) \tag{3.4.34}$$

Proof: Equation (3.4.34) follows from the equation

$$\tilde{\mathbf{w}}\tilde{\mathbf{u}}\mathbf{v} = \mathbf{w} \times (\mathbf{u} \times \mathbf{v})$$

$$= \mathbf{u}(\mathbf{w}\cdot\mathbf{v}) - \mathbf{v}(\mathbf{w}\cdot\mathbf{u})$$

and the linearity of the dual operator. □

Equation (3.4.34), for $\mathbf{w} = \mathbf{u}$, becomes $dual(\tilde{\mathbf{u}}\tilde{\mathbf{u}}\mathbf{v}) = \tilde{\mathbf{u}}(\mathbf{u}\cdot\mathbf{v}) - \tilde{\mathbf{v}}(\mathbf{u}\cdot\mathbf{u})$ which can be further manipulated to give

$$dual(\tilde{\mathbf{u}}\tilde{\mathbf{u}}\mathbf{v}) = \tilde{\mathbf{u}}(\mathbf{u}\cdot\mathbf{v}) + \tilde{\mathbf{v}}(\mathbf{u}\cdot\mathbf{u}) - 2\tilde{\mathbf{v}}(\mathbf{u}\cdot\mathbf{u})$$

$$= -[\tilde{\mathbf{v}}\tilde{\mathbf{u}}\tilde{\mathbf{u}} + \tilde{\mathbf{u}}\tilde{\mathbf{u}}\tilde{\mathbf{v}}] - 2\tilde{\mathbf{v}}(\mathbf{u}\cdot\mathbf{u}) \qquad \text{[by (3.4.18)]}$$

$$= -[\tilde{\mathbf{v}}\tilde{\mathbf{u}}\tilde{\mathbf{u}} + \tilde{\mathbf{u}}\tilde{\mathbf{u}}\tilde{\mathbf{v}}] + tr[\tilde{\mathbf{u}}\tilde{\mathbf{u}}]\tilde{\mathbf{v}} \qquad \text{[by (3.4.16)]} \qquad (3.4.35)$$

Now, based on equation (3.4.35), we can show the following:

Proposition 3.11: Let $\mathbf{I}$ be a symmetric tensor. Then the dual tensor $dual(\mathbf{I}\cdot\mathbf{v})$, where $\mathbf{v}$ is any vector, satisfies the following equation

$$dual(\mathbf{I}\cdot\mathbf{v}) = -[\mathbf{I}\tilde{\mathbf{v}} + \tilde{\mathbf{v}}\mathbf{I}] + tr[\mathbf{I}]\tilde{\mathbf{v}} \qquad (3.4.36)$$

Proof: Since $\mathbf{I}$ is a symmetric tensor, let

$$\mathbf{I} = \lambda_1\mathbf{x}\otimes\mathbf{x} + \lambda_2\mathbf{y}\otimes\mathbf{y} + \lambda_3\mathbf{z}\otimes\mathbf{z} \qquad (3.4.37)$$

be its spectral decomposition, where λ_i, $i = 1, 2, 3$, are the eigenvalues of $\mathbf{I}$ and $\mathbf{x}$, $\mathbf{y}$ and $\mathbf{z}$ are its normalized eigenvectors. Now, using equations (3.4.14), we can write the spectral decomposition of $\mathbf{I}$ in terms of the dual tensor of its eigenvectors, i.e., we can write

$$\mathbf{I} = \lambda_1\tilde{\mathbf{x}}\tilde{\mathbf{x}} + \lambda_2\tilde{\mathbf{y}}\tilde{\mathbf{y}} + \lambda_3\tilde{\mathbf{z}}\tilde{\mathbf{z}} + tr[\mathbf{I}]\mathbf{1} \qquad (3.4.38)$$

Then, the right dot product $\mathbf{I}\cdot\mathbf{v}$ becomes

$$\mathbf{I}\cdot\mathbf{v} = \lambda_1\tilde{\mathbf{x}}\tilde{\mathbf{x}}\cdot\mathbf{v} + \lambda_2\tilde{\mathbf{y}}\tilde{\mathbf{y}}\cdot\mathbf{v} + \lambda_3\tilde{\mathbf{z}}\tilde{\mathbf{z}}\cdot\mathbf{v} + tr[\mathbf{I}]\mathbf{v}$$

Now, based on the linearity of the dual operator and using equation (3.4.35), we can write

$$dual(\mathbf{I}\cdot\mathbf{v}) = -\left\{\tilde{\mathbf{v}}\,[\lambda_1\tilde{\mathbf{x}}\tilde{\mathbf{x}} + \lambda_2\tilde{\mathbf{y}}\tilde{\mathbf{y}} + \lambda_3\tilde{\mathbf{z}}\tilde{\mathbf{z}}] + [\lambda_1\tilde{\mathbf{x}}\tilde{\mathbf{x}} + \lambda_2\tilde{\mathbf{y}}\tilde{\mathbf{y}} + \lambda_3\tilde{\mathbf{z}}\tilde{\mathbf{z}}]\tilde{\mathbf{v}}\right\}$$

$$+ \left[\lambda_1 tr[\tilde{\mathbf{x}}\cdot\tilde{\mathbf{x}}] + \lambda_2 tr[\tilde{\mathbf{y}}\cdot\tilde{\mathbf{y}}] + \lambda_3 tr[\tilde{\mathbf{z}}\cdot\tilde{\mathbf{z}}]\right]\tilde{\mathbf{v}} + tr[\mathbf{I}]\tilde{\mathbf{v}}$$

Moreover, since for unit vectors we have

$$tr[\tilde{\mathbf{x}}\tilde{\mathbf{x}}] = tr[\tilde{\mathbf{y}}\tilde{\mathbf{y}}] = tr[\tilde{\mathbf{z}}\tilde{\mathbf{z}}] = -2,$$

we can write

$$dual(\mathbf{I}\cdot\mathbf{v}) = -\left\{\tilde{\mathbf{v}}\left[\lambda_1\tilde{\mathbf{x}}\tilde{\mathbf{x}} + \lambda_2\tilde{\mathbf{y}}\tilde{\mathbf{y}} + \lambda_3\tilde{\mathbf{z}}\tilde{\mathbf{z}} + tr[\mathbf{I}]\mathbf{1}\right]\right.$$

$$\left. + \left[\lambda_1\tilde{\mathbf{x}}\tilde{\mathbf{x}} + \lambda_2\tilde{\mathbf{y}}\tilde{\mathbf{y}} + \lambda_3\tilde{\mathbf{z}}\tilde{\mathbf{z}} + tr[\mathbf{I}]\mathbf{1}\right]\tilde{\mathbf{v}}\right\} + tr[\mathbf{I}]\tilde{\mathbf{v}}$$

from where using (3.4.38) we get (3.4.36). □

Proposition 3.12: Let **I** be a symmetric tensor, and let $\tilde{\mathbf{u}}$ and $\tilde{\mathbf{v}}$ be skew-symmetric tensors. Then, the following equation holds:

$$\mathbf{u}\cdot\mathbf{I}\cdot\mathbf{v} = -\,tr[\tilde{\mathbf{u}}\,\mathbf{J}\,\tilde{\mathbf{v}}] \qquad (3.4.39)$$

where $\mathbf{J} = -\mathbf{I} + \frac{1}{2}tr[\mathbf{I}]\mathbf{1}$.

Proof: Using equation (3.4.16), we can write the left-hand side of equation (3.4.39) as follows

$$\mathbf{u}\cdot\mathbf{I}\cdot\mathbf{v} = -\frac{1}{2}tr\left[\tilde{\mathbf{u}}\cdot dual(\mathbf{I}\cdot\mathbf{v})\right]$$

$$= -\frac{1}{2}tr\left[\tilde{\mathbf{u}}\left\{-\mathbf{I}\tilde{\mathbf{v}} - \tilde{\mathbf{v}}\mathbf{I} + tr[\mathbf{I}]\tilde{\mathbf{v}}\right\}\right]$$

$$= -\frac{1}{2}tr\left[-\tilde{\mathbf{u}}\mathbf{I}\tilde{\mathbf{v}} - \tilde{\mathbf{u}}\tilde{\mathbf{v}}\mathbf{I} + tr[\mathbf{I}]\tilde{\mathbf{u}}\tilde{\mathbf{v}}\right]$$

Moreover, since $tr[\mathbf{S}] = tr[\mathbf{S}^T]$ and $tr[\mathbf{S}\,\mathbf{T}] = tr[\mathbf{T}\,\mathbf{S}]$ for any tensors $\mathbf{S}$ and $\mathbf{T}$, we have

$$\begin{aligned} tr[\tilde{\mathbf{u}}(\tilde{\mathbf{v}}\mathbf{I})] &= tr[(\tilde{\mathbf{v}}\mathbf{I})\tilde{\mathbf{u}}] \\ &= tr[(\tilde{\mathbf{v}}\mathbf{I}\tilde{\mathbf{u}})^T]. \\ &= tr[\tilde{\mathbf{u}}\mathbf{I}\tilde{\mathbf{v}}]. \end{aligned}$$

Therefore, we can write

$$\begin{aligned} \mathbf{u}\cdot\mathbf{I}\cdot\mathbf{v} &= -\,tr\left[-\,\tilde{\mathbf{u}}\mathbf{I}\tilde{\mathbf{v}} + \frac{1}{2}tr[\mathbf{I}]\tilde{\mathbf{u}}\tilde{\mathbf{v}}\right] \\ &= -\,tr\left[\tilde{\mathbf{u}}\left(-\mathbf{I} + \frac{1}{2}tr[\mathbf{I}]\mathbf{1}\right)\tilde{\mathbf{v}}\right] \\ &= -\,tr[\tilde{\mathbf{u}}\mathbf{J}\tilde{\mathbf{v}}] \end{aligned}$$

where $\mathbf{J} = -\mathbf{I} + \frac{1}{2}tr[\mathbf{I}]\mathbf{1}$, completing the proof. □

With these tensor-vector identities at our disposal, we proceed in the next chapter to rewrite the classical vector equations of the Newtonian rigid body dynamics in a Cartesian tensor formulation. Moreover, in the next chapter we shall show that the Cartesian tensor formulation of the equations of motion can be implemented in a computationally efficient manner.

3.5 References

[1] G. Ricci and T. Levi-Civita, *Methodes de calcul differentiel absolu et leurs applications*, (Paris, 1923). (Reprinted from *Mathematische Annalen*, tome 54, 1900).

[2] D. Lovelock and H. Rund, *Tensors, Differential Forms and Variational Principles*, John Wiley & Sons, New York, 1965.

[3] R. L. Bishop and S. I. Goldberg *Tensor Analysis on Manifolds*, The Macmillan Company, New York, 1968.

[4] E. A. Lord, *Tensors, Relativity and Cosmology*, Tata McGraw-Hill Publishing Co. Ltd., New Delhi, 1976.

[5] I. S. Sokolnikoff, *Tensor Analysis: Theory and Applications to Geometry and Mechanics of Continua*, John Wiley & Sons, New York, 1965.

[6] S. F. Borg, *Matrix-Tensor Methods in Continuum Mechanics*, D. Van Nostrand Company, Princeton, New Jersey, 1963.

[7] P. Chadwick, *Continuum Mechanics: Concise Theory and Problems*, George Allen & Unwin Ltd., London, 1976.

[8] C. Truesdell, *A First Course in Rational Continuum Mechanics*, Vol. 1, Academic Press, New York, 1977.

[9] W. M. Lai, D. Rubin and E. Krempl, *Introduction to Continuum Mechanics*, Pergamon Press, New York, 1978.

[10] M. E. Gurtin, *An Introduction to Continuum Mechanics*, Academic Press, New York, 1981.

[11] A. J. McConnell, *Applications of Tensor Analysis*, Dover Publications, New York, 1947.

[12] J. L. Synge and A. Schild, *Tensor Calculus*, University of Toronto Press, Toronto, 1949.

[13] M. S. Smith, *Principles & Applications of Tensor Analysis*, Howard W. Sams & Co., New York, 1963.

[14] J. L. Mercier, *An Introduction to Tensor Calculus*, Wolters-Noordhoff Publishing, Groningen, 1971.

[15] J. G. Simmonds, *A Brief on Tensor Analysis*, Springer-Verlag, New York, 1982.

[16] L. Brillouin, *Tensors in Mechanics and Elasticity*, Academic Press, New York, 1964.

[17] W. Gibbs *Vector Analysis*, Dover Publications, New York, 1960.

[18] L. Brand, *Vector and Tensor Analysis*, John Wiley & Sons, New York, 1947.

[19] W. G. Bickley and R. E. Gibson, *Via Vector to Tensor*, The English Universities Press, London, 1962.

[20] A. I. Borisenko and I. E. Tarapov, *Vector and Tensor Analysis With Applications*, Prentice-Hall, Englewood Cliffs, New Jersey, 1968.

[21] D. E. Bourne and P. C. Kendall, *Vector Analysis and Cartesian Tensors*, Thomas Nelson & Sons, England, 1977.

[22] A. Lichnerowicz, *Elements of Tensor Calculus*, Methuen, London, 1962.

[23] G. Temple, *Cartesian Tensors: An Introduction*, John Wiley & Sons, New York, 1960.

[24] H. Jeffreys, *Cartesian Tensors*, Cambridge University Press, Cambridge, 1961.

[25] A. M. Goodbody, *Cartesian Tensors: With Applications to Mechanics, Fluid Mechanics and Elasticity*, Ellis Horwood, England, 1982.

[26] I. J. Wittenburg, *Dynamics of Systems of Rigid Bodies*, B. G. Teubner, Stuttgart, 1977.

[27] J. Angeles, *Rational Kinematics*, Springer-Verlag, New York, 1988.

[28] A. Graham, *Kronecker Products and Matrix Calculus with Applications*, Ellis Horwood. London, 1981.

[29] R. Gilmore, *Lie Groups, Lie Algebras, and Some of Their Applications*, John Wiley & Sons, New York, 1974.

[30] G. W. Stewart, *Introduction to Matrix Computations*, Academic Press, New York, 1973.

[31] D. Hestenes, *New Foundations of Classical Mechanics*, D. Reidel Publishing Company, Dordrecht, Holland, 1986.

Chapter 4

Cartesian Tensors and Rigid Body Motion

4.1 Introduction

As we have mentioned in Chapter 1, the representation of the physical quantities involved in the formulation of the equations of motion of a rigid body system determines the kind of mathematical analysis that will be used in deriving these equations. In the classical Newtonian formulation of rigid body dynamics, vectors are usually used [1-5] to represent basic physical quantities and therefore vector analysis is used for deriving the equations of rigid body motion. Vector analysis is usually imposed on classical Newtonian dynamics by the consideration that *angular rates* (i.e., linearly independent rates of change of a rigid body orientation) constitute the components of a vector quantity, the *angular velocity vector*. This consideration also assigns a vector character to other physical quantities which are defined in terms of the angular velocity vector such as *angular acceleration, angular momentum* and *external torque*. However, as is well known [9], basic physical quantities in rigid body motion such as angular velocity, angular acceleration, angular or rotational momentum and resultant torque can be described by using second order Cartesian tensors. Therefore, based on the Cartesian tensor representation of these quantities, we can use Cartesian tensor analysis for the study of rigid body motion.

Applications of Cartesian tensor analysis to rigid body dynamics can be usually found within the framework of continuum mechanics. This is natural since, as is well known [8-12], tensor analysis has been universally adopted as the mathematical system upon which continuum mechanics is built. Thus, within the framework of continuum mechanics, an elegant theoretical treatment of rigid body motion based on Cartesian tensor analysis has been given by Truesdell [9]. Following an abstract axiomatic approach in his analysis, Truesdell treated all physical quantities involved in rigid body dynamics as second order Cartesian tensors rather than vectors. Without considering 3-dimensional vector analysis and based only on the tensor representation of the quantities involved in rigid body dynamics, Truesdell derived a very simple Cartesian tensor formulation for the equations of rigid body motion. Therefore, as shown by Truesdell, Cartesian tensor analysis alone suffices for a comprehensive study of rigid body dynamics.

However, in continuum mechanics most of the time attention is centered on rigorous theoretical foundations and concise formulations for the equations of motion. Thus, important aspects concerning the various formulations of the equations of motion (such as their computational efficiency) are often overlooked and not properly addressed. This is in particular true in [9] where, since Truesdell is not concerned with applications, he does not provide any computational complexity analysis for the tensor formulation of the equations of rigid body motion. Thus, although this tensor formulation for the equations of rigid body motion is conceptually simple and, as we shall see later, has computational advantages over the classical vector formulation, to our knowleage, it has not been used in practical applications where the vector formulation is still much more popular.

In this chapter, we shall not be concerned with the rigorous theoretical foundations of the tensor formulation of the equations of rigid body motion. This, has been done already by Truesdell in [9]. Here, we shall consider some practical aspects of these equations. In particular, first, starting from the classical vector description for the equations of rigid body motion and using the tensor-vector identities developed in Chapter 3, we shall provide an equivalent tensor formulation for these vector equations. This tensor formulation is obviously similar to that used by Truesdell [9]. However, the approach is different and, we believe, easier to follow for students or engineers who are familiar with the classical vector formulation of the equations of rigid body motion. Then, we shall examine the important issue (in practical applications) of computational complexity in implementing the equations of rigid body motion. In particular, we shall examine the computational complexity of implementing the vector and the tensor formulations of these equations. It will then became clear that the tensor formulation of these equations can be implemented far more efficiently. Therefore, in practical applications (e.g., in robotics) where computational efficiency is of importance, the tensor formulation of the equations of rigid body motion is to be preferred over the classical vector formulation.

The outline of this chapter is as follows: Section 4.2 deals with kinematic aspects of rigid body motion. In particular, we show that by using the angular velocity and angular acceleration tensors, the velocity and acceleration of any point on the rigid body can by computed very easily. Section 4.3 deals with dynamic aspects of rigid body motion. In particular, an analysis of the rigid body inertia tensor is given and the vector formulation of rigid body motion is reviewed. From this, the angular momentum and torque tensors surface naturally and they lead to a tensor formulation for Euler's equation of rotational rigid body motion.

4.2 On Kinematic Analysis of Rigid Body Motion

Kinematic analysis of rigid body motion deals with motion without regard to forces or moments that cause that motion. In particular, kinematic analysis of rigid body motion is concerned with the body's *configuration* and *motion* analysis. Configuration kinematic analysis deals with possible descriptions of the rigid body spatial configuration as a function of time, and motion kinematic analysis deals with the first and second time derivatives of these configuration functions. We dealt with configuration kinematic analysis in Chapter II. In this section, we shall be concerned mainly with the kinematic analysis of rigid body rotational motion.

As we have outlined in section 2.2, the spatial configuration of a rigid body, relative to a Cartesian orthogonal coordinate system, is defined by considering a position vector and a rotation tensor which defines the orientation of the rigid body. Moreover, as is well known (Theorem 2.2) a general rigid body motion can be considered as the superposition of a purely translational and a purely rotational rigid body motion. Therefore, the decomposition of the rigid body spatial configuration, as outlined above, implies that the position vector and its time derivatives describe purely translational rigid body motion and, similarly, the rotation tensor and its time derivatives describe purely rotational rigid body motion about a fixed point. The latter follows also from the fact that a rotation tensor can be used to describe a finite displacement about a fixed point (see Chapter II), which in the 3-D physical space is equivalent to finite rigid body displacement about a fixed axis. Hence, a rotation tensor in the 3-D physical space, when it is considered to be a continuous function of time, and its time derivatives are sufficient to study pure rotational rigid body motion about a fixed point.

Therefore, for the kinematic analysis of a rigid body pure rotational motion we need only to consider a rotation tensor $\mathbf{R} \equiv \mathbf{R}(t)$, which defines the rigid body orientation as a function of time, and its first and second time derivatives. For the absolute time derivatives (i.e., the time derivatives relative to an inertia frame) of **R** we shall use the classical Newtonian notation, i.e., we shall write $\frac{d\mathbf{R}}{dt} \equiv \dot{\mathbf{R}}$ and $\frac{d^2\mathbf{R}}{dt^2} \equiv \ddot{\mathbf{R}}$.

4.2.1 The Angular Velocity Tensor

Perhaps the easiest way to introduce the angular velocity tensor is by using a corollary of the following theorem [13].

Theorem 4.1 : Any differentiable orthogonal† tensor $\mathbf{Q} \equiv \mathbf{Q}(t)$ satisfies the following first order differential equation

$$\dot{\mathbf{Q}} = \mathbf{\Phi}\mathbf{Q} \tag{4.2.1}$$

where, $\mathbf{\Phi}$ is a second order skew-symmetric tensor.

Proof : Since for any orthogonal tensor we have $\mathbf{Q}^T\mathbf{Q} = \mathbf{1}$, we can write

$$\dot{\mathbf{Q}} = \dot{\mathbf{Q}}\mathbf{Q}^T\mathbf{Q}. \tag{4.2.2}$$

Now, let

$$\mathbf{\Phi} = \dot{\mathbf{Q}}\mathbf{Q}^T. \tag{4.2.3}$$

Then, (4.2.2) can be written as (4.2.1). Therefore, **Q** satisfies a first order differential equation. Now, we shall show that $\mathbf{\Phi}$, as defined by (4.2.3), is skew-symmetric. It follows from the orthogonality of **Q** that

$$\dot{\mathbf{Q}}\mathbf{Q}^T + \mathbf{Q}\dot{\mathbf{Q}}^T = 0$$

or

$$\mathbf{\Phi} + \mathbf{\Phi}^T = 0$$

i.e.,

---

† A tensor **Q**, defined in an inner-product space, is *orthogonal* if it preserves the inner product, i.e., if $\mathbf{Q}(\mathbf{u})\cdot\mathbf{Q}(\mathbf{v}) = \mathbf{u}\cdot\mathbf{v}$ for all vector **u** and **v**.

$$\mathbf{\Phi} = -\mathbf{\Phi}^T$$

and this completes the proof. □

Now, since a rotation tensor is an orthogonal tensor, it satisfies Theorem 4.1 and we can state the following corollary.

Corollary 4.1 : A rotation tensor **R** satisfies a first order differential equation given by

$$\dot{\mathbf{R}} = \mathbf{\Phi}\mathbf{R} \tag{4.2.4}$$

where, $\mathbf{\Phi}$ is a second order skew-symmetric tensor.

From this Corollary we can derive the definition of the angular velocity tensor as follows: When the rotation tensor **R** is defined in a 3-D Euclidean space, it describes the orientation of a rigid body. In this case, we denote the skew-symmetric tensor $\mathbf{\Phi}$ by $\tilde{\boldsymbol{\omega}}$, i.e., we write equation (4.2.4) as

$$\dot{\mathbf{R}} = \tilde{\boldsymbol{\omega}}\mathbf{R}. \tag{4.2.5}$$

and refer to $\tilde{\boldsymbol{\omega}}$ as the *angular velocity tensor* (or *spin tensor* [9]). Equation (4.2.5), which is sometimes referred [9] to as *Poisson's equation*, can also be written in the form

$$\tilde{\boldsymbol{\omega}} = \dot{\mathbf{R}}\mathbf{R}^T \tag{4.2.6}$$

and can be used as the definition of the angular velocity tensor. Moreover, since the skew-symmetric tensor $\tilde{\boldsymbol{\omega}}$ is defined in a 3-D Euclidean space, it has a unique dual (or axial) vector The dual vector $\boldsymbol{\omega}$, of $\tilde{\boldsymbol{\omega}}$, is the familiar *angular velocity vector*. Therefore, equation (4.2.6) and the dual operator provide a simple definition for the angular velocity vector.

A tensor representation for angular velocity has many advantages over the classical vector representation (which, however, may claim superiority over a tensor representation when pictures are to be drawn). First of all, from a practical point of view, a tensor treatment for angular rates allows us to relate the angular velocity to the derivative of another tensor quantity, namely, the derivative of the orientation tensor $\mathbf{R}$ as expressed by equation (4.2.5). This is not possible when we describe the angular rates with the angular velocity vector $\boldsymbol{\omega}$, since it is well known [2,3] that there is no vector quantity whose derivative is related to the angular velocity vector. Also from a theoretical point of view, the tensor representation for angular rates is to be preferred as is obvious from the work of Truesdell on continuum mechanics [9]. Finally, the angular velocity tensor is not restricted to a 3-D Euclidean space. As we can see, by generalizing equation (4.2.5) from a 3-D to an n-D Euclidean space, the tensor $\tilde{\boldsymbol{\omega}}$ becomes $\boldsymbol{\Phi}$. Therefore, by analogy, we can refer to the skew-symmetric tensor $\boldsymbol{\Phi}$ as the angular velocity tensor in an n-D Euclidean space. From the foregoing, the angular velocity tensor $\boldsymbol{\Phi}$ is well defined in any n-D Euclidean space. This unfortunately is not true for the angular velocity vector which exists only in a 3-D Euclidean space.

4.2.2 The Angular Acceleration Tensor

Let us consider now the second time derivative $\ddot{\mathbf{R}}$, relative to an inertial frame, of a rotation tensor $\mathbf{R}$. The functional relationship between $\ddot{\mathbf{R}}$ and $\mathbf{R}$ follows from the following theorem.

Theorem 4.2 : Any differentiable orthogonal tensor $\mathbf{Q} \equiv \mathbf{Q}(t)$ satisfies the following second order differential equation

$$\ddot{\mathbf{Q}} = \boldsymbol{\Psi}\mathbf{Q} \tag{4.2.7}$$

where, $\boldsymbol{\Psi}$ is a second order tensor defined by

$$\boldsymbol{\Psi} = \dot{\boldsymbol{\Phi}} + \boldsymbol{\Phi}^2, \tag{4.2.8}$$

with $\boldsymbol{\Phi}$ the angular velocity tensor defined by (4.2.3) and $\boldsymbol{\Phi}^2 = \boldsymbol{\Phi}\cdot\boldsymbol{\Phi} \equiv \boldsymbol{\Phi}\boldsymbol{\Phi}$.

Proof : Differentiation of (4.2.1) gives

$$\ddot{\mathbf{Q}} = \dot{\boldsymbol{\Phi}}\mathbf{Q} + \boldsymbol{\Phi}\dot{\mathbf{Q}}$$

and substituting for $\dot{\mathbf{Q}}$ we get

$$\ddot{\mathbf{Q}} = (\dot{\mathbf{\Phi}} + \mathbf{\Phi\Phi})\mathbf{Q}$$
$$= \mathbf{\Psi Q}$$

where $\mathbf{\Psi} = \dot{\mathbf{\Phi}} + \mathbf{\Phi}^2$. This complete the proof. □

Corollary 4.2 : A second order rotation tensor **R** satisfies the following second order differential equation

$$\ddot{\mathbf{R}} = \mathbf{\Psi R} \tag{4.2.9}$$

where $\mathbf{\Psi}$ is a second order tensor defined by

$$\mathbf{\Psi} = \dot{\mathbf{\Phi}} + \mathbf{\Phi}^2, \tag{4.2.10}$$

with $\mathbf{\Phi}$ the angular velocity tensor which corresponds to **R** and $\mathbf{\Phi}^2 = \mathbf{\Phi\Phi}$.

Theorem 4.2 and its Corollary are obviously valid for orthogonal tensors defined in an n-D Euclidean space, but here we shall be concerned only with the particular case where $n = 3$. In this case, we introduce the notation $\mathbf{\Omega}$ to denote the tensor $\mathbf{\Psi}$ and write equation (4.2.9) as

$$\ddot{\mathbf{R}} = \mathbf{\Omega R}. \tag{4.2.11}$$

Also, using the corresponding notation for the tensor $\mathbf{\Phi}$, we write equation (4.2.10) as

$$\mathbf{\Omega} = \dot{\tilde{\boldsymbol{\omega}}} + \tilde{\boldsymbol{\omega}}\tilde{\boldsymbol{\omega}} \tag{4.2.12}$$

and we refer to the tensor $\mathbf{\Omega}$ as the *angular acceleration tensor*. Obviously, using equation (4.2.9), we can also define the angular acceleration tensor $\mathbf{\Omega}$ by the equation

$$\mathbf{\Omega} = \ddot{\mathbf{R}}\mathbf{R}^T \tag{4.2.13}$$

where **R** is a rotation tensor in a 3-D Euclidean space.

The tensor $\mathbf{\Omega}$ is neither symmetric nor skew-symmetric. Actually, since $\dot{\tilde{\boldsymbol{\omega}}}$ is skew-symmetric and $\tilde{\boldsymbol{\omega}}\tilde{\boldsymbol{\omega}}$ is symmetric, equation (4.2.12) represents the *Cartesian decomposition* of $\mathbf{\Omega}$. Moreover, since the tensor $\tilde{\boldsymbol{\omega}}\tilde{\boldsymbol{\omega}}$ is symmetric we have from equations (3.3.28) and (3.3.30) that

$$vect\,(\mathbf{\Omega}) = vect\,(\dot{\tilde{\boldsymbol{\omega}}}) \equiv dual\,(\dot{\tilde{\boldsymbol{\omega}}}) = \dot{\boldsymbol{\omega}} \tag{4.2.14}$$

i.e., the vector invariant of the angular acceleration tensor $\mathbf{\Omega}$ is the familiar *angular acceleration vector*. This obviously justifies the name given to the tensor $\mathbf{\Omega}$.

As is obvious from the extensive literature on rigid bodies, the angular acceleration tensor is rarely used in the kinematic and dynamic analysis of rigid body motion (a few exceptions may be found in [9,16,17,21]). This is probably due to the fact that a vector approach to analyzing rigid body motion fails to establish a clear relationship between vector invariants and tensors other than skew-symmetric tensors. In the case of the angular velocity vector and angular velocity tensor, a 1–1 relationship between them is obvious, since the angular velocity tensor is skew-symmetric. Thus, in this case, one representation is the dual of the other. A similar dual relationship between the angular acceleration vector and angular acceleration tensor does not exist because the angular acceleration tensor is not skew-symmetric. The dual tensor of the angular acceleration vector (which is known as *Euler's acceleration* [9]) defines only the skew-symmetric part of the angular acceleration tensor and not the whole tensor. As we can see from equation (4.2.12), to define the angular acceleration tensor we need to use not only the angular acceleration vector but also the angular velocity vector. Therefore, the transition from vector analysis to tensor analysis is not straightforward.

We shall conclude this section with applications of the angular velocity and angular acceleration tensors in the computation of linear velocity and acceleration of points on a moving rigid body.

4.2.3 Linear Velocity and Acceleration in Rigid Body Motion

The concepts of angular velocity and angular acceleration tensors provide powerful tools for describing the motion of a rigid body since they enable us to derive equations with very simple structure. To see this, let us consider some arbitrary vector involved in a mechanical problem, such as the position vector **r** of a point on the rigid body. Usually such a vector will vary in time as the body moves. Therefore, it is important that its linear velocity and acceleration relative to an inertial coordinate system be determined in a computationally efficient manner. To solve problems of this type, we proceed as follows.

As is usually the practice in rigid body motion, we consider two coordinate systems: an inertial coordinate system and a body coordinate system, which we denote by { **e** } and { **e**′ } respectively. The body coordinate system is rigidly attached to the rigid body and so moves with it. Suppose now that a point **p** on the body has a position vector $\mathbf{r}_1$ relative to the origin **o** of the inertial coordinate system and a position vector $\mathbf{r}_2$ relative to the origin **o**′ of the moving coordinate system. Also, let **s** be the vector from **o** to **o**′. Then, as we can see in Figure 4.1 the three vectors are related by the equation

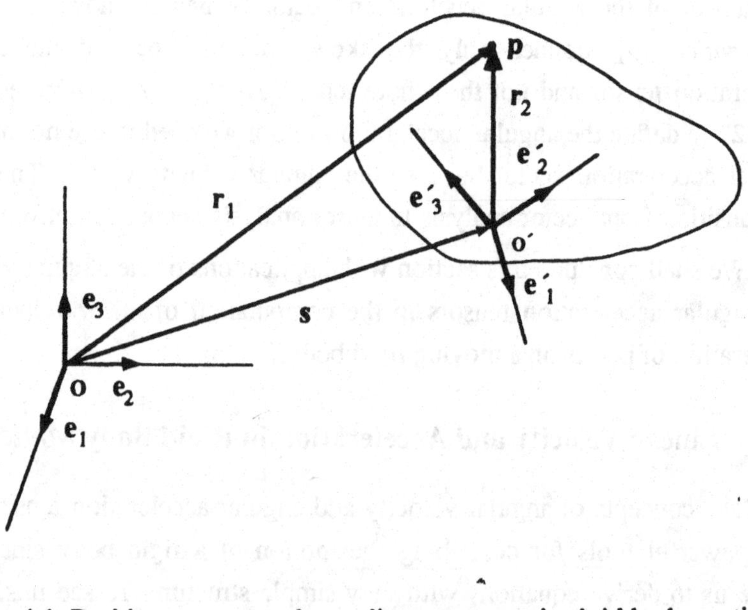

Figure 4.1: Position vectors and coordinate systems in rigid body motion.

$$\mathbf{r}_1 = \mathbf{s} + \mathbf{r}_2. \tag{4.2.15}$$

Moreover, let $\mathbf{R}$ be the rotation tensor which specifies the orientation of the moving coordinate system relative to the inertial one. Then, if $\mathbf{r}'_2$ denotes the position vector of the point $\mathbf{p}$ relative to $\mathbf{o}'$ when it is expressed in the moving coordinate system, premultiplication by the rotation tensor $\mathbf{R}$ expresses it relative to the inertial coordinate system, i.e., we can write

$$\mathbf{r}_2 = \mathbf{R}\mathbf{r}'_2 \tag{4.2.16}$$

Therefore, equation (4.2.15) can also be written as

$$\mathbf{r}_1 = \mathbf{s} + \mathbf{R}\mathbf{r}'_2 \tag{4.2.17}$$

Now, as is well known, the vector of the absolute linear velocity of the point $\mathbf{p}$ is defined to be the first time derivative of its position vector relative to the inertial coordinate system. In other words the absolute linear velocity of the point $\mathbf{p}$ is given by the vector $\dot{\mathbf{r}}_1$. To compute this derivative we shall use equation (4.2.17). Thus, we have

$$\dot{\mathbf{r}}_1 = \dot{\mathbf{s}} + \dot{\mathbf{R}}\mathbf{r}'_2, \tag{4.2.18}$$

since the vector $\mathbf{r}'_2$ is time independent relative to the body coordinate system. Now, substituting for $\dot{\mathbf{R}}$ from equation (4.2.5) and using equation (4.2.16), we finally have for the velocity of the point $\mathbf{p}$

$$\dot{\mathbf{r}}_1 = \dot{\mathbf{s}} + \tilde{\boldsymbol{\omega}}\mathbf{r}_2 \tag{4.2.19}$$

where $\tilde{\boldsymbol{\omega}}$ is the angular velocity tensor of the moving body. Also, the second time derivative of equation (4.2.17) defines the vector of the absolute linear acceleration of the point $\mathbf{p}$ relative to the inertial coordinate system. Therefore, for the absolute linear acceleration of the point $\mathbf{p}$ we have

$$\ddot{\mathbf{r}}_1 = \ddot{\mathbf{s}} + \ddot{\mathbf{R}}\mathbf{r}'_2$$

which can be simplified to

$$\ddot{\mathbf{r}}_1 = \ddot{\mathbf{s}} + \boldsymbol{\Omega}\mathbf{r}_2 \tag{4.2.20}$$

where $\mathbf{\Omega}$ is the angular acceleration tensor of the moving body. Moreover, as we can see from equation (4.2.16), the time derivatives for vectors which are constant relative to the body coordinate system are computed by using the simple equations

$$\dot{\mathbf{r}}_2 = \tilde{\boldsymbol{\omega}}\mathbf{r}_2 \tag{4.2.21}$$

and

$$\ddot{\mathbf{r}}_2 = \mathbf{\Omega}\mathbf{r}_2 \tag{4.2.22}$$

Let us recall at this point, that in the classical vector description of rigid body dynamics, the vectors of the linear velocity and acceleration of the same point $\mathbf{p}$ are computed from the following vector equations

$$\dot{\mathbf{r}}_1 = \dot{\mathbf{s}} + \boldsymbol{\omega} \times \mathbf{r}_2 \tag{4.2.23}$$

and

$$\ddot{\mathbf{r}}_1 = \ddot{\mathbf{s}} + \dot{\boldsymbol{\omega}} \times \mathbf{r}_2 + \boldsymbol{\omega} \times (\boldsymbol{\omega} \times \mathbf{r}_2) \tag{4.2.24}$$

respectively. Obviously, equations (4.2.19) and (4.2.20) are equivalent to equations (4.2.23) and (4.2.24) respectively. However, as we can see, the introduction of the angular velocity and angular acceleration tensors enables us to derive a simple and compact representation for the linear velocity and acceleration of points on a moving rigid body. In particular, equations (4.2.20) and (4.2.22) enable us to manipulate very effectively, equations involving the linear acceleration of various position vectors on the same rigid body. From the foregoing, we see that the introduction of the angular velocity and angular acceleration tensors provides a more efficient means for the analysis of motion kinematics. Moreover, as we shall see in the following section, the angular velocity and angular acceleration tensors can be used to simplify motion dynamics as well.

4.3 On Dynamic Analysis of Rigid Body Motion

In the dynamic analysis of motion, we deal with relationships between the motion of a body and the forces and/or torques which cause or result from that motion. As is well known from classical dynamics [1-5], a number of schemes have been developed over the years for the dynamic analysis of rigid body motion such as those based on the equations of *d'Alembert, Newton-Euler, Euler-Lagrange* and *Hamilton*. In this section we shall be concerned with the Newtonian formulation of the equations of rigid body motion.

As we have mentioned before, a general motion of a rigid body can be considered as resulting from the superposition of two independent motions: a pure translational motion of a point (usually its center of mass) and a pure rotational motion about that point. The Newton-Euler procedure uses exactly this decomposition. In particular, in the Newtonian formulation of the equations of rigid body motion, the translational motion is described by Newton's equation (or Newton's second law) which symbolically is stated as follows

$$\mathbf{F}_c = m\ddot{\mathbf{r}}_c \tag{4.3.1}$$

where $\mathbf{F}_c$ is the total external (or resultant) force acting on the rigid body, m is the mass of the rigid body and $\ddot{\mathbf{r}}_c$ is the absolute linear acceleration of its center of mass. The rotational motion is described by Euler's equation which is symbolically stated as

$$\mathbf{M}_c = \mathbf{I}_c \cdot \dot{\boldsymbol{\omega}} + \boldsymbol{\omega} \times \mathbf{I}_c \cdot \boldsymbol{\omega} \tag{4.3.2}$$

where $\mathbf{M}_c$ is the total external (or resultant) torque about the center of mass, $\mathbf{I}_c$ is the inertia tensor of the body about its center of mass and $\boldsymbol{\omega}$ ($\dot{\boldsymbol{\omega}}$) is the vector of angular velocity (acceleration) of the body.

Equations (4.3.1) and (4.3.2) are the fundamental equations which describe the rigid body motion in the classical Newtonian formulation. As we can see, these two vector equations provide six differential scalar equations which, when the external force and torque (with appropriate initial conditions) are given, can be solved to determine the six degrees of freedom of a rigid body in three dimensional physical space, i.e., the position of its center of mass and its orientation.

It is obvious from equations (4.3.1) and (4.3.2) that the dynamic analysis of rigid body motion in the classical Newtonian formulation, for both translational and rotational rigid body motion, is based on vector analysis. In this section, as an alternative to vector analysis, we shall use Cartesian tensors to analyze the rotational rigid body motion. Since, in rigid body rotational motion, the inertia tensor of the body plays an important role, we first review some relevant facts about the rigid body inertia tensor.

4.3.1 The Rigid Body Inertia Tensor

As is well known (e.g., see [1]), the *inertia tensor* of a rigid body characterizes the mass distribution of the body relative to a point, and is usually defined by the equation

$$\mathbf{I}_o = \int_m (\mathbf{r} \cdot \mathbf{r}\, \mathbf{1} - \mathbf{r} \otimes \mathbf{r})\, dm \tag{4.3.3}$$

where $\mathbf{o}$ denotes a point of the body and $\mathbf{r}$ denotes the position vector of a point mass relative to the point $\mathbf{o}$.

The rigid body inertia tensor, as defined by equation (4.3.3), is used extensively in the dynamic analysis of rigid body motion. Actually, it is often the only definition provided for the inertia tensor, especially when a vector treatment of the Newtonian dynamic analysis of rigid body motion is used. In order to treat the dynamic analysis of rigid body motion based on Cartesian tensors, this definition of the inertia tensor needs to be modified. As we shall see later, the proper definition for a Cartesian tensor formulation of rigid body rotational dynamics is provided by the equation

$$\mathbf{J}_o = \int_m \mathbf{r} \otimes \mathbf{r}\, dm \tag{4.3.4}$$

where o is a point on the rigid body and $\mathbf{r}$ is the position vector of a point mass relative to point o. We shall refer to the inertia tensor $\mathbf{J}_o$ as the *Euler tensor* of a rigid body [9].

Obviously, the two inertia tensors $\mathbf{I}_o$ and $\mathbf{J}_o$ describe the same physical property of a rigid body, and thus they have to be equivalent. To see this, we proceed as follows [24]. First we note that equation (4.3.3) can be written in the form

$$\mathbf{I}_o = -\int_m \tilde{\mathbf{r}}\tilde{\mathbf{r}}\, dm \tag{4.3.5}$$

if one uses the tensor equation (3.4.10). Then, starting from equation (4.3.4) and using the tensor equation (3.4.14) we can write :

$$\begin{aligned}
\mathbf{J}_o &= \int_m (\tilde{\mathbf{r}}\tilde{\mathbf{r}} + \mathbf{r}\cdot\mathbf{r}\,\mathbf{1})dm \\
&= \int_m \tilde{\mathbf{r}}\tilde{\mathbf{r}}\, dm + \mathbf{1}\int_m -\frac{1}{2} tr[\tilde{\mathbf{r}}\tilde{\mathbf{r}}]dm \qquad \text{[by (3.4.16)]} \\
&= \int_m \tilde{\mathbf{r}}\tilde{\mathbf{r}}\, dm + \frac{1}{2} tr[-\int_m \tilde{\mathbf{r}}\tilde{\mathbf{r}}\, dm]\mathbf{1} \\
&= -\mathbf{I}_o + \frac{1}{2} tr[\mathbf{I}_o]\mathbf{1} \qquad \text{[by (4.3.5)]}
\end{aligned}$$

i.e., we have

$$\mathbf{J}_o = \frac{1}{2} tr[\mathbf{I}_o]\mathbf{1} - \mathbf{I}_o \tag{4.3.6}$$

Therefore, equation (4.3.6) provides the equivalence relationship between the two tensors, $\mathbf{J}_o$ and $\mathbf{I}_o$. Similarly, it can be shown that the equation

$$\mathbf{I}_o = tr[\mathbf{J}_o]\mathbf{1} - \mathbf{J}_o, \tag{4.3.7}$$

is also valid.

Now, as is often the case with most mathematical definitions, in practical applications we cannot use these definitions for the computation of the inertia tensor. In practice, the inertia tensor of a rigid body is computed experimentally. Moreover, even in experimental measurements, the direct computation of the rigid body inertia tensor about any point o other than the center of mass, is in general very difficult. Therefore, the body center of mass c is used when the inertia tensor of a body is evaluated. Then, in applications where the inertia tensor relative to points other than the center of mass is required and the inertia tensor about the center of mass is known, the *parallel axis theorem* is used. The parallel axis theorem for the inertia tensor which is defined by equation (4.3.3) is usually stated in the following form

$$\mathbf{I}_o = \mathbf{I}_c + m[\mathbf{r}_c \cdot \mathbf{r}_c \mathbf{1} - \mathbf{r}_c \otimes \mathbf{r}_c] \tag{4.3.8}$$

where $\mathbf{I}_c$ is the rigid body inertia tensor about its center of mass, $\mathbf{r}_c$ is the position vector of the center of mass relative to point o, and m is the mass of the body. Equation (4.3.8) can also be written in a compact tensor form as follows:

$$\mathbf{I}_o = \mathbf{I}_c - m\tilde{\mathbf{r}}_c \tilde{\mathbf{r}}_c \tag{4.3.9}$$

where equation (4.3.9) is derived from (4.3.8) using equation (3.4.10).

The parallel axis theorem for the inertia tensor $\mathbf{I}_o$ is a basic theorem in rigid body dynamics and its proof can be found in any book on classical dynamics (e.g. [1]). Obviously, the parallel axis theorem is also valid for the Euler tensor $\mathbf{J}_o$, which is defined by (4.3.5). Since the application of the parallel axis theorem for the tensor $\mathbf{J}_o$ is not well known, we provide here a formulation and a proof for it.

Theorem 4.3 (*parallel axis theorem*) : When the Euler tensor $\mathbf{J}_c$ of a rigid body about its center of mass is known, then the Euler tensor, $\mathbf{J}_o$, about any other point o, is given by

$$\mathbf{J}_o = \mathbf{J}_c + m\mathbf{r}_c \otimes \mathbf{r}_c \tag{4.3.10}$$

where $\mathbf{r}_c$ is the position vector of the center of mass relative to point o and m is the total mass of the body.

Proof : From the formulation of the parallel axis theorem in terms of the inertia tensor $\mathbf{I}_c$, i.e., from equation (4.3.8), we have

$$tr\,[\mathbf{I}_o] = tr\,[\mathbf{I}_c] + (3\mathbf{r}_c \cdot \mathbf{r}_c - \mathbf{r}_c \cdot \mathbf{r}_c)m$$

or

$$tr\,[\mathbf{I}_o] = tr\,[\mathbf{I}_c] + 2\mathbf{r}_c \cdot \mathbf{r}_c m. \tag{4.3.11}$$

Now, using (4.3.8) and (4.3.11) we can rewrite (4.3.6) as follows

$$\mathbf{J}_o = \frac{1}{2}\left\{tr\,[\mathbf{I}_c] + 2\mathbf{r}_c \cdot \mathbf{r}_c m\right\}\mathbf{1} - \mathbf{I}_c - (\mathbf{r}_c \cdot \mathbf{r}_c \mathbf{1} - \mathbf{r}_c \otimes \mathbf{r}_c)m$$

$$= \frac{1}{2} tr\,[\mathbf{I}_c]\mathbf{1} - \mathbf{I}_c + m\mathbf{r}_c \otimes \mathbf{r}_c.$$

Further, since equation (4.3.6) is valid for any point o, it is valid for the center of mass, i.e., we have

$$\mathbf{J}_c = \frac{1}{2} tr\,[\mathbf{I}_c]\mathbf{1} - \mathbf{I}_c \tag{4.3.12}$$

Therefore, by substituting equation (4.3.12) into the last expression for $\mathbf{J}_o$, we get equation (4.3.10). □

The inertia tensor, like any other tensor, is described relative to a coordinate system by a set of components which are known as the *moments of inertia* and *products of inertia.* These components define the coordinate matrix I_o for the tensor $\mathbf{I}_o$. If J_o is the coordinate matrix of the Euler tensor $\mathbf{J}_o$ relative to the same coordinate system { **e** }, then the equivalence which is established above by equation (4.3.6) or equation (4.3.7) between $\mathbf{J}_o$ and $\mathbf{I}_o$ leads us to a component-wise relationship between the coordinate matrices J_o and I_o. This component-wise relationship is expressed as follows:

$$J_o = \begin{bmatrix} J_{11} & J_{12} & J_{13} \\ J_{12} & J_{22} & J_{23} \\ J_{13} & J_{23} & J_{33} \end{bmatrix}$$

$$= \begin{bmatrix} \frac{-I_{11} + I_{22} + I_{33}}{2} & -I_{12} & -I_{13} \\ -I_{12} & \frac{I_{11} - I_{22} + I_{33}}{2} & -I_{23} \\ -I_{13} & -I_{23} & \frac{I_{11} + I_{22} - I_{33}}{2} \end{bmatrix} \tag{4.3.13}$$

or

$$I_o = \begin{bmatrix} I_{11} & I_{12} & I_{13} \\ I_{12} & I_{22} & I_{23} \\ I_{13} & I_{23} & I_{33} \end{bmatrix}$$

$$= \begin{bmatrix} J_{22} + J_{33} & -J_{12} & -J_{13} \\ -J_{12} & J_{11} + J_{33} & -J_{23} \\ -J_{13} & -J_{23} & J_{11} + J_{22} \end{bmatrix} \tag{4.3.14}$$

Now, depending on the coordinate system, the components of the inertia tensor can be time-dependent or time-independent. In particular, relative to a coordinate system which is rigidly attached to the rigid body, the components of the inertia tensor are always time independent. However, when the rigid body is moving in space, the components of the inertia tensor relative to an inertial coordinate system will be time-dependent, and in this case, calculation of their time derivatives may be required. For example, in practical applications it is usually necessary to know the first time derivative of the inertia tensor $\mathbf{I}_c$ (or $\mathbf{J}_c$) relative to an inertial coordinate system. Therefore, in the following, we provide a simple formulation for this derivative.

Let us consider the coordinate system { $\mathbf{e}$ } to be an inertial coordinate system. Also, let us consider a body coordinate system { $\mathbf{e}'$ } whose orientation relative to the inertial coordinate system is described by the rotation tensor $\mathbf{R}$ which is assumed to be a continuous differentiable function of time. Moreover, let us denote by $\mathbf{I}_c$ and $\mathbf{I}'_c$ the rigid body inertia tensors relative to the inertial and body coordinate systems respectively. From the foregoing, it is obvious that the inertia tensor $\mathbf{I}'_c$ is time independent whereas the inertia tensor $\mathbf{I}_c$ is time dependent. We express the time dependence of $\mathbf{I}_c$ by writing

$$\mathbf{I}_c = \mathbf{R}\mathbf{I}'_c\mathbf{R}^T. \tag{4.3.15}$$

Equation (4.3.15) allows us to derive the time derivative (i.e., the time derivative in an inertial reference frame) of the inertia tensor $\mathbf{I}_c$ of a rigid body in a simple and concise manner as follows:

$$\begin{aligned}\dot{\mathbf{I}}_c &= \dot{\mathbf{R}}\mathbf{I}'_c\mathbf{R}^T + \mathbf{R}\mathbf{I}'_c\dot{\mathbf{R}}^T \\ &= \tilde{\boldsymbol{\omega}}\mathbf{R}\mathbf{I}'_c\mathbf{R}^T + \mathbf{R}\mathbf{I}'_c\mathbf{R}^T\tilde{\boldsymbol{\omega}}^T \qquad \text{[by (4.2.5)]} \\ &= \tilde{\boldsymbol{\omega}}\mathbf{I}_c + \mathbf{I}_c\tilde{\boldsymbol{\omega}}^T\end{aligned}$$

or, since $\tilde{\boldsymbol{\omega}}$ is skew-symmetric, we can write

$$\dot{\mathbf{I}}_c = \tilde{\boldsymbol{\omega}}\mathbf{I}_c - \mathbf{I}_c\tilde{\boldsymbol{\omega}}. \tag{4.3.16}$$

Equation (4.3.16) is also valid if we consider the inertia tensor of the rigid body about any other point $\mathbf{o}$ instead of that about the center of mass $\mathbf{c}$. We can show this as follows:

Using the parallel axis theorem (equation (4.3.9)), we have

$$\dot{\mathbf{I}}_o = \dot{\mathbf{I}}_c - m\left[\dot{\tilde{\mathbf{r}}}_c\tilde{\mathbf{r}}_c + \tilde{\mathbf{r}}_c\dot{\tilde{\mathbf{r}}}_c\right]$$

which, by equation (4.2.21), can be written as

$$\dot{\mathbf{I}}_o = \dot{\mathbf{I}}_c - m\left[dual(\tilde{\boldsymbol{\omega}}\mathbf{r}_c)\tilde{\mathbf{r}}_c + \tilde{\mathbf{r}}_c dual(\tilde{\boldsymbol{\omega}}\mathbf{r}_c\right].$$

Now, using equation (3.4.26b), after a few manipulations we get

$$\begin{aligned}\dot{\mathbf{I}}_o &= \dot{\mathbf{I}}_c - \tilde{\boldsymbol{\omega}}[m\tilde{\mathbf{r}}_c\tilde{\mathbf{r}}_c] + [m\tilde{\mathbf{r}}_c\tilde{\mathbf{r}}_c]\tilde{\boldsymbol{\omega}} \\ &= \tilde{\boldsymbol{\omega}}[\mathbf{I}_c - m\tilde{\mathbf{r}}_c\tilde{\mathbf{r}}_c] - [\mathbf{I}_c - m\tilde{\mathbf{r}}_c\tilde{\mathbf{r}}_c]\tilde{\boldsymbol{\omega}}\end{aligned}$$

or, finally

$$\dot{\mathbf{I}}_o = \tilde{\boldsymbol{\omega}}\mathbf{I}_o - \mathbf{I}_o\tilde{\boldsymbol{\omega}}. \qquad (4.3.17)$$

Equations (4.3.16) and (4.3.17) are also valid if we use the Euler tensor $\mathbf{J}_o$ instead of $\mathbf{I}_o$. To see this, we need only to notice that

$$\dot{\mathbf{I}}_o = -\dot{\mathbf{J}}_o \qquad (4.3.18)$$

for any point $\mathbf{o}$. Equation (4.3.18) follows from either (4.3.6) or (4.3.7), since the trace of a tensor is a scalar invariant (see section 3.3.3) and thus is time independent. Then a simple substitution in equation (4.3.17) shows that the derivative of $\mathbf{J}_o$ is given by

$$\dot{\mathbf{J}}_o = \tilde{\boldsymbol{\omega}}\mathbf{J}_o - \mathbf{J}_o\tilde{\boldsymbol{\omega}}. \qquad (4.3.19)$$

In the following, we shall use the two inertia tensors $\mathbf{I}$ and $\mathbf{J}$ to compute other basic physical quantities in rigid body motion such as the angular momentum and the external (or resultant) torque.

4.3.2 The Angular Momentum Tensor

One of the most important physical quantities in rigid body dynamics is *angular* (or *rotational*) *momentum* or *moment of momentum.* In the classical vectorial treatment of rigid body dynamics, angular momentum is represented by a vector which is defined [1] by the equation

$$\mathbf{L}_o = \mathbf{s} \times [\dot{\mathbf{s}} + \boldsymbol{\omega} \times \mathbf{r}_c]m + \mathbf{r}_c \times \dot{\mathbf{s}}m + \mathbf{I}_{o'}\boldsymbol{\omega} \tag{4.3.20}$$

where o is the origin of the inertial coordinate system, o′ is a point fixed on the rigid body, **s** is the position vector of o′ relative to o, $\mathbf{r}_c$ is the position vector of the center of mass relative to o′, $\mathbf{I}_{o'}$ is the rigid body inertia tensor about the point o′ and $\boldsymbol{\omega}$ is the vector of the angular velocity. The expression for $\mathbf{L}_o$ in equation (4.3.20) becomes particularly simple if either the body fixed point o′ is also fixed in inertial space ($\dot{\mathbf{s}} = 0$) or the center of mass is used as the reference point o′ ($\mathbf{r}_c = 0$). In both cases the term $\mathbf{r}_c \times \dot{\mathbf{s}}m$ vanishes. The first term then represents the angular momentum with respect to o, due to the translation of the center of mass, and the last term represents the angular momentum caused by the rotation of the rigid body. In the following, we shall assume that there is no translational motion ($\dot{\mathbf{s}} = 0$) and so equation (4.3.20) takes the form

$$\mathbf{L}_o = \mathbf{s} \times [\boldsymbol{\omega} \times \mathbf{r}_c]m + \mathbf{I}_{o'}\boldsymbol{\omega}. \tag{4.3.21}$$

Moreover, we shall assume that the inertial coordinate system has its origin at the point o′ ($\mathbf{s} = 0$) and in this case we shall write

$$\mathbf{L}_o = \mathbf{I}_o\boldsymbol{\omega}. \tag{4.3.22}$$

Obviously, when the center of rotation is at the center of mass, equation (4.3.22) becomes

$$\mathbf{L}_c = \mathbf{I}_c\boldsymbol{\omega}. \tag{4.3.23}$$

However, even when the center of rotation is different from the center of mass, it is useful to write $\mathbf{L}_o$ in terms of $\mathbf{L}_c$. An expression for $\mathbf{L}_o$ in terms of

$\mathbf{L}_c$ can be easily derived by using Cartesian tensor analysis as follows.

Using the parallel axis theorem, equation (4.3.22) can be modified as shown

$$\begin{aligned}\mathbf{L}_o &= [\mathbf{I}_c - m\tilde{\mathbf{r}}_c\tilde{\mathbf{r}}_c]\boldsymbol{\omega} \\ &= \mathbf{L}_c - m\tilde{\mathbf{r}}_c\tilde{\mathbf{r}}_c\boldsymbol{\omega} \\ &= \mathbf{L}_c + m\tilde{\mathbf{r}}_c\dot{\mathbf{r}}_c.\end{aligned} \tag{4.3.24}$$

In a pure vector notation, equation (4.3.24) takes the form

$$\mathbf{L}_o = \mathbf{L}_c + \mathbf{r}_c \times \dot{\mathbf{r}}_c m. \tag{4.3.25}$$

Angular momentum can also be defined in terms of the Euler tensor $\mathbf{J}_o$. To see this we need only to substitute $\mathbf{I}_o$ in equation (4.3.22) by $\mathbf{J}_o$. For this, we use equation (4.3.7) and get

$$\mathbf{L}_o = -\mathbf{J}_o\boldsymbol{\omega} + tr[\mathbf{J}_o]\boldsymbol{\omega}. \tag{4.3.26}$$

In addition to its vector description, the angular momentum can also be described [6,9] by a second order skew-symmetric Cartesian tensor. To see this, we need only apply the dual operator on the angular momentum vector $\mathbf{L}_o$. This gives a dual skew-symmetric tensor $\tilde{\mathbf{L}}_o$ which we express by writing

$$\tilde{\mathbf{L}}_o = dual(\mathbf{L}_o). \tag{4.3.27}$$

We refer to the dual skew-symmetric tensor $\tilde{\mathbf{L}}_o$ as the *angular momentum tensor* about the point o.

The dual operator provides an indirect definition for the angular momentum tensor. However, as the following theorem shows, it is possible to define the angular momentum tensor $\tilde{\mathbf{L}}_o$ directly in terms of the inertia tensor $\mathbf{I}_o$ and the angular velocity tensor $\tilde{\boldsymbol{\omega}}$, i.e., without the need to first compute the angular momentum vector.

Theorem 4.4 : The angular momentum tensor of a rotating rigid body about a point o, satisfies the equation

$$\tilde{\mathbf{L}}_o = [\tilde{\omega}\mathbf{I}_o]^T - [\tilde{\omega}\mathbf{I}_o] + tr[\mathbf{I}_o]\tilde{\omega} \tag{4.3.28}$$

where $\mathbf{I}_o$ is the inertia tensor of the rigid body about the center of rotation o and $\tilde{\omega}$ is the angular velocity tensor.

Proof : Using equation (4.3.22) we can write equation (4.3.27) as

$$\tilde{\mathbf{L}}_o = dual(\mathbf{I}_o\tilde{\omega}).$$

Further, since the inertia tensor $\mathbf{I}_o$ is symmetric, by using Proposition (3.11) we get

$$\begin{aligned}\tilde{\mathbf{L}}_o &= -[\mathbf{I}_o\tilde{\omega} + \tilde{\omega}\mathbf{I}_o] + tr[\mathbf{I}_o]\tilde{\omega} \\ &= \mathbf{I}_o\tilde{\omega}^T - \tilde{\omega}\mathbf{I}_o + tr[\mathbf{I}_o]\tilde{\omega} \\ &= [\tilde{\omega}\mathbf{I}_o]^T - \tilde{\omega}\mathbf{I}_o + tr[\mathbf{I}_o]\tilde{\omega}\end{aligned}$$

and this completes the proof. □

Theorem 4.4 can also be written in terms of the Euler tensor $\mathbf{J}_o$. First we notice that from equation (4.3.7) we have

$$tr[\mathbf{I}_o] = 2tr[\mathbf{J}_o]. \tag{4.3.29}$$

Therefore, if we substitute equation (4.3.7) and (4.3.29) into equation (4.3.28) we have

$$\begin{aligned}\tilde{\mathbf{L}}_o &= \left[tr[\mathbf{J}_o]\tilde{\omega} - \tilde{\omega}\mathbf{J}_o\right]^T - \left[tr[\mathbf{J}_o]\tilde{\omega} - \tilde{\omega}\mathbf{J}_o\right] + 2tr[\mathbf{J}_o]\tilde{\omega} \\ &= -tr[\mathbf{J}_o]\tilde{\omega} + \mathbf{J}_o\tilde{\omega} - tr[\mathbf{J}_o]\tilde{\omega} + \tilde{\omega}\mathbf{J}_o + 2tr[\mathbf{J}_o]\tilde{\omega} \\ &= \mathbf{J}_o\tilde{\omega} + \tilde{\omega}\mathbf{J}_o\end{aligned}$$

or, finally

$$\tilde{\mathbf{L}}_o = \mathbf{J}_o\tilde{\omega} - [\mathbf{J}_o\tilde{\omega}]^T. \tag{4.3.30}$$

Obviously, when the rigid body is rotating about its center of mass, the angular momentum tensor is defined by the equation

$$\tilde{\mathbf{L}}_{\mathbf{C}} = [\tilde{\omega}\mathbf{I}_{\mathbf{C}}]^T - [\tilde{\omega}\mathbf{I}_{\mathbf{C}}] + tr\,[\mathbf{I}_{\mathbf{C}}]\tilde{\omega} \tag{4.3.31}$$

or the equation

$$\tilde{\mathbf{L}}_{\mathbf{C}} = \mathbf{J}_{\mathbf{C}}\tilde{\omega} - [\mathbf{J}_{\mathbf{C}}\tilde{\omega}]^T. \tag{4.3.32}$$

As we can see from equations (4.3.22) and (4.3.26), the angular momentum vector has a simpler expression when it is written in terms of the inertia tensor $\mathbf{I}_{\mathbf{o}}$. But this is not true for the angular momentum tensor. Equations (4.3.28) and (4.3.30) show that the angular momentum tensor has a simpler expression when it is expressed in terms of the Euler tensor $\mathbf{J}_{\mathbf{o}}$.

In the following, we shall use angular momentum in its vector or tensor representation to describe the dynamic behavior of a rotating rigid body.

4.3.3 The Torque Tensor

As is well known (e.g., [1]), the time derivative of angular momentum equals the *resultant torque* or *moment of force*. This derivative expresses the basic axiom (*Euler's axiom*) governing the rotational motion of a rigid body and in vector form is written as

$$\mathbf{M}_{\mathbf{o}} = \dot{\mathbf{L}}_{\mathbf{o}}. \tag{4.3.33}$$

When the point of rotation is the center of mass of the rigid body, equation (4.3.33) takes the form

$$\mathbf{M}_{\mathbf{C}} = \dot{\mathbf{L}}_{\mathbf{C}}. \tag{4.3.34}$$

To show that equation (4.3.34) is equivalent to equation (4.3.2), we proceed as follows

$$\mathbf{M}_{\mathbf{C}} = \frac{d}{dt}(\mathbf{I}_{\mathbf{C}}\omega)$$

$$= \dot{\mathbf{I}}_c \boldsymbol{\omega} + \mathbf{I}_c \dot{\boldsymbol{\omega}})$$

$$= [\tilde{\boldsymbol{\omega}} \mathbf{I}_c - \mathbf{I}_c \tilde{\boldsymbol{\omega}}]\boldsymbol{\omega} + \mathbf{I}_c \dot{\boldsymbol{\omega}} \quad [\text{by}\,(4.3.16)]$$

$$= \mathbf{I}_c \dot{\boldsymbol{\omega}} + \tilde{\boldsymbol{\omega}} \mathbf{I}_c \boldsymbol{\omega} \quad [\text{by}\,(3.4.4)]$$

$$= \mathbf{I}_c \dot{\boldsymbol{\omega}} + \boldsymbol{\omega} \times \mathbf{I}_c \boldsymbol{\omega}$$

Sometimes, in the literature on classical dynamics, equation (4.3.2) or equation (4.3.34) is referred to as the *generalized Euler equation* for rigid body rotational motion. Here, we shall refer to equation (4.3.2) as the *vector* formulation of the generalized Euler equation.

Now, as in the case of $\mathbf{M}_c$, it can be shown that the vector $\mathbf{M}_o$ satisfies the equation

$$\mathbf{M}_o = \mathbf{I}_o \dot{\boldsymbol{\omega}} + \tilde{\boldsymbol{\omega}} \mathbf{I}_o \boldsymbol{\omega} \tag{4.3.35}$$

when the center of rotation, i.e., point $\mathbf{o}$, is any point other than the center of mass. Moreover, the torque vector $\mathbf{M}_o$ can be written in terms of the torque vector $\mathbf{M}_c$ as follows

$$\mathbf{M}_o = \mathbf{M}_c + \mathbf{r}_c \times \mathbf{F}_c \tag{4.3.36}$$

where $\mathbf{F}_c$ is the total force caused at the center of the mass of the rigid body due to its rotational motion. To see that equation (4.3.36) is equivalent to equation (4.3.35), we write from equation (4.3.36),

$$\mathbf{M}_o = \mathbf{M}_c + m \tilde{\mathbf{r}}_c \ddot{\mathbf{r}}_c$$

$$= \mathbf{M}_c + m \tilde{\mathbf{r}}_c \boldsymbol{\Omega} \mathbf{r}_c$$

$$= \mathbf{M}_c + m [\tilde{\mathbf{r}}_c (\dot{\tilde{\boldsymbol{\omega}}} + \tilde{\boldsymbol{\omega}}\tilde{\boldsymbol{\omega}}) \mathbf{r}_c]$$

$$= \mathbf{M}_c - m [\tilde{\mathbf{r}}_c \tilde{\mathbf{r}}_c \dot{\boldsymbol{\omega}} + \tilde{\mathbf{r}}_c \tilde{\boldsymbol{\omega}} \tilde{\mathbf{r}}_c \boldsymbol{\omega}] \tag{4.3.37}$$

where equation (3.4.6) has been used in the last step. Now, from equation (3.4.19b) we have

$$\tilde{\mathbf{r}}_c \tilde{\boldsymbol{\omega}} \tilde{\mathbf{r}}_c = \tilde{\boldsymbol{\omega}} \tilde{\mathbf{r}}_c \tilde{\mathbf{r}}_c + \tilde{\mathbf{r}}_c \tilde{\mathbf{r}}_c \tilde{\boldsymbol{\omega}} - \frac{1}{2} tr [\tilde{\mathbf{r}}_c \tilde{\mathbf{r}}_c] \tilde{\boldsymbol{\omega}}$$

and since $\tilde{\omega}\omega = 0$, equation (4.3.37) becomes

$$\mathbf{M}_o = \mathbf{M}_c - m[\tilde{\mathbf{r}}_c\tilde{\mathbf{r}}_c\dot{\omega} + \tilde{\omega}\tilde{\mathbf{r}}_c\tilde{\mathbf{r}}_c\omega]$$

$$= [\mathbf{I}_c - m\tilde{\mathbf{r}}_c\tilde{\mathbf{r}}_c]\dot{\omega} + \tilde{\omega}[\mathbf{I}_c - m\tilde{\mathbf{r}}_c\tilde{\mathbf{r}}_c]\omega. \qquad \text{(by (4.3.2))}$$

Finally, using the parallel axis theorem, we can see that the last equation is equivalent to equation (4.3.35).

Equation (4.3.35) has been stated in terms of the inertia tensor $\mathbf{I}_o$. If we use equation (4.3.7) to substitute for $\mathbf{I}_o$ in terms of $\mathbf{J}_o$, equation (4.3.35) becomes

$$\mathbf{M}_o = -[\mathbf{J}_o\dot{\omega} + \tilde{\omega}\mathbf{J}_o\omega] + tr[\mathbf{J}_o]\dot{\omega}. \tag{4.3.38}$$

Similarly, for $\mathbf{M}_c$ we can write

$$\mathbf{M}_c = -[\mathbf{J}_c\dot{\omega} + \tilde{\omega}\mathbf{J}_c\omega] + tr[\mathbf{J}_c]\dot{\omega}. \tag{4.3.39}$$

Now, as with angular momentum, the torque can be described by a second order skew-symmetric Cartesian tensor, which we shall refer to as the *torque tensor*. Using the dual operator we define the torque tensor as follows

$$\tilde{\mathbf{M}}_o = dual(\mathbf{M}_o). \tag{4.3.40}$$

This definition of the torque tensor requires the computation of the torque vector $\mathbf{M}_o$. Another definition of the torque tensor in terms of the inertia and the angular acceleration tensors is given by the following theorem.

Theorem 4.5 : The torque tensor about the center of rotation o, is defined as the time derivative of the angular momentum tensor about the point o and satisfies the equation

$$\tilde{\mathbf{M}}_o = [\Omega\mathbf{I}_o]^T - \Omega\mathbf{I}_o + tr[\mathbf{I}_o]\dot{\tilde{\omega}} \tag{4.3.41}$$

where Ω is the angular acceleration tensor of rotation and $\mathbf{I}_o$ is the rigid body inertia tensor about the point o.

Proof : From Euler's axiom, we have

$$\tilde{\mathbf{M}}_o = \frac{d}{dt}\tilde{\mathbf{L}}_o.$$

Further, using equation (4.3.28), we get

$$\tilde{\mathbf{M}}_o = \left[\dot{\tilde{\omega}}\mathbf{I}_o + \tilde{\omega}\dot{\mathbf{I}}_o\right]^T - \left[\dot{\tilde{\omega}}\mathbf{I}_o + \tilde{\omega}\dot{\mathbf{I}}_o\right] + tr\,[\mathbf{I}_o]\dot{\tilde{\omega}},$$

since the time derivative of the scalar invariant $tr\,[\mathbf{I}_o]$ is zero. Now, using equation (4.3.17), we have

$$\begin{aligned}
\tilde{\mathbf{M}}_o &= \left[\dot{\tilde{\omega}}\mathbf{I}_o + \tilde{\omega}\tilde{\omega}\mathbf{I}_o - \tilde{\omega}\mathbf{I}_o\tilde{\omega}\right]^T - \left[\dot{\tilde{\omega}}\mathbf{I}_o + \tilde{\omega}\tilde{\omega}\mathbf{I}_o - \tilde{\omega}\mathbf{I}_o\tilde{\omega}\right] + tr\,[\mathbf{I}_o]\dot{\tilde{\omega}} \\
&= \left[\dot{\tilde{\omega}}\mathbf{I}_o + \tilde{\omega}\tilde{\omega}\mathbf{I}_o\right]^T - \left[\dot{\tilde{\omega}}\mathbf{I}_o + \tilde{\omega}\tilde{\omega}\mathbf{I}_o\right] + tr\,[\mathbf{I}_o]\dot{\tilde{\omega}} \\
&= [\Omega\mathbf{I}_o]^T - \Omega\mathbf{I}_o + tr\,[\mathbf{I}_o]\dot{\tilde{\omega}}
\end{aligned}$$

□

The torque tensor $\tilde{\mathbf{M}}_o$ can also be defined in terms of the Euler tensor $\mathbf{J}_o$ and the angular acceleration tensor Ω. To see this, we can consider the time derivative of equation (4.3.30) or use equation (4.3.7) to substitute for $\mathbf{I}_o$ in equation (4.3.41). In both cases after a few manipulations we arrive at the following equation

$$\tilde{\mathbf{M}}_o = \Omega\mathbf{J}_o - [\Omega\mathbf{J}_o]^T. \tag{4.3.42}$$

Obviously, when the rigid body is rotated about its center of mass equations (4.3.41) and (4.3.42) are written as

$$\tilde{\mathbf{M}}_c = [\Omega\mathbf{I}_c]^T - \Omega\mathbf{I}_c + tr\,[\mathbf{I}_c]\dot{\tilde{\omega}} \tag{4.3.43}$$

and

$$\tilde{\mathbf{M}}_c = \Omega\mathbf{J}_c - [\Omega\mathbf{J}_c]^T \tag{4.3.44}$$

respectively. We shall refer to equation (4.3.44) as the *tensor* formulation of the generalized Euler equation of a rigid body rotational motion.

As we can see from equations (4.3.2) and (4.3.39) the torque vector $\mathbf{M}_c$ has a simpler expression when it is defined in terms of the inertia tensor $\mathbf{I}_c$. But, as in the case of the angular momentum tensor, the Euler tensor $\mathbf{J}_o$ leads to a simpler equation for the definition of the torque tensor $\tilde{\mathbf{M}}_c$. This implies that for a vector formulation of the equations of rotational rigid body motion, the proper definition of the inertia tensor is given by equation (4.3.3). But, when a tensor formulation for the equations of rotational rigid body motion is required, then the proper definition of the inertia tensor is given by equation (4.3.4).

From the foregoing, to describe rotational rigid body motion within the Newtonian formulation, we can use either vector analysis or Cartesian tensor analysis. The two approaches are equivalent in the sense that the torque vector $\mathbf{M}_c$ is the dual vector or vector invariant of the torque tensor $\tilde{\mathbf{M}}_c$. Therefore, we can use either approach for describing the resultant (or external) torque of rigid body motion. But as we shall see in the following, in practical applications a tensor description of the resultant torque is to be preferred since it is computationally more efficient.

4.3.4 Computational Considerations

In the following, we shall assume that the angular velocity vector, $\boldsymbol{\omega}$, the angular acceleration vector, $\dot{\boldsymbol{\omega}}$, and the angular acceleration tensor $\boldsymbol{\Omega}$ are available (their computation will be consider in Chapter 5) and we shall examine the computational cost of evaluating the vectors $\mathbf{F}_c$ and $\mathbf{M}_c$ which describe rigid body motion.

To compute the vector $\mathbf{F}_c$ we need to evaluate equation (4.3.1). It is obvious that the computational burden of evaluating this equation results mainly from the computation of the vector $\ddot{\mathbf{r}}_c$. From a computational point of view, in general computing the vector $\ddot{\mathbf{r}}_c$ is similar to computing the vector $\ddot{\mathbf{r}}_1$ which is defined by either equation (4.2.20) or equation (4.2.24). Thus, to compute the vector $\ddot{\mathbf{r}}_c$ we can use equation (4.2.20) or equation (4.2.24). In the latter case we need to perform three vector cross product operations and two vector additions and this requires a total of 18 scalar multiplications and 15 scalar additions. In the former case we need to perform a matrix-vector multiplication and a vector addition requiring a total of 9 scalar multiplications and 9 scalar additions. To compute the torque vector $\mathbf{M}_c$ we can use equation (4.3.2), or equation (4.3.44) which computes the torque tensor $\tilde{\mathbf{M}}_c$ and then from that skew-symmetric tensor we can extract the vector $\mathbf{M}_c$ by using the correspondence (3.3.31). In the first approach, i.e., using equation (4.3.2), we need to perform two matrix-vector multiplications, a vector cross product operation and a vector addition requiring a total of 24 scalar multiplications and 18 scalar additions. In the second approach, we can evaluate the torque vector $\mathbf{M}_c$ with only 18 scalar multiplications (which can be reduced to 15, if the symmetry of $\mathbf{J}_c$ is taken into account) and 15 scalar additions. This is so, because there is no need to compute the complete matrix-matrix multiplication, which is involved in equation (4.3.44), since the tensor $\tilde{\mathbf{M}}_c$ is skew-symmetric, and the extraction of its dual vector requires no computations.

From the foregoing, when vectors are used to describe the Newtonian formulation of rigid body motion, i.e., the vectors $\mathbf{F}_c$ and $\mathbf{M}_c$ are computed using equations (4.3.1), (4.2.24) and (4.3.2), we require a total of 45 scalar multiplications and 33 scalar additions. On the other hand, when Cartesian tensors are used to describe the Newtonian formulation of rigid body motion, i.e., equations (4.3.1), (4.2.20), (4.3.44) and (3.3.31) are used, a total of only 30 (or even 27) scalar multiplications and 24 scalar additions are need to evaluate the same vectors. Therefore, the tensor treatment of rigid body motion which is presented in this chapter reduces the computational cost of evaluating the equations of rigid body motion. This, obviously, has very important consequences for many practical problems of mechanics where the equations of rigid body motion is needed to be computed a number of times.

In the following chapters, we shall use this tensor treatment of rigid body motion to solve, in a computationally efficient manner, the three problems of robot dynamics: the problem of inverse dynamics, the problem of forward dynamics, and the linearization of the equations of motion of rigid-link open-chain robot manipulators.

4.4 References

[1] I. J. Wittenburg, *Dynamics of Systems of Rigid Bodies*, B. G. Teubner, Stuttgart, 1977.

[2] H. Goldstein, *Classical Mechanics*. 2nd Ed., Addison-Wesley, Reading, MA : 1980.

[3] J. B. Marion, *Classical Dynamics of Particles and Systems*, 2nd Ed., Academic Press, New York, 1970.

[4] S. N. Rasband, *Dynamics*, John Wiley & Sons, New York, 1983.

[5] L. A. Pars, *A Treatise on Analytical Dynamics*, Heinemann, London, 1965.

[6] H. Jeffreys, *Cartesian Tensors*, Cambridge University Press, Cambridge, 1961.

[7] J. Casey, "A Treatment of Rigid Body Dynamics", *J. Appl. Mech.*, pp. 905-907, Vol. 50, 1983.

[8] C. Truesdell and R. Toupin, "The Classical Field Theories", *Encyclopedia of Physics*, Voll. III/1, (S. Flugge Ed.), Springer-Verlag, Berlin, 1960.

[9] C. Truesdell, *A First Course in Rational Continuum Mechanics*, Vol. 1, Academic Press, New York, 1977.

[10] P. Chadwick, *Continuum Mechanics: Concise Theory and Problems*, George Allen & Unwin Ltd., London, 1976.

[11] W. M. Lai, D. Rubin and E. Krempl, *Introduction to Continuum Mechanics*, Pergamon Press, New York, 1978.

[12] M. E. Gurtin, *An Introduction to Continuum Mechanics*, Academic Press, New York, 1981.

[13] O. Bottema and B. Roth, *Theoretical Kinematics*, North-Holland Publishing Co., Amsterdam, 1978.

[14] E. J. Konopinski, *Classical Descriptions of Motion*, W. H. Freeman and Company, San Francisco, 1969.

[15] J. Angeles, *Spatial Kinematic Chains : Analysis, Synthesis, Optimization*, Springer-Verlag, New York, 1982.

[16] J. Casey and V. C. Lam, "A Tensor Method for the Kinematical Analysis of Systems of Rigid Bodies", *Mechanism and Machine Theory*, pp 87-97, Vol. 21, No. 1, 1986.

[17] J. Angeles, *Rational Kinematics*, Springer-Verlag, New York, 1989.

[18] J. S. Beggs, *Kinematics*, Hemisphere Publishing Corporation, Washington, 1983.

[19] T. Crouch, *Matrix Methods Applied to Engineering Rigid Body Mechanics*, Pergamon Press, Oxford, 1981.

[20] S. F. Borg, *Matrix-Tensor Methods in Continuum Mechanics*, D. Van Nostrand Company, Princeton, New Jersey, 1963.

[21] C. A. Balafoutis, R. V. Patel and P. Misra, "Efficient Modeling and Computation of Manipulator Dynamics Using Orthogonal Cartesian Tensors" *IEEE J. Robotics and Automation*, pp 665-676, Vol. 4, No. 6, 1988.

[22] C. G. Atkeson, C. H. An and J. M. Hollerbach, "Rigid Body Load Identification for Manipulators", *24th IEEE Conf. on Decision and Control*, pp. 996-1002, December 1985.

[23] W. M. Silver, ''On the Equivalence of Lagrangian and Newton-Euler Dynamics for Manipulators'', *Int. Journal of Robotics Research*, pp. 60-70, Vol. 1, No. 2, 1982.

[24] C. A. Balafoutis, R. V. Patel and J. Angeles, ''A Comparative Study of Newton-Euler, Euler-Lagrange and Kane's Formulation for Robot Manipulator Dynamics'', *Robotics and Manufacturing : Recent Trends in Research, Education and Applications*, M. Jamsidi, J. Y. S. Luh, H. Seraji, and G. P. Starr, *Eds.*,, ASME Press, New York, 1988.

Chapter 5

Manipulator Inverse Dynamics

5.1 Introduction

Manipulator inverse dynamics, or simply inverse dynamics, is the calculation of the forces and/or torques required at a robot's joints in order to produce a given motion trajectory consisting of a set of joint positions, velocities and accelerations. The principal uses of inverse dynamics are in robot control and trajectory planning. In control applications computation of inverse dynamics is usually incorporated as an element of the feedback or feedforward path to convert positions, velocities and accelerations, computed according to some desired trajectory, into the joint generalized forces which will achieve those accelerations (e.g. see [1-4]). In trajectory planning, inverse dynamics can be used to check or ensure that a proposed trajectory can be executed without exceeding the actuators' limits [5-7]. Also, using certain time-scaling properties of inverse dynamics, we can facilitate minimum-time (or near minimum-time) trajectory planning [8]. Moreover, inverse dynamics are also taken into consideration in defining manipulability measures of robot arms. (Manipulability is usually expressed as a quantitative measure of a robot arm's manipulating ability in positioning and orienting its end-effector [9]). Finally, as we shall see in the next chapter, computation of inverse dynamics is also used as a building block for constructing forward dynamics algorithms which are useful in performing dynamic simulations of robot arms.

Mathematically, the inverse dynamics problem (IDP) can be described by a vector equation of the form

$$\boldsymbol{\tau} = \mathbf{f}\,(\mathbf{q}\,,\dot{\mathbf{q}}\,,\ddot{\mathbf{q}}\,,\text{manipulator parameters}) \tag{5.1.1}$$

where $\boldsymbol{\tau}$ is the vector of the unknown generalized forces, $(\mathbf{q}\,,\dot{\mathbf{q}}\,,\ddot{\mathbf{q}})$ denotes a given manipulator trajectory and the "manipulator parameters" are all those parameters which characterize the particular geometry and dynamics of a robot manipulator. Equation (5.1.1) is referred to as the *dynamic model* or the *dynamic equation* of a robot manipulator.

To derive equation (5.1.1), we can use well-established procedures from classical mechanics such as those based on the equations of *Newton* and *Euler*, *Euler* and *Lagrange*, *Kane*, etc. Intuitively, one would expect that all the different approaches of formulating this dynamic model should result in the same or equivalent equations. This, is in fact true. However, it is understandable that the choice of a particular procedure is important because it determines the nature of the analysis and the amount of effort needed to obtain these equations. But, otherwise, it is not important which procedure we choose because, as we shall see in this Chapter, different procedures can be formulated to lead to the same algorithm for solving the IDP.

It is known (from experience) that independent of which procedure we use to derive the dynamic model of a *simple* mechanical system, the vector function $\mathbf{f}$ is usually simple and thus it is possible to express it explicitly in terms of the system kinematic parameters (generalized positions, velocities and accelerations). But, unfortunately, this is extremely difficult for mechanical systems of the complexity of a robot manipulator for which the vector function $\mathbf{f}$ is known to be highly nonlinear and coupled. Although, based on symbolic manipulations, some explicit formulations for the function $\mathbf{f}$ have been proposed [10,11], usually the function **f** is obtained via a structured algorithm, i.e., the dynamic model of a robot manipulator is usually evaluated in stages. The results of each stage are a set of values for intermediate variables which are used in subsequent stages to evaluate other variables (or expressions). Depending on what we consider as intermediate variables and how we represent them, we can derive different algorithms for evaluating the dynamic model of a robot manipulator.

The computational complexity of different algorithms varies enormously and these differences are accounted for by the amount of calculation involved in evaluating the equations of motion via a prescribed set of intermediate variables. The key to efficient dynamics calculation is to find a set of common sub-expressions which will effectively be the intermediate variables to be calculated by the algorithm. This eliminates most of the repetition inherent in the equations of motion. Also, the representation (and thus the description) of the intermediate variables is important because, based on their representation other subsequent intermediate quantities may be formulated more efficiently. Furthermore, the structure of the computations is an other important factor, since they can lead us to *closed-form* or *recursive* algorithms. Finally, particular implementations (e.g., tabularization or customization) can be used to improve the computational efficiency of an algorithm which solves the IDP for a certain class of robotic manipulators.

In this Chapter, after a review of some basic methods proposed so far for solving the IDP, we shall apply Cartesian tensor theory for obtaining more efficient solutions for this fundamental problem of robotics. In particular, the outline of this Chapter is as follows : Section 5.2 contains a review of existing methods for solving the IDP; some "classical" algorithms which have been derived from these methods are presented. Also, some observations are made about various issues concerning the computational efficiency of these algorithms. Section 5.3, presents two new algorithms for solving the IDP which are based on Cartesian tensors and use two different modeling schemes. The computational complexity of these algorithms is analyzed and compared with that of other algorithms in the literature. In Section 5.4 it is demonstrated that the computational efficiency of an algorithm which solves the IDP is actually independent of the particular formulation from classical mechanics that is used for its derivation. Finally, Section 5.5 concludes this chapter.

5.2 Previous Results and General Observations on Manipulator Inverse Dynamics

An extensive literature exists on the subject of rigid-link manipulator dynamics in general and on manipulator inverse dynamics in particular. In this section, basic contributions on this subject will be mentioned and the "classical" computational algorithms for solving inverse dynamics will be presented.

The methods for computing inverse dynamics are usually classified with respect to the laws of mechanics on the basis of which the equations of motion are formed. Thus, one may distinguish methods based on the *Euler-Lagrange, Newton-Euler, Kane's, d'Alembert's,* and other equations. Among them, methods based on the Euler-Lagrange and Newton-Euler equations have gained popularity and, as a consequence, there are many algorithms available today that have been derived using these equations to obtain inverse dynamics. The Euler-Lagrange equations are popular because they are conceptually simple and the methods based on them can lead easily to closed-form algorithms which are attractive from both the dynamic modeling and control points of view. On the other hand, the Newton-Euler equations became popular because they led to computationally efficient recursive algorithms which could be used for real-time control applications and simulation. Besides these two commonly used formulations, in the past few years, researchers have also successfully used Kane's dynamical equations in deriving efficient recursive algorithms for computing inverse dynamics.

In the following, we present a brief survey of the methods for solving the IDP based on the Euler-Lagrange, Newton-Euler and Kane's equations.

5.2.1 Formulations Based on Euler-Lagrange Equations

In the Lagrangian approach, to derive the dynamic equations of motion of a robot manipulator we first express the *Lagrangian*

$$\mathbf{L} = \mathbf{\Phi} - \mathbf{P}, \tag{5.2.1}$$

where $\mathbf{\Phi}$ is the kinetic and $\mathbf{P}$ is the potential energy of the robot manipulator, in terms of the joint positions q_i and velocities $\dot{q}_i$ (which are the generalized coordinates and their derivatives). In the early formulations, this was usually done in terms of homogeneous coordinates and the analysis was based on the following modeling scheme: The robot manipulator is considered to be an ideally connected, open-loop, serial-chain of rigid bodies. When frictional forces at the joints are to be considered, we compute them based on the joint velocities and add them directly to the joint generalized forces. Based on this modeling scheme and using the principle of superposition, the kinetic energy of a robotic manipulator can be computed as follows:.

The kinetic energy $\mathbf{\Phi}_i$ of the i-th link of a manipulator, is written in the form

$$\mathbf{\Phi}_i = \frac{1}{2} \sum_{j=1}^{i} \sum_{k=1}^{i} tr \left[\frac{\partial \mathbf{W}_i}{\partial q_j} {}^i\mathbf{J}_i \frac{\partial \mathbf{W}_i^T}{\partial q_k} \dot{q}_j \dot{q}_k \right] \tag{5.2.2}$$

where ${}^i\mathbf{J}_i$ is the *pseudo-inertia* tensor of the i-th link with respect to the origin of the i-th link coordinate system expressed in i-th coordinates. The pseudo-tensor ${}^i\mathbf{J}_i$ is defined by the equation

$${}^i\mathbf{J}_i = \int_{link\ i} {}^i\mathbf{r}_i \, {}^i\mathbf{r}_i^T \, dm \qquad \left(\equiv \int_{link\ i} {}^i\mathbf{r}_i \otimes {}^i\mathbf{r}_i \, dm \right) \tag{5.2.3a}$$

where $\mathbf{r}_i^i$ is a four dimensional homogeneous vector (see Chapter 2) which denotes the position of a point mass *dm* relative to the origin of the i-th link coordinate system expressed in i-th coordinates. In terms of three dimensional vectors equation (5.2.3a) can be written as

$$\mathbf{J}_i^i = \left[\begin{array}{c|c} \mathbf{J}_{O_i}^i & m_i \mathbf{r}_{i,i}^i \\ \hline m_i \mathbf{r}_{i,i}^{i^T} & m_i \end{array}\right] = \left[\begin{array}{c|c} \mathbf{J}_{C_i}^i + m_i \mathbf{r}_{i,i}^i \otimes \mathbf{r}_{i,i}^i & m_i \mathbf{r}_{i,i}^i \\ \hline m_i \mathbf{r}_{i,i}^{i^T} & m_i \end{array}\right] \tag{5.2.3b}$$

where $\mathbf{J}_{O_i}^i$ is the Euler inertia tensor (see Chapter 4) of the i-th link.

Using the above formulation for the kinetic energy of the i-th link and the principle of superposition, the total kinetic energy of a manipulator is computed by using the equation

$$\Phi = \frac{1}{2} \sum_{i=1}^{n} \sum_{j=1}^{i} \sum_{k=1}^{i} tr \left[\frac{\partial \mathbf{W}_i}{\partial q_j} \mathbf{J}_i^i \frac{\partial \mathbf{W}_i^T}{\partial q_k} \dot{q}_j \dot{q}_k \right]. \tag{5.2.4}$$

The potential energy $\mathbf{P}$ of a manipulator, which is equal to the work required to transport the mass center of each link from a reference plane to a given position, can be written as

$$\mathbf{P} = \text{constant} - \sum_{j=1}^{n} m_j \mathbf{g}^T \mathbf{r}_{0,j} \tag{5.2.5a}$$

or

$$\mathbf{P} = \text{constant} - \sum_{j=1}^{n} m_j \mathbf{g}^T \mathbf{W}_j \mathbf{r}_{j,j}^j \tag{5.2.5b}$$

where $\mathbf{g}$, $\mathbf{W}_j$ and $\mathbf{r}_{j,j}^j$ are all expressed in homogeneous coordinates. Now, based on these expressions for the kinetic and potential energy, we can define the Lagrangian $\mathbf{L}$ of a robot manipulator in terms of the generalized coordinates. Then, the Lagrangian is substituted into the Euler-Lagrange equation

$$\tau_i = \frac{d}{dt}\frac{\partial \mathbf{L}}{\partial \dot{q}_i} - \frac{\partial \mathbf{L}}{\partial q_i} \tag{5.2.6}$$

which is expanded by symbolic differentiation to give the generalized forces (joint forces and torques) τ_i, $i = 1, 2, \cdots, n$, in terms of the generalized joint positions, velocities and accelerations.

The first results on dynamic analysis of robotic mechanisms based on the Lagrangian formulation were reported by J. J. Uicker [12]. He was concerned with the dynamic analysis of joint-connected systems of arbitrary structure and with an arbitrary number of closed kinematic chains. Uicker's method was later modified by Kahn [13] to include open-loop mechanisms. The Uicker/Kahn method leads to the following set of equations

$$\tau_i = \sum_{j=i}^{n}\left\{\sum_{k=1}^{j}\left[tr\left(\frac{\partial \mathbf{W}_j}{\partial q_i}{}^j\mathbf{J}_j^j\frac{\partial \mathbf{W}_j^T}{\partial q_i}\right)\right]\ddot{q}_k\right.$$
$$\left.+\sum_{k=1}^{j}\sum_{l=1}^{j}\left[tr\left(\frac{\partial \mathbf{W}_j}{\partial q_i}{}^j\mathbf{J}_j^j\frac{\partial^2 \mathbf{W}_j^T}{\partial q_k \partial q_l}\right)\dot{q}_k\dot{q}_l\right] - m_j\mathbf{g}^T\frac{\partial \mathbf{W}_j}{\partial q_i}{}^j\mathbf{r}_{jj}^j\right\} \tag{5.2.7}$$

where $i = 1, 2, \cdots, n$. These equations can be put into a more compact vector-tensor notation as

$$\boldsymbol{\tau} = \mathbf{D}(\mathbf{q})\ddot{\mathbf{q}} + \mathbf{C}(\mathbf{q}, \dot{\mathbf{q}}) + \mathbf{G}(\mathbf{q}) \tag{5.2.8}$$

where

$\boldsymbol{\tau}$	is the vector of generalized forces
$\mathbf{q}\ (\dot{\mathbf{q}}, \ddot{\mathbf{q}})$	is the vector of joint positions (velocities, accelerations)
$\mathbf{D}(\mathbf{q})$	is the generalized inertia tensor of the manipulator

$\mathbf{C}(\mathbf{q}, \dot{\mathbf{q}})$ is a vector containing the Coriolis and centrifugal forces, and

$\mathbf{G}(\mathbf{q})$ is the gravity force vector

Equation (5.2.8) is obviously non-recursive and is referred to as the *closed-form* dynamic robot model. As has been estimated in [14], equation (5.2.7) (or equation (5.2.8)) has $o(n^4)$ computational complexity, where n is the number of links or the number of degrees-of-freedom for a serial-type manipulator to which the algorithm is applied. In particular, for a 6 degrees-of-freedom manipulator the evaluation of the generalized forces at a trajectory point using the Uicker/Kahn algorithm requires 66,271 scalar multiplications and 51,548 scalar additions. This led some researchers to consider simplifications in the dynamical equations, namely, ignoring the Coriolis and centrifugal forces (Paul [15], Bejczy [16]). However, since simplifications of this nature are justifiable only for slow movements of the manipulator [17], this approach was soon abandoned. Another approach for reducing the computational complexity was considered by Albus [18] and Raibert [19]. They proposed a table look-up method, whereby all the configuration dependent terms in the dynamical equations were computed in advance and tabulated for discrete points on the trajectory. Because of the large memory requirements involved in this approach, Horn and Raibert [20] proposed yet another method in which only the position dependent terms where tabulated. However, besides memory requirements, tabular methods have other serious limitations such as poor accuracy of the trajectory, due to interpolation between the stored discrete points. Moreover, the requirement that the trajectory has to be known in advance makes such methods unattractive because this obviously prevents their applicability to robots working in a dynamically changing environment.

Soon it was realized that inverse dynamics for open-loop manipulators with a simple kinematic chain structure, could be analyzed more effectively using recursive methods. In this approach, the task is first broken down into a number of partially ordered steps. In each step a number of intermediate variables are evaluated. The value of each variable is determined by the application of a formula to each link in turn. Where possible and appropriate, the formula defines the quantity of interest for the link in question in terms of that quantity for one or more of the link's immediate neighbors, and in this case the formula is known [36] as a *recurrence relation.* At the end, these steps are stated in the form of a *recursive algorithm* which solves the problem of inverse dynamics.

Waters [21] was the first to notice that the equation (5.2.7) of the generalized forces could be written in the following form

$$\tau_i = \sum_{j=i}^{n} \left[tr\left(\frac{\partial \mathbf{W}_j}{\partial q_i} \mathbf{J}_j {}^j\ddot{\mathbf{W}}_j^T \right) - m_j \mathbf{g}^T \frac{\partial \mathbf{W}_j}{\partial q_i} {}^j\mathbf{r}_{j\,j} \right] , \; i = 1, \cdots, n. \tag{5.2.9}$$

where

$$\ddot{\mathbf{W}}_j = \sum_{k=1}^{j} \frac{\partial \mathbf{W}_j}{\partial q_k} \ddot{\mathbf{q}}_k + \sum_{k=1}^{j} \sum_{l=1}^{j} \frac{\partial^2 \mathbf{W}_j}{\partial q_k \partial q_l} \dot{\mathbf{q}}_k \dot{\mathbf{q}}_l .$$

Equation (5.2.9) allows one to take advantage of the following kinematic recurrence relations

$$\mathbf{W}_i = \mathbf{W}_{i-1} \mathbf{A}_i \tag{5.2.10a}$$

$$\dot{\mathbf{W}}_i = \dot{\mathbf{W}}_{i-1} \mathbf{A}_i + \mathbf{W}_{i-1} \frac{\partial \mathbf{A}_i}{\partial q_i} \dot{q}_i \tag{5.2.10b}$$

$$\ddot{\mathbf{W}}_i = \ddot{\mathbf{W}}_{i-1} \mathbf{A}_i + 2\, \dot{\mathbf{W}}_{i-1} \frac{\partial \mathbf{A}_i}{\partial q_i} \dot{q}_i + \mathbf{W}_{i-1} \frac{\partial^2 \mathbf{A}_i}{\partial q_i^2} \dot{q}_i^2 + \mathbf{W}_{i-1} \frac{\partial \mathbf{A}_i}{\partial q_i} \ddot{q}_i \tag{5.2.10c}$$

for efficiently computing the homogeneous transformations and their time derivatives. Therefore, based on these equations, Waters proposed a recursive algorithm for solving inverse dynamics which has computational complexity $o(n^2)$. In terms of scalar multiplications and additions, for $n = 6$, Waters' algorithm [14] requires 7051 multiplications and 5652 additions. Later on, by introducing the following dynamic recurrence relations,

$$\mathbf{D}_i = \mathbf{J}_i\,{}^i\ddot{\mathbf{W}}_i^T + \mathbf{A}_{i+1}\mathbf{D}_{i+1} \qquad \left[= \sum_{j=i}^{n} {}^i\mathbf{W}_j \mathbf{J}_j\,{}^j\ddot{\mathbf{W}}_j^T\right] \tag{5.2.11a}$$

and

$$\mathbf{c}_i^i = \mathbf{c}_i^i + \mathbf{A}_{i+1}\mathbf{c}_{i+1}^{i+1} \qquad \left[= \sum_{j=i}^{n} m_j\,{}^i\mathbf{W}_j \mathbf{r}_{j\,j}^j\right] \tag{5.2.11b}$$

Hollerbach [14] proposed modifications to Water's algorithm which led to a complete recursive algorithm for implementing equation (5.2.9). Hollerbach's algorithm, formulated in the notation of this monograph, can be stated as follows:

ALGORITHM 5.1

Step 0: Initialization

$$\mathbf{W}_0 = \mathbf{I}\,,\quad \dot{\mathbf{W}}_0 = 0\,,\quad \ddot{\mathbf{W}}_0 = 0\,,\quad \mathbf{A}_{n+1} = 0$$

Step 1: Outward Recursion:- $i = 1, n$

$$\mathbf{W}_i = \mathbf{W}_{i-1}\mathbf{A}_i \tag{5.2.12a}$$

$$\dot{\mathbf{W}}_i = \dot{\mathbf{W}}_{i-1}\mathbf{A}_i + \mathbf{W}_{i-1}\frac{\partial \mathbf{A}_i}{\partial q_i}\dot{q}_i \tag{5.2.12b}$$

$$\ddot{\mathbf{W}}_i = \ddot{\mathbf{W}}_{i-1}\mathbf{A}_i + 2\,\dot{\mathbf{W}}_{i-1}\frac{\partial \mathbf{A}_i}{\partial q_i}\dot{q}_i + \mathbf{W}_{i-1}\frac{\partial^2 \mathbf{A}_i}{\partial q_i^2}\dot{q}_i^2 + \mathbf{W}_{i-1}\frac{\partial \mathbf{A}_i}{\partial q_i}\ddot{q}_i \tag{5.2.12c}$$

Step 2 : Inward Recursion :- $i = n, 1$

$$\mathbf{D}_i = \mathbf{J}_i \, {}^{i}\ddot{\mathbf{W}}_i^T + \mathbf{A}_{i+1}\mathbf{D}_{i+1} \tag{5.2.13a}$$

$$\mathbf{c}_i^i = m_i \mathbf{r}_{i,i}^i + \mathbf{A}_{i+1}\mathbf{c}_{i+1}^{i+1} \tag{5.2.13b}$$

$$\tau_i = tr\left[\frac{\partial \mathbf{W}_i}{\partial q_i}\mathbf{D}_i\right] - \mathbf{g}^T \frac{\partial \mathbf{W}_i}{\partial q_i}\mathbf{c}_i^i \tag{5.2.13c}$$

end

For its implementation, Algorithm 5.1 requires $830n - 592$ scalar multiplications and $675n - 464$ scalar additions. Hence, Algorithm 5.1 has $o(n)$ computational complexity, but is still computationally inefficient for real-time applications, since, for $n = 6$ it requires 4388 multiplications and 3586 additions. However, it was noticed by Hollerbach that the computational inefficiency of this algorithm resulted from the fact that homogeneous transformations are used to describe general rigid body motion. Therefore, using rotation tensors and three dimensional vectors to describe rigid body displacements, equation (5.2.9) can be written as follows† :

$$\tau_i = \sum_{j=i}^{n}\left\{ tr\left[m_j \frac{\partial \mathbf{s}_{0,j}}{\partial q_i}\ddot{\mathbf{s}}_{0,j}^T + \frac{\partial \mathbf{s}_{0,j}}{\partial q_i}(\mathbf{n}_j^j)^T \, {}^{j}\ddot{\mathbf{W}}_j^T + \frac{\partial \mathbf{W}_j}{\partial q_i}\mathbf{n}_j^j \, \ddot{\mathbf{s}}_{0,j}^T \right.\right.$$
$$\left.\left. + \frac{\partial \mathbf{W}_j}{\partial q_i}\mathbf{J}_{0_j}^j \, {}^{j}\ddot{\mathbf{W}}_j^T \right] - m_j \mathbf{g}^T \frac{\partial \mathbf{W}_i}{\partial q_i}\mathbf{r}_{i,j}^i \right\}, \qquad i = 1, \cdots, n \tag{5.2.14}$$

---

† This form is different from equation (15) in Hollerbach's formulation [14] and leads to a slightly modified analysis which is given in Appendix A. A consequence of this is that equation (5.2.16c) in Algorithm 5.2 is different from the corresponding equation (equation (13)) in Ref. [14].

where $\mathbf{n}_j^j = m_j \mathbf{r}_{j,j}^j$ and $\mathbf{W}_j$ now denotes a rotational tensor and not a homogeneous transformation. Based on this equation, Hollerbach proposed another algorithm which has basically the same recursive structure as Algorithm 5.1 and can be stated as follows:

ALGORITHM 5.2

Step 0: Initialization

$$\mathbf{W}_0 = \mathbf{I}\ ,\quad \dot{\mathbf{W}}_i = 0\ ,\quad \ddot{\mathbf{W}}_i = 0\ ,\quad \ddot{\mathbf{s}}_{0,i} = 0\ ,\quad \mathbf{n}_i^i = m_i \mathbf{r}_{i,i}^i\ ,\quad \mathbf{A}_{n+1} = 0\ ,\quad \mathbf{e}_{n+1} = 0$$

Step 1: Outward Recursion:- $i = 1, n$

$$\mathbf{W}_i = \mathbf{W}_{i-1}\mathbf{A}_i \tag{5.2.15a}$$

$$\dot{\mathbf{W}}_i = \dot{\mathbf{W}}_{i-1}\mathbf{A}_i + \mathbf{W}_{i-1}\frac{\partial \mathbf{A}_i}{\partial q_i}\dot{q}_i \tag{5.2.15b}$$

$$\ddot{\mathbf{W}}_i = \ddot{\mathbf{W}}_{i-1}\mathbf{A}_i + 2\,\dot{\mathbf{W}}_{i-1}\frac{\partial \mathbf{A}_i}{\partial q_i}\dot{q}_i + \mathbf{W}_{i-1}\frac{\partial^2 \mathbf{A}_i}{\partial q_i^2}\dot{q}_i^2 + \mathbf{W}_{i-1}\frac{\partial \mathbf{A}_i}{\partial q_i}\ddot{q}_i \tag{5.2.15c}$$

$$\ddot{\mathbf{s}}_{0,i} = \ddot{\mathbf{s}}_{0,i-1} + \ddot{\mathbf{W}}_{i-1}\mathbf{s}_{i-1,i}^{i-1} \tag{5.2.15d}$$

Step 2 : Inward Recursion :- $i = n, 1$

$$\mathbf{e}_i = \mathbf{e}_{i+1} + m_i \ddot{\mathbf{s}}_{0,i}^T + \mathbf{n}_i^{i^T}\ddot{\mathbf{W}}_i^T \tag{5.2.16a}$$

$$\mathbf{D}_i = \mathbf{A}_{i+1}\mathbf{D}_{i+1} + \mathbf{s}_{i,i+1}^i\mathbf{e}_{i+1} + \mathbf{n}_i^i\ddot{\mathbf{s}}_{0,i}^T + \mathbf{J}_{0_i}^i\ddot{\mathbf{W}}_i^T \tag{5.2.16b}$$

$$\mathbf{c}_i^i = m_i \mathbf{r}_{i,i}^i + \bar{m}_{i+1}\mathbf{s}_{i,i+1}^i + \mathbf{A}_{i+1}\mathbf{c}_{i+1}^{i+1} \tag{5.2.16c}$$

$$\tau_i = tr\left[\frac{\partial \mathbf{W}_i}{\partial q_i}\mathbf{D}_i\right] - \mathbf{g}^T\frac{\partial \mathbf{W}_i}{\partial q_i}\mathbf{c}_i^i \tag{5.2.16d}$$

end

For its implementation† , Algorithm 5.2 requires $412n - 277$ scalar multiplications and $675n - 201$ scalar additions, which for $n = 6$ gives 2195 scalar multiplications and 1719 scalar additions, and this is a significant improvement over Algorithm 5.1.

Besides the effects of homogeneous descriptions, it was soon realized that the structure of the kinematic and dynamic recurrence relations have a direct effect on the computational complexity of recursive as well as closed-form manipulator dynamics algorithms. In particular, it was realized that the structure of the kinematic and dynamic recurrence relations depends on the particular modeling scheme which is used for deriving the equations of motion. As result of this, modeling schemes which are based on the ideas of *augmented* and *generalized links* [22,23] were proposed. The ideas of generalized and augmented links surface quite naturally in the modeling schemes of joint connected mechanisms and were first introduced (in 1906) in the underlying modeling scheme of mechanical systems with a tree topology [22].

One of the advantages of using the concepts of augmented and generalized links in robotics is that they lead to efficient recurrent relations for computing the coefficients of the generalized inertia tensor $\mathbf{D}(\mathbf{q})$ of a robot manipulator. Therefore, in solving the IDP of a robot manipulator based on the Lagrangian approach, the concepts of augmented and generalized links are best utilized when the equations of motion are described by the closed-form equation

$$\tau_i(t) = \sum_{j=1}^{n} d_{i,j}\ddot{q}_j + \sum_{j=1}^{n}\sum_{k=1}^{n} \dot{q}_j c_{j,k}(i)\dot{q}_k + G_i \qquad i = 1, 2, \cdots, n \tag{5.2.17}$$

† Note that the difference in equation (5.2.16c) does not change the computational requirements of the original equation (equation (13) in Ref. [14]) when the $(i+1)$-th joint is revolute. When the $(i+1)$-th joint is prismatic the implementation of equation (5.2.16c) requires a few extra computations.

where $d_{i,j}$ are the coefficients of the generalized inertia tensor,

$$c_{j,k}(i) = \frac{1}{2}\left[\frac{\partial d_{i,j}}{\partial q_k} + \frac{\partial d_{i,k}}{\partial q_j} - \frac{\partial d_{j,k}}{\partial q_i}\right] \tag{5.2.18}$$

are the centrifugal and Coriolis coefficients, defined using Christoffel symbols, and G_i are the gravitational coefficients. Renaud [24] was among the first to use these ideas for solving the IDP. In particular, he proposed [24] an iterative analytical procedure for computing the generalized inertia tensor $\mathbf{D}(\mathbf{q})$ but, he did not work out the partial derivatives which are involved in the Christoffel symbols for the case of general manipulators. Instead, to demonstrate his approach, Renaud proposed a customized algorithm for computing the IDP for a simple 6 degrees-of-freedom revolute joint manipulator. Following Renaud's approach, Vukobratovic et al., [25] proposed the following analytical expressions for computing the coefficients of equation (5.2.17)

$$d_{i,j} = \mathbf{z}_i^T \mathbf{E}_{\mathrm{o}_i} \mathbf{z}_j + \mathbf{z}_i^T (\mathbf{U}_{\mathrm{o}_i} \times (\mathbf{z}_j \times \mathbf{s}_{j,i})) \tag{5.2.19a}$$

$$c_{i,k}(j) = \mathbf{z}_j^T (\mathbf{z}_i \times (\mathbf{E}_{\mathrm{o}_i} - (\frac{1}{2} tr\, \mathbf{E}_{\mathrm{o}_i} 1)\mathbf{z}_k) + \mathbf{z}_j^T (\mathbf{s}_{j,i} \times (\mathbf{z}_k \times (\mathbf{z}_i \times \mathbf{U}_{\mathrm{o}_i})) \tag{5.2.19b}$$

$$G_i = \mathbf{z}_i^T (\mathbf{r}_{i,i} \times (\overline{m}_i \mathbf{g})) \tag{5.2.19c}$$

where, $\mathbf{E}_{\mathrm{o}_i}$ denotes the inertia tensor of the i-th generalized link and is computed recursively using the equation

$$\mathbf{E}_{\mathrm{o}_i} = \mathbf{E}_{\mathrm{o}_{i+1}} + (2\mathbf{U}_{\mathrm{o}_{i+1}}^T \mathbf{s}_{i,i+1})1 - \mathbf{s}_{i,i+1}\mathbf{U}_{\mathrm{o}_{i+1}}^T - \mathbf{U}_{\mathrm{o}_{i+1}}\mathbf{s}_{i,i+1}^T$$
$$+ \mathbf{I}_{\mathrm{o}_i} + \overline{m}_{i+1}(\mathbf{s}_{i,i+1}^T \mathbf{s}_{i,i+1} 1 - \mathbf{s}_{i,i+1}\mathbf{s}_{i,i+1}^T), \qquad i = n-1, \cdots, 1 \tag{5.2.20a}$$

and $\mathbf{U}_{\mathrm{o}_i}$ is the first moment of the i-th generalized link and is defined recursively by the equation

$$\mathbf{U}_{\mathrm{o}_i} = \mathbf{U}_{\mathrm{o}_{i+1}} + \overline{m}_{i+1}\mathbf{s}_{i,i+1} + m_i \mathbf{r}_{i,i} \tag{5.2.20b}$$

Based on these equations, they proposed a recursive algorithm for solving the IDP of a robot manipulator with all joints of revolute type which for its evaluation requires $3/2n^3 + 35/2n^2 + 9n - 36$ scalar multiplications and $7/6n^3 + 23/2n^2 + 64/2n - 28$ scalar additions. However, for its implementation when $n = 6$, this algorithm requires 992 scalar multiplications and 776 scalar additions, and thus it is more efficient that Hollerbach's algorithms.

5.2.2 Formulations Based on Newton-Euler Equations

In the Newton-Euler approach, a general motion of a rigid body is considered to result from the superposition of two independent motions; namely a pure translational motion of a point (usually its center of mass) and a pure rotational motion about that point. The translational motion is then described by Newton's equation (or Newton's second law) which is symbolically stated as

$$\mathbf{F}_C = m\ddot{\mathbf{r}}_C \tag{5.2.21}$$

where $\mathbf{F}_C$ is the total external (or resultant) force acting on the rigid body, m is the mass of the rigid body and $\ddot{\mathbf{r}}_C$ is the absolute linear acceleration of its center of mass. The rotational motion is described by Euler's equation which is symbolically stated as

$$\mathbf{M}_C = \mathbf{I}_C \cdot \dot{\boldsymbol{\omega}} + \boldsymbol{\omega} \times \mathbf{I}_C \cdot \boldsymbol{\omega} \tag{5.2.22}$$

where $\mathbf{M}_C$ is the total external (or resultant) torque about the center of mass, $\mathbf{I}_C$ is the inertia tensor of the body about its center of mass and $\boldsymbol{\omega}$ ($\dot{\boldsymbol{\omega}}$) is the angular velocity (acceleration) of the body. Therefore, in solving the IDP for rigid-link robot manipulators following the Newton-Euler approach, these two equations are applied to each link and the resulting equations are combined with constraint equations from the joints in such a way as to give the joint generalized forces in terms of the joint acceleration. Methods based on this approach were originally developed to describe multi-body satellite and spacecraft dynamics [27]. One of the earliest applications of the Newton-

Euler dynamic equations to robotic systems may be found, among others, in the work of Stepanenko and Vucobratovic [28], Vucobratovic [29], Ho [30] and Hughes [31]. However, algorithms based on these methods, as in the case of the Uicker/Kahn method of the Lagrangian formulation, are computationally very inefficient. These early formulations led to closed-form algorithms which have computational complexity $o(n^3)$ or in some cases even $o(n^4)$ [14,25].

A more efficient method was proposed by Orin et al. [32] by introducing link coordinate systems. Using relationships between moving coordinate systems, they were able to achieve more efficient kinematic recurrence relations for computing velocities and accelerations and dynamic recurrence relations for computing forces and torques. Based on these relations, they derived an algorithm which has computational complexity $o(n^2)$. However, in this method, the basic equations of motion for each link are expressed in the inertial coordinate system and this involves unnecessary coordinate transformations. Luh, Walker and Paul [33], modified this method by expressing the equations of motion in link coordinate systems instead of the inertial coordinate system. The approach proposed by Luh, Walker and Paul is probably the best one for deriving recursive algorithms to compute manipulator inverse dynamics. This approach can be outlined as follows: Based on moving coordinate systems, kinematic recurrence relations are used to compute velocities and accelerations from the base of the manipulator to the end-effector, link-by-link. Then dynamic recurrence relations are used to compute forces and torques from the end-effector back to the base of the manipulator. In this process, because of the nature of these recurrence relations, the generalized forces are computed by simple projections of vector quantities onto the joint axes. From this outline, it is obvious that only the information needed to characterize rigid-body movements are computed and many duplications in the computations are avoided. Based on this approach, Luh, Walker and Paul proposed a recursive algorithm which for its implementation requires $150n - 48$ scalar multiplications and $131n - 48$ scalar additions, so that for $n = 6$ this algorithm requires 852 scalar multiplications and 738 scalar additions [14]. A modified version of this algorithm is as follows:

ALGORITHM 5.3

Step 0: Initialization

$$\sigma_i = \begin{cases} 1 \text{ revolute } i\text{–th joint} \\ 0 \text{ prismatic } i\text{–th joint} \end{cases}, \quad q_i = \begin{cases} \theta_i \text{ revolute } i\text{–th joint} \\ d_i \text{ prismatic } i\text{–th joint} \end{cases}$$

$$\boldsymbol{\omega}_0^0 = 0, \quad \dot{\boldsymbol{\omega}}_0^0 = 0, \quad \ddot{\mathbf{s}}_{0,0}^0 = -\mathbf{g}, \quad \mathbf{A}_{n+1} = 0, \quad \mathbf{z}_i^i = [\,0\,0\,1\,]^T$$

Step 1 : Forward recursion :- For *i = 0, n-1* do

$$\boldsymbol{\omega}_{i+1}^{i+1} = \mathbf{A}_{i+1}^T \boldsymbol{\omega}_i^i + \sigma_{i+1} \mathbf{z}_{i+1}^{i+1} \dot{q}_{i+1} \tag{5.2.23a}$$

$$\dot{\boldsymbol{\omega}}_{i+1}^{i+1} = \mathbf{A}_{i+1}^T \dot{\boldsymbol{\omega}}_i^i + \sigma_{i+1} [\mathbf{A}_{i+1}^T \boldsymbol{\omega}_i^i \times \mathbf{z}_{i+1}^{i+1} \dot{q}_{i+1} + \mathbf{z}_{i+1}^{i+1} \ddot{q}_{i+1}] \tag{5.2.23b}$$

$$\begin{aligned} \ddot{\mathbf{s}}_{0,i+1}^{i+1} &= \mathbf{A}_{i+1}^T [\ddot{\mathbf{s}}_{0,i}^i + \dot{\boldsymbol{\omega}}_i^i \times \mathbf{s}_{i,i+1}^i + \boldsymbol{\omega}_i^i \times (\boldsymbol{\omega}_i^i \times \mathbf{s}_{i,i+1}^i)] \\ &\quad + (1-\sigma_{i+1}) [2\boldsymbol{\omega}_i^{i+1} \times \mathbf{z}_{i+1}^{i+1} \dot{q}_{i+1} + \mathbf{z}_{i+1}^{i+1} \ddot{q}_{i+1}] \end{aligned} \tag{5.2.23c}$$

$$\ddot{\mathbf{r}}_{0,i+1}^{i+1} = \dot{\boldsymbol{\omega}}_{i+1}^{i+1} \times \mathbf{r}_{i+1,i+1}^{i+1} + \boldsymbol{\omega}_{i+1}^{i+1} \times (\boldsymbol{\omega}_{i+1}^{i+1} \times \mathbf{r}_{i+1,i+1}^{i+1}) + \ddot{\mathbf{s}}_{0,i+1}^{i+1} \tag{5.2.23d}$$

$$\mathbf{F}_{\mathrm{C}_{i+1}}^{i+1} = m_{i+1} \ddot{\mathbf{r}}_{0,i+1}^{i+1} \tag{5.2.23e}$$

$$\mathbf{M}_{\mathrm{C}_{i+1}}^{i+1} = \mathbf{I}_{\mathrm{C}_{i+1}}^{i+1} \dot{\boldsymbol{\omega}}_{i+1}^{i+1} + \boldsymbol{\omega}_{i+1}^{i+1} \times \mathbf{I}_{\mathrm{C}_{i+1}}^{i+1} \boldsymbol{\omega}_{i+1}^{i+1} \tag{5.2.23f}$$

Step 2 : Backward recursion :- For *i = n, 1* do

$$\mathbf{f}_i^i = \mathbf{A}_{i+1} \mathbf{f}_{i+1}^{i+1} + \mathbf{F}_{\mathrm{C}_i}^i \tag{5.2.24a}$$

$$\boldsymbol{\eta}_i^i = \mathbf{M}_{\mathrm{C}_i}^i + \mathbf{s}_{i,i+1}^i \times \mathbf{A}_{i+1} \mathbf{f}_{i+1}^{i+1} + \mathbf{r}_{i,i}^i \times \mathbf{F}_{\mathrm{C}_i}^i + \mathbf{A}_{i+1} \boldsymbol{\eta}_{i+1}^{i+1} \tag{5.2.24b}$$

$$\tau_i = \sigma_i (\boldsymbol{\eta}_i^i \cdot \mathbf{z}_i^i) + (1-\sigma_i)(\mathbf{f}_i^i \cdot \mathbf{z}_i^i) \tag{5.2.24c}$$

end

The algorithm by Luh,Walker and Paul has $o(n)$ computational complexity and is far more efficient than Algorithm 5.2 which also has $o(n)$ computational complexity, but is derived using the Lagrangian formulation. The difference in the computational complexity of these two algorithms sparked a debate about which of the two formulations, i.e., the Lagrangian or the Newton-Euler, leads to more efficient computational algorithms for solving the IDP. For some time it was believed, due to a lack of deeper understanding of the mathematical representations used to describe the equations of motion, that the algorithms derived from the Newton-Euler formulation were computationally more efficient than those derived using the Lagrangian formulation. Finally, Silver [35] resolved the issue by showing that both formulations are equivalent, in the sense that the Lagrangian formulation will yield a similar algorithm to that obtained using the Newton-Euler formulation, if an equivalent representation of angular velocity is employed.

In the following, we shall briefly review two other methods which are based on Newton-Euler equations but are somehow different from those reviewed so far. The main differences in these two methods, which have been proposed by Featherstone [36] and Rodriguez [37], are that in the kinematic and dynamic analysis of the equations of motion, *spatial notation* and *spatial algebra* have been used. The spatial notation is based on the use of 6-dimensional vectors, called *spatial vectors*, to represent the combined linear and angular components of physical quantities involved in rigid body dynamics, and 6×6 matrices to represent the inertia tensors of rigid bodies. The basic advantage of using spatial notation in the dynamic analysis of rigid body systems is that it reduces the number of quantities and the number of equations required to express and solve various problems associated with the motion of such systems. In the following (and only for this section), symbols with a hat ($\hat{}$) over them denote spatial quantities. Spatial algebra is based on spatial operations (defined on spatial quantities) which are usually implemented using operations of standard vector and matrix algebra.

In Featherstone's analysis [36], spatial vectors are defined in terms of Plucker coordinates and thus spatial algebra is similar to that of screw and motors. Also, spatial tensors, say the spatial tensor of the i-th link about the origin o_i of the i-th coordinate system, are defined by an equation of the form

$$\hat{\mathbf{I}}_{o_i} = \begin{bmatrix} -m_i \tilde{\mathbf{r}}_{i,i} & m_i \mathbf{1} \\ \mathbf{I}_{C_i} - m_i \tilde{\mathbf{r}}_{i,i} \tilde{\mathbf{r}}_{i,i} & m_i \tilde{\mathbf{r}}_{i,i} \end{bmatrix} = \begin{bmatrix} -m_i \tilde{\mathbf{r}}_{i,i} & m_i \mathbf{1} \\ \mathbf{I}_{o_i} & m_i \tilde{\mathbf{r}}_{i,i} \end{bmatrix} \tag{5.2.25}$$

Based on Newton-Euler equations, and using spatial quantities, Featherstone proposed [36] the following algorithm for computing the IDP which has the same structure as that of the ones mentioned so far.

ALGORITHM 5.4

Step 0 : Initialization

$$\hat{\mathbf{v}}_0 = \hat{\mathbf{a}}_0 = \hat{\mathbf{f}}^{n+1}_{n+1} = \hat{0}$$

Step 1 : Forward recursion :- For $i = 1, n$ do

$$\hat{\mathbf{v}}^i_i = \hat{\mathbf{X}}^i_{i-1} \hat{\mathbf{v}}^{i-1}_{i-1} + \hat{\mathbf{s}}^i_i \dot{q}_i \tag{5.2.26a}$$

$$\hat{\mathbf{a}}^i_i = \hat{\mathbf{X}}^i_{i-1} \hat{\mathbf{a}}^{i-1}_{i-1} + \hat{\mathbf{v}}^i_i \hat{\times} \hat{\mathbf{s}}^i_i \dot{q}_i + \hat{\mathbf{s}}^i_i \ddot{q}_i \tag{5.2.26b}$$

$$\hat{\mathbf{F}}^i_i = \hat{\mathbf{I}}^i_{o_i} \hat{\mathbf{a}}^i_i + \hat{\mathbf{v}}^i_i \hat{\times} \hat{\mathbf{I}}^i_{o_i} \hat{\mathbf{v}}^i_i \tag{5.2.26c}$$

Step 2 : Backward recursion :- For $i = n, 1$ do

$$\hat{\mathbf{f}}^i_i = \hat{\mathbf{X}}^i_{i+1} \hat{\mathbf{f}}^{i+1}_{i+1} + \hat{\mathbf{F}}^i_i \tag{5.2.27a}$$

$$\tau_i = \hat{\mathbf{s}}^i_i \hat{\cdot} \hat{\mathbf{f}}^i_i \tag{5.2.27b}$$

end

The spatial operations which are involved in this algorithm are defined as follows. The spatial dot and cross product operations are defined by the equations

$$\begin{bmatrix} \mathbf{a} \\ \mathbf{b} \end{bmatrix} \hat{\cdot} \begin{bmatrix} \mathbf{c} \\ \mathbf{d} \end{bmatrix} \equiv \mathbf{a} \cdot \mathbf{d} + \mathbf{b} \cdot \mathbf{c} \tag{5.2.28}$$

$$\begin{bmatrix} \mathbf{a} \\ \mathbf{b} \end{bmatrix} \hat{\times} \equiv \begin{bmatrix} \tilde{\mathbf{a}} & 0 \\ \tilde{\mathbf{b}} & \tilde{\mathbf{a}} \end{bmatrix} \tag{5.2.29}$$

respectively, and the spatial rotational transformation $\hat{\mathbf{X}}_{i+1}^{i}$ is defined by the equation

$$\hat{\mathbf{X}}_{i+1}^{i} = \begin{bmatrix} \mathbf{A}_{i+1} & 0 \\ 0 & \mathbf{A}_{i+1} \end{bmatrix} \tag{5.2.30}$$

To implement this algorithm, Featherstone used a special spatial arithmetic package and, according to his estimates [36], this algorithm requires $130n - 68$ scalar multiplications and $101n - 56$ scalar additions, so that, for $n = 6$ it requires 712 scalar multiplications and 550 scalar additions.

In the approach by Rodriguez [37], spatial vectors are defined from 3-dimensional vectors. For example, a spatial acceleration is defined to be a six-dimensional vector formed by an angular acceleration and a linear acceleration. Also, the spatial inertia tensor of the i-th link about the origin of the i-th coordinate system is defined by the equation

$$\hat{\mathbf{I}}(i) = \begin{bmatrix} \mathbf{I}_o(i) & m(i)\tilde{\mathbf{r}}(i) \\ -m(i)\tilde{\mathbf{r}}(i) & m(i)\mathbf{1} \end{bmatrix} \tag{5.2.31}$$

The differences from Featherstone's definitions have been introduced to simplify the operations of spatial algebra. Note that the differences have been incorporated in the notation. To transfer the new notation to the notation used so far, one needs to write $m(i) \equiv m_i$, $\mathbf{r}(i) \equiv \mathbf{r}_{i,i}$, $\mathbf{I}_c(i) \equiv \mathbf{I}_{c_i}$ and so on. However, besides the differences in the definitions of spatial quantities and in the spatial notation, the main difference in Rodriguez's approach is to be found in the analysis of the equations of motion. Basic to this analysis, is a spatial quantity of the form

$$\hat{\Phi}(i,j) = \begin{bmatrix} 1 & \tilde{s}(i,j) \\ 0 & 1 \end{bmatrix} \tag{5.2.32}$$

where $\tilde{s}(i,j) \equiv \tilde{s}_{i,j}$. As can be seen, the spatial quantity $\hat{\Phi}(i,j)$ satisfies the properties

$$\hat{\Phi}(i,j) = \hat{\Phi}(i,k)\hat{\Phi}(k,j) \tag{5.2.33a}$$

$$\hat{\Phi}(i,i) = 1 \tag{5.2.33b}$$

$$\hat{\Phi}^{-1}(i,j) = \hat{\Phi}(j,i) \tag{5.2.33c}$$

which are usually associated with a *transition* matrix of a (discrete) linear state space system. The idea of the "transition matrix" $\hat{\Phi}(i,j)$ in Rodriguez's approach is not the only one which is associated with the linear system theory. There are other ideas too which are obvious from the following outline.

The equations of translational and rotational motion (derived from Newton and Euler's equations) for each link are written as linear difference equations that allow the spatial force at the proximal joint to be computed from the spatial force at the distal joint and the spatial acceleration of the link. These difference equations are similar to those describing the evolution of the state of a discrete-time state space system. Here the state is defined by the vector of spatial force which is "propagated" in space (instead of time) from link to link. Thus, in this state-space equation, the role of the time interval between discrete-time samples is played by the spatial interval which is defined as the vector from the proximal to the distal joint. An "output" equation is also associated with this "state" equation in order to generate the scalar joint moments/forces which form the output vector. Also, in Rodriguez's approach, complementary to the state equation is another difference equation which propagates spatial acceleration within the link and across the joint and plays the role of the co-states in his analysis. From the foregoing, in Rodriguez's approach the IDP is formulated as a two-point boundary-value problem, with boundary conditions ensuring that the state (spatial force) vanishes at the tip of the manipulator and the co-state (spatial acceleration) vanishes at the base. These boundary conditions reflect the assumptions that the tip of a manipulator is free while the base is immobile. Obviously, different boundary conditions may also be considered.

Based on these ideas, the following algorithm has been proposed in [38].

ALGORITHM 5.5

Step 0 : Initialization

$$\hat{\mathbf{v}}(0) = \hat{\mathbf{a}}(0) = \hat{\mathbf{f}}(n) = \hat{0}$$

Step 1 : Forward recursion :- For $i = 0, n\text{-}1$ do

$$\hat{\mathbf{v}}(i+1) = \hat{\mathbf{\Phi}}^T(i,i+1)\hat{\mathbf{v}}(i) + \hat{\mathbf{z}}^T(i+1)\dot{q}_{i+1} \tag{5.2.34a}$$

$$\hat{\mathbf{a}}(i+1) = \hat{\mathbf{\Phi}}^T(i,i+1)\hat{\mathbf{a}}(i) + \hat{\mathbf{z}}^T(i+1)\ddot{q}_{i+1} + \hat{\alpha}(i+1) \tag{5.2.34b}$$

Step 2 : Backward recursion :- For $i = n, 1$ do

$$\hat{\mathbf{f}}(i) = \hat{\mathbf{\Phi}}(i,i+1)\hat{\mathbf{f}}(i+1) + \hat{\mathbf{I}}(i)\hat{\mathbf{a}}(i) + \hat{\beta}(i) \tag{5.2.35a}$$

$$\tau(i) = \hat{\mathbf{z}}^T(i)\hat{\mathbf{f}}(i) \tag{5.2.35b}$$

end

In this algorithm, the spatial vectors $\hat{\alpha}(i)$ and $\hat{\beta}(i)$, which are not defined here, are "bias" quantities which in the i-th iteration can be computed (see [38]) from other known quantities. For the implementation of this algorithm, a computational complexity analysis has not been given in [38]. However, since the spatial algebra used in [38] can be implemented based on straightforward vector-matrix algebra and the structure of Algorithm 5.5 is similar to that of Algorithm 5.4 it should be expected that these two algorithms have similar computational complexity.

5.2.3 Formulations Based on Kane's Equations

In Kane's approach, one first describes the generalized *active* force and the generalized *inertia* force of a system in terms of generalized coordinates, generalized speeds, partial linear velocities, and partial angular velocities. Then, the dynamic equations of motion (Kane's equations) are obtained by setting the sum of these two forces equal to zero according to the d'Alembert principle. Kane's equations were first introduced for general nonholonomic mechanical systems [39] and were used, as were the Newton-Euler and Euler-Lagrange equations, in rigid multi-body satellite and spacecraft dynamic analysis [40,41]. Huston and Kelly [42], were among the first to apply Kane's equations to robotics. However, they presented neither an explicit algorithm for inverse dynamics nor a complexity analysis of their method. Kane's equations were also used by Faessler [43], who presented a method which led to a closed-form algorithm for evaluating inverse dynamics. Using analytical procedures, Faessler expressed the entries of the coefficient matrices in symbolic form, but did not provide a complexity analysis of his method. Kane and Levinson [44] also presented a customized algorithm for solving inverse dynamics for the Stanford arm. But, although their procedure is conceptually simple, it requires considerable experience with handling complex dynamical systems and involves an extensive manual analysis for setting up a large number of intermediate variables, which are not defined via recurrence relations.

A recursive algorithm for computing inverse dynamics, based on Kane's equations has been proposed by Angeles, Ma and Rojas [45]. Using analytical procedures, they were able to derive from Kane's equations the following equations for the i-th component ($i = 1, 2, \cdots, n$) of the generalized force τ:

i) When the i-th joint is revolute,

$$\tau_i = \mathbf{z}_i^i \cdot \left\{ \sum_{j=i}^{n} \left[\mathbf{M}_{C_j}^i + m_j \mathbf{r}_{i,j}^i \times \ddot{\mathbf{r}}_{0,j}^i \right] \right\} \tag{5.2.36}$$

ii) When the i-th joint is prismatic,

$$\tau_i = \mathbf{z}_i^i \cdot \left\{ \sum_{j=i}^{n} m_j \ddot{\mathbf{r}}_{0,j}^i \right\} \tag{5.2.37}$$

Equations (5.2.36) and (5.2.37) are identical to the equations derived by Silver [35] using the Euler-Lagrange equations and therefore, as Silver has shown, they can also be derived from the Newton-Euler formulation. Based on equations (5.2.36) and (5.2.37) and kinematic and dynamic recurrence relations, like those introduced by Luh, Walker and Paul [33], Angeles, Ma and Rojas proposed a recursive algorithm which is similar in structure to Algorithm 5.4. For a "semi-customized" implementation, this algorithm requires $105n - 109$ scalar multiplications and $90n - 105$ scalar additions (where $n \geq 2$).

Finally, we conclude this survey on inverse manipulator dynamics with the following note. In an effort to reduce further the computational cost for solving the IDP the particular kinematic and dynamic structures of the manipulator were taken into consideration by some researchers. This effort resulted in a class of dynamic algorithms which are referred to in the literature as *customized* algorithms. Customized robot dynamics algorithms, which can be derived based on any formulation (the Lagrangian, the Newton-Euler or Kane's) are usually based on mathematical models of individual manipulators. Hence, customized algorithms can be systematically organized [46-53] and therefore such algorithms exhibit a significant increase in computational efficiency over general-purpose algorithms. Also, some aspects of better manipulator designs for reduced dynamic complexity have been considered [54,55]. In this approach, the kinematic structure and mass distribution of a manipulator arm are designed so that the inertia matrix of the manipulator becomes diagonal and/or invariant for an arbitrary arm configuration. However, this approach leads to algorithms which are applicable to particular classes of manipulators with two and three degrees of freedom only. Finally, to facilitate real-time implementation of advanced robot control strategies, parallel processing techniques have been used [56-60] to implement many of the existing algorithms which compute inverse dynamics.

5.2.4 Observations Concerning Computational Issues in the IDP

Based on this brief survey, we can make the following observations concerning various computational issues in evaluating manipulator inverse dynamics.

It is clear that for solving manipulator inverse dynamics, recursive algorithms are computationally more efficient than closed-form ones, since in recursive algorithms unnecessary computations (usually duplications) are avoided. Also, it can be shown that the computational efficiency of an algorithm for solving the IDP is independent of the particular equations of motion (Newton-Euler, Euler-Lagrange or Kane's) used to derive it. The computational efficiency of these recursive algorithms, as Silver [35] has pointed out, depends mainly on "the structure of the computation and choice of representation'' and this survey on inverse dynamics confirms Silver's remark. To emphasize this important remark, we restate it as follows : The computational efficiency of a recursive algorithm for evaluating inverse dynamics depends mainly on the following factors:

(a) the particular representation of various physical quantities appearing in the equations of motion;

(b) the underlying modeling scheme used for the manipulator;

(c) the organization of the computations and the degree of customization involved in its numerical or symbolic implementation.

From the brief survey presented above, one can see that the organization of the computations and the degree of customization is actually the point on which most of the existing recursive algorithms differ. The large number of these algorithms reveals that many alternatives have been considered in reducing the computational cost. However, particular analytical organization procedures and customization are usually used for the implementation of an algorithm and not for deriving it. Our basic objective in this monograph is to improve the computational efficiency of algorithms for solving the IDP through a better understanding of the mathematical representations used to describe the equations of motion and not through better implementations of existing formulations. Therefore, we shall not consider the latter aspect here. More information on organization procedures and customization can be found, for instance, in [47] and in the extensive list of references cited therein.

For the class of robot manipulators we are dealing with, namely, rigid-link, open-chain manipulators, the modeling scheme is simple and common to most of the existing algorithms. Usually, each link is considered to be a rigid body and the manipulator is modeled as an ideally connected (i.e, without friction or any backlash) open-chain of rigid bodies. This chain is assumed to be a rigid structure when static force analysis is required, as in the case of the Newton-Euler formulation. This modeling scheme, as well as another one which utilizes the ideas of augmented and generalized links, will be used in the next section to demonstrate the effects of the underlying modeling scheme on the structure of a recursive algorithm. In particular, we shall show that the modeling scheme which is based on augmented and generalized links leads to algorithms which have computational advantages, because they allow for some quantities to be computed *off-line*.

Finally, the choice of representation is the most important factor affecting the computational efficiency of the algorithms we are dealing with. From the survey, it is clear that the debate about the Euler-Lagrange and Newton-Euler formulations, mentioned above, was actually an indirect debate about the proper representation (as far as computational efficiency is concerned) of angular velocity. Theoretically, it is known that the two formulations are equivalent. Therefore, the real question, although never stated explicitly as such, was which representation for angular velocity describes the angular motion of a rigid body more efficiently. As noted by Silver [35], the angular motion of a rigid body can be described equally well by either angular velocity vector, which is used in the Newton-Euler formulation, or the derivative of a rotation tensor which is used in the Lagrangian formulation. At the time that Silver's work was published, the algorithm by Luh, Walker and Paul clearly indicated that the angular motion of a rigid body was described more efficiently by angular velocity vector. Thus, the vector representation of angular velocity became standard in the dynamic analysis of robotic systems. Following the vector representation of angular velocity, all the other quantities which are defined in terms of angular velocity were also represented by vectors. Thus, vector representations and vector analysis have been used almost exclusively for deriving computationally efficient algorithms for solving the IDP.

However, as we have shown in Chapter 4, the angular motion of a rigid body is described more efficiently by using a Cartesian tensor representation for the angular velocity. Obviously, this provides another possibility for describing efficiently the dynamic equations of a system of rigid bodies, in general, and solving the manipulator IDP in particular. Therefore, the question which arises at this point is the following : Does the tensor representation of the angular velocity lead us to recursive algorithms which compute inverse dynamics more efficiently than those algorithms which are derived using the traditional vector representation of angular velocity ?

The answer to this question is in the affirmative as will be shown in the next section.

5.3 A Cartesian Tensor Approach for Solving the IDP

As we mentioned in the previous section, the method proposed by Luh, Walker and Paul is the most suitable one for deriving computationally efficient recursive algorithms for solving the IDP. In this section, following this method and using Cartesian tensor analysis, we shall derive recursive algorithms for computing inverse dynamics, which are computationally far more efficient than similar algorithms derived using vector analysis. In particular, employing the methodology and basic theorems introduced in Chapter 4, we shall reformulate Algorithm 5.3, which was presented in Section 5.2. We shall do this by rewriting the basic vector equations of this algorithm in an equivalent, but computationally more efficient, tensor formulation. Moreover, to increase the computational efficiency of this algorithm further, we shall examine the underlying modeling scheme for the class of robot manipulators that we are dealing with. To this end, we shall derive a second algorithm by using a modeling scheme which employs the ideas of augmented and generalized links. Finally, the numerical implementation and computational complexity of these algorithms are considered and compared with similar algorithms that can be found in the literature.

5.3.1 New Algorithms for Solving the IDP

In the Newton-Euler formulation, the dynamic equations for robot manipulators are obtained by evaluating recursively the velocities and accelerations for each link and then applying Newton's and Euler's equations to each link. In the first step, the recursions are performed from link 1 to link n. Then, in the second step, using static force and torque analysis, the joint actuator forces/torques are computed with the recursions applied from joint n to joint 1. These recursions can be stated in an algorithmic form as was done for Algorithm 5.3.

In Algorithm 5.3, the absolute linear acceleration of the center of mass of each link is computed following the classical vector approach in which the absolute angular velocity and acceleration are represented by vectors. Also, in this algorithm, the generalized Euler equation is stated in its classical vector formulation in terms of the vector angular velocity and vector angular acceleration. However, as we have seen in Chapter 4, the absolute linear acceleration of a point on a moving rigid body as well as the Euler equation can be described by using the angular acceleration tensor instead of using the vectors of angular velocity and acceleration. Therefore, in an effort to improve the computational efficiency of Algorithm 5.3, we shall use Cartesian tensor analysis to reformulate most of the recurrence relations in this algorithm. The basic tensor-vector identities, proven in Chapters 3 and 4, will make the process here straightforward and simple. We need only to note that the equations of motion in Algorithm 5.3 are written with reference to link coordinate systems as opposed to Chapter 4 where the equations of motion are written with reference to an inertial coordinate system. However, as we mentioned in Chapter 3, Cartesian tensor equations are invariant under orthogonal coordinate transformations. Therefore, all the equations which describe the rigid body motion and which, in Chapter 4, are written relative to an inertial coordinate system, will be written here relative to link coordinate systems by using appropriate orthogonal coordinate transformations.

To derive a tensor formulation for Algorithm 5.3, we obviously have to abandon Gibb's classical vector cross product operation. As was shown in Chapter 4, we have to use the tensors $\tilde{\omega}$ and $\Omega = \dot{\tilde{\omega}} + \tilde{\omega}\tilde{\omega}$ written here with reference to appropriate link coordinate systems. Using these two tensors, the equations of Algorithm 5.3 can be modified as follows :

Equation (5.2.23b) can be expressed as

$$\dot{\omega}_{i+1}^{i+1} = \mathbf{A}_{i+1}^{T}\dot{\omega}_{i}^{i} + \sigma_{i+1}[\tilde{\omega}_{i}^{i+1}\mathbf{z}_{i+1}^{i+1}\dot{q}_{i+1} + \mathbf{z}_{i+1}^{i+1}\ddot{q}_{i+1}]. \tag{5.3.1}$$

Also, equation (5.2.23c) can be written as

$$\ddot{\mathbf{s}}_{0,i+1}^{i+1} = \mathbf{A}_{i+1}^{T}[\ddot{\mathbf{s}}_{0,i}^{i} + \Omega_{i}^{i}\mathbf{s}_{i,i+1}^{i}] + (1-\sigma_{i+1})[2\tilde{\omega}_{i}^{i+1}\mathbf{z}_{i+1}^{i+1}\dot{q}_{i+1} + \mathbf{z}_{i+1}^{i+1}\ddot{q}_{i+1}] \tag{5.3.2}$$

and equation (5.2.23d) can be written as

$$\ddot{\mathbf{r}}_{0,i+1}^{i+1} = \Omega_{i+1}^{i+1}\mathbf{r}_{i+1,i+1}^{i+1} + \ddot{\mathbf{s}}_{0,i+1}^{i+1}. \tag{5.3.3}$$

Newton's equation, i.e., equation (5.2.23e) is very simple in its vector form and therefore we do not modify it. However, Euler's equation, i.e., equation (5.2.23f), assumes a simpler structure when it is written in tensor form and, in particular, when it is expressed in terms of the Euler tensor $\mathbf{J}_{C_{i+1}}^{i+1}$. Therefore, using equation (4.3.6) to translate the inertia tensor $\mathbf{I}_{C_{i+1}}^{i+1}$ to the Euler tensor $\mathbf{J}_{C_{i+1}}^{i+1}$, we first write equation (5.2.23f) in the following tensor form

$$\tilde{\mathbf{M}}_{C_{i+1}}^{i+1} = \Omega_{i+1}^{i+1}\mathbf{J}_{C_{i+1}}^{i+1} - [\Omega_{i+1}^{i+1}\mathbf{J}_{C_{i+1}}^{i+1}]^{T} \tag{5.3.4}$$

and then we recover the vector $\mathbf{M}_{C_{i+1}}^{i+1}$ from the skew-symmetric tensor $\tilde{\mathbf{M}}_{C_{i+1}}^{i+1}$ by using the dual operator. The dual operator has been introduced in Chapter 3 by equation (3.3.31). Thus, in a tensor formulation the vector $\mathbf{M}_{C_{i+1}}^{i+1}$ is computed by the following equations,

$$\mathbf{J}_{C_{i+1}}^{i+1} = \frac{1}{2}tr[\mathbf{I}_{C_{i+1}}^{i+1}]\mathbf{1} - \mathbf{I}_{C_{i+1}}^{i+1} \tag{5.3.5a}$$

$$\tilde{\mathbf{M}}_{C_{i+1}}^{i+1} = \Omega_{i+1}^{i+1}\mathbf{J}_{C_{i+1}}^{i+1} - [\Omega_{i+1}^{i+1}\mathbf{J}_{C_{i+1}}^{i+1}]^{T} \tag{5.3.5b}$$

$$\mathbf{M}_{C_{i+1}}^{i+1} = dual(\tilde{\mathbf{M}}_{C_{i+1}}^{i+1}). \tag{5.3.5c}$$

In the second step of Algorithm 5.3, we need only to reformulate equation (5.2.24b) which can be written in the form

$$\eta_i^i = \mathbf{M}_{C_i}^i + \tilde{\mathbf{s}}_{i,i+1}^i \mathbf{f}_{i+1}^i + \tilde{\mathbf{r}}_{i,i}^i \mathbf{F}_{C_i}^i + \mathbf{A}_{i+1} \eta_{i+1}^{i+1}. \tag{5.3.6}$$

Now, using equations (5.3.1)-(5.3.6), we can state Algorithm 5.3 in a new formulation as follows :

ALGORITHM 5.6

Step 0 : Initialization

$$\sigma_i = \begin{cases} 1 & \text{revolute } i\text{-th joint} \\ 0 & \text{prismatic } i\text{-th joint} \end{cases}, \quad q_i = \begin{cases} \theta_i & \text{revolute } i\text{-th joint} \\ d_i & \text{prismatic } i\text{-th joint} \end{cases}$$

$$\omega_0^0 = 0, \; \dot{\omega}_0^0 = 0, \; \Omega_0^0 = 0, \; \ddot{\mathbf{s}}_{0,0}^0 = -\mathbf{g}, \; \mathbf{A}_{n+1} = 0, \; \mathbf{z}_i^i = [\,0\,0\,1\,]^T$$

$$\mathbf{J}_{C_{i+1}}^{i+1} = \frac{1}{2} tr[\mathbf{I}_{C_{i+1}}^{i+1}]\mathbf{1} - \mathbf{I}_{C_{i+1}}^{i+1}$$

Step 1 : Forward recursion :- For $i = 0, n\text{-}1$ do

$$\omega_{i+1}^{i+1} = \mathbf{A}_{i+1}^T \omega_i^i + \sigma_{i+1} \mathbf{z}_{i+1}^{i+1} \dot{q}_{i+1} \tag{5.3.7a}$$

$$\dot{\omega}_{i+1}^{i+1} = \mathbf{A}_{i+1}^T \dot{\omega}_i^i + \sigma_{i+1}[\tilde{\omega}_i^{i+1} \mathbf{z}_{i+1}^{i+1} \dot{q}_{i+1} + \mathbf{z}_{i+1}^{i+1} \ddot{q}_{i+1}] \tag{5.3.7b}$$

$$\Omega_{i+1}^{i+1} = \tilde{\dot{\omega}}_{i+1}^{i+1} + \tilde{\omega}_{i+1}^{i+1} \tilde{\omega}_{i+1}^{i+1} \tag{5.3.7c}$$

$$\ddot{\mathbf{s}}_{0,i+1}^{i+1} = \mathbf{A}_{i+1}^T [\ddot{\mathbf{s}}_{0,i}^i + \Omega_i^i \mathbf{s}_{i,i+1}^i] + (1-\sigma_{i+1})[2\tilde{\omega}_i^{i+1} \mathbf{z}_{i+1}^{i+1} \dot{q}_{i+1} + \mathbf{z}_{i+1}^{i+1} \ddot{q}_{i+1}] \tag{5.3.7d}$$

$$\ddot{\mathbf{r}}_{0,i+1}^{i+1} = \Omega_{i+1}^{i+1} \mathbf{r}_{i+1,i+1}^{i+1} + \ddot{\mathbf{s}}_{0,i+1}^{i+1} \tag{5.3.7e}$$

$$\mathbf{F}_{C_{i+1}}^{i+1} = m_{i+1} \ddot{\mathbf{r}}_{0,i+1}^{i+1} \tag{5.3.7f}$$

$$\tilde{\mathbf{M}}_{\mathrm{C}_{i+1}}^{i+1} = \boldsymbol{\Omega}_{i+1}^{i+1}\mathbf{J}_{\mathrm{C}_{i+1}}^{i+1} - [\boldsymbol{\Omega}_{i+1}^{i+1}\mathbf{J}_{\mathrm{C}_{i+1}}^{i+1}]^T \tag{5.3.7g}$$

$$\mathbf{M}_{\mathrm{C}_{i+1}}^{i+1} = dual(\tilde{\mathbf{M}}_{\mathrm{C}_{i+1}}^{i+1}) \tag{5.3.7h}$$

Step 2 : Backward recursion :- For *i = n, 1* do

$$\mathbf{f}_i^i = \mathbf{A}_{i+1}\mathbf{f}_{i+1}^{i+1} + \mathbf{F}_{\mathrm{C}_i}^i \tag{5.3.8a}$$

$$\boldsymbol{\eta}_i^i = \mathbf{M}_{\mathrm{C}_i}^i + \tilde{\mathbf{s}}_{i,i+1}^i\mathbf{f}_{i+1}^i + \tilde{\mathbf{r}}_{i,i}^i\mathbf{F}_{\mathrm{C}_i}^i + \mathbf{A}_{i+1}\boldsymbol{\eta}_{i+1}^{i+1} \tag{5.3.8b}$$

$$\tau_i = \sigma_i(\boldsymbol{\eta}_i^i \cdot \mathbf{z}_i^i) + (1-\sigma_i)(\mathbf{f}_i^i \cdot \mathbf{z}_i^i) \tag{5.3.8c}$$

end

We shall be concerned with the numerical implementation of Algorithm 5.6 in the next section. However, as we can notice here, the structure of this algorithm reveals that for its implementation all the quantities (with the exception of the Euler tensors) have to be computed *on-line*. Obviously, from a computational point of view it is desirable to devise algorithms which allow us to compute *off-line* as many quantities as possible and at the same time, to keep the on-line computations as simple as possible. To see if this is feasible, we have to examine the underlying modeling scheme for the class of robot manipulators we are dealing with, since the structure of an algorithm obviously depends on it.

As we mentioned in Section 5.2, the robot manipulator is modeled, in general, as an ideally connected, open-loop, serial-chain of rigid bodies. When frictional forces at the joints are to be considered we compute them based on the joint velocities and add them directly to the joint generalized forces. This clearly justifies the idealization in the connections of the rigid links. Now, utilizing this modeling scheme, kinematic and dynamic recurrence relations are defined, based on which the recursive Algorithm 5.3 (or Algorithm 5.6) has been derived. As is well known [34], the kinematic recurrence relations are defined by analyzing the velocity "propagation" from link to link starting from the base of the manipulator to the end-effector. The dynamic recurrence relations are defined based on a static force and moment analysis. In particular, in deriving the dynamic recurrence relations in Algorithm 5.3, or in Algorithm 5.6, it has been assumed that the manipulator is locked at the joints so that it becomes a structure which is "rigid" in static equilibrium, and that static force and moment analysis has been performed for each link. However, in defining these dynamic recurrence relations, the analysis for the static forces and moments can be modified. For the static analysis, as long as we do not disturb the static equilibrium position of the manipulator, we are free to merge links and thus to generate hypothetical "generalized" links or even to assume the presence of "fictitious" links. In the following, using this "unconventional" static analysis, we shall modify the dynamic recurrence relations of Algorithm 5.6. Then, based on these modified dynamic recurrence relations, we shall state a new algorithm which, when applied to most industrial robot manipulators in use today, allows us to compute some quantities off-line.

To proceed, we first need to introduce the concepts of augmented and generalized links, which are shown in Figure 5.1.

Definition 5.1 : An augmented link i is a *fictitious* link composed of link i and the mass of links $i+1, i+2, \cdots, n$, attached to the origin of the $(i+1)$-th coordinate system.

This definition, which can be applied regardless of the type of joint (i.e., revolute or prismatic), is slightly different from the one presented in [23] in that the mass of the augmented link is not the total mass of the system (here the robot manipulator). Note that an augmented link is "rigid" (i.e., has fixed geometry) if and only if the (i+1)th joint is revolute because, when the $(i+1)$-th joint is prismatic, the position vector of the origin of the $(i+1)$-th coordinate system relative to the origin of the i-th coordinate system, i.e., the vector $\mathbf{s}_{i,i+1}^{i}$, is not constant.

Definition 5.2 : A generalized link i is a composite link, consisting of links i through n treated as a single rigid body structure.

In order to modify the dynamic recurrence relations of Algorithm 5.6, we need to define the following moments :

(a) The 0-th moment or mass of the i-th augmented or the i-th generalized link :

$$\overline{m}_i = m_i + \overline{m}_{i+1} \tag{5.3.9}$$

where m_i is the mass of the i-th link.

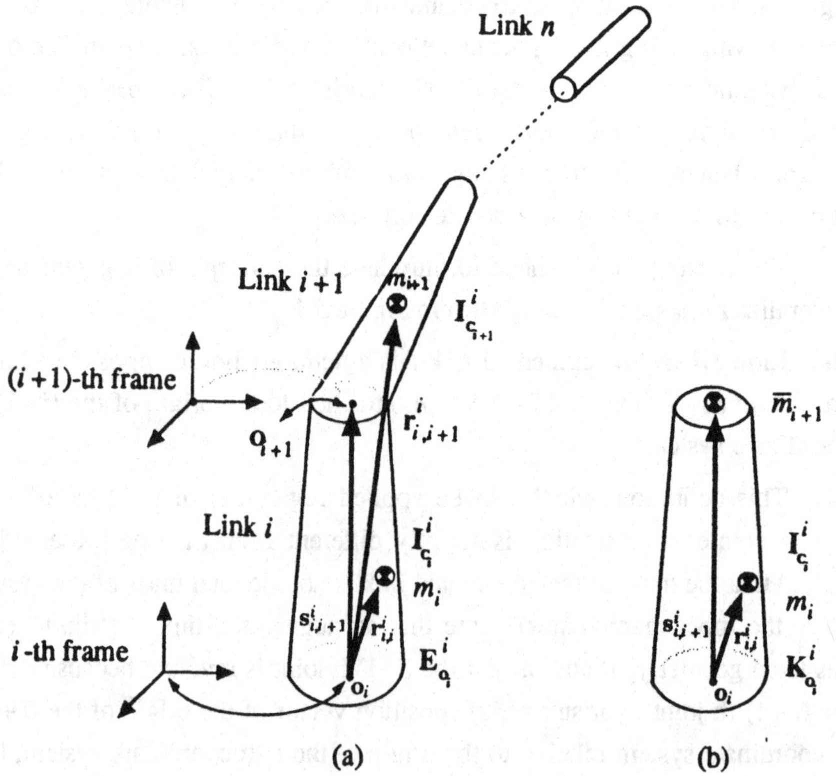

Figure 5.1: (a) The i-th generalized link, (b) The i-th augmented link .

Also, the first and second moments of the augmented link i with respect to the origin of the i-th coordinate system, expressed in the i-th link frame, are defined as follows:

(b) First moment of the augmented link i :

$$\mathbf{u}^i_{\mathrm{o}_i} = m_i \mathbf{r}^i_{i,i} + \bar{m}_{i+1} \mathbf{s}^i_{i,i+1}. \tag{5.3.10}$$

(c) Second moment or inertia tensor of the augmented link i :

$$\mathbf{K}^i_{\mathrm{o}_i} = \mathbf{I}^i_{\mathrm{c}_i} - m_i \tilde{\mathbf{r}}^i_{i,i} \tilde{\mathbf{r}}^i_{i,i} - \bar{m}_{i+1} \tilde{\mathbf{s}}^i_{i,i+1} \tilde{\mathbf{s}}^i_{i,i+1} \tag{5.3.11}$$

where $\mathbf{I}^i_{\mathrm{c}_i}$ is the inertia tensor of link i with respect to its center of mass expressed in the i-th coordinate frame. Note that when the $(i+1)$-th joint is revolute, the first and second moments are independent of the configuration of the manipulator and can be computed off-line. We also need the first moment about the origin of the i-th coordinate frame of the generalized link i. This is obtained as

(d) First moment of the generalized link i :

$$\mathbf{U}^i_{\mathrm{o}_i} = \sum_{j=i}^{n} m_j \mathbf{r}^i_{i\,j}. \tag{5.3.12}$$

In the equation above, $\mathbf{U}^i_{\mathrm{o}_i}$, expressed in the i-th link frame, is configuration dependent, and therefore must be calculated on-line. However, we can compute $\mathbf{U}^i_{\mathrm{o}_i}$ recursively as the following lemma shows:

Lemma 5.1 : The first moment of the i-th generalized link satisfies the following recursive equation

$$\mathbf{U}^i_{\mathrm{o}_i} = \mathbf{u}^i_{\mathrm{o}_i} + \mathbf{A}_{i+1} \mathbf{U}^{i+1}_{\mathrm{o}_{i+1}}. \tag{5.3.13}$$

Proof: From equation (5.3.12) we have

$$\mathbf{U}^i_{\mathrm{o}_i} = \sum_{j=i}^{n} m_j \mathbf{r}^i_{i\,j}$$

$$= m_i \mathbf{r}^i_{i,i} + \sum_{j=i+1}^{n} m_j \mathbf{r}^i_{i,j}.$$

Since for $j > i$, $\mathbf{r}^i_{i,j} = \mathbf{s}^i_{i,i+1} + \mathbf{A}_{i+1} \mathbf{r}^{i+1}_{i+1,j}$, we have

$$\mathbf{U}^i_{o_i} = m_i \mathbf{r}^i_{i,i} + \sum_{j=i+1}^{n} m_j \mathbf{s}^i_{i,i+1} + \sum_{j=i+1}^{n} \mathbf{A}_{i+1} m_j \mathbf{r}^{i+1}_{i+1,j}$$

$$= m_i \mathbf{r}^i_{i,i} + \bar{m}_{i+1} \mathbf{s}^i_{i,i+1} + \mathbf{A}_{i+1} \mathbf{U}^{i+1}_{o_{i+1}}$$

$$= \mathbf{u}^i_{o_i} + \mathbf{A}_{i+1} \mathbf{U}^{i+1}_{o_{i+1}}$$ □

In the following, we shall analyze the rotational motion of an augmented link, say the i-th one. For the sake of simplicity we assume first that the i-th augmented link has rigid body characteristics, i.e., the $(i+1)$-th joint is assumed to be revolute. Later we shall extend the analysis to include augmented links for which the $(i+1)$-th joint is prismatic. We begin by reviewing the rotational motion of the i-th link about the i-th joint since both the i-th link and the i-th augmented link require similar dynamic analysis. In particular, both have the same angular velocity and angular acceleration. However, since the i-th augmented link has different mass from the i-th link, it has obviously different dynamic characteristics.

When the i-th link experiences a rotational motion with center of rotation at the origin of the i-th coordinate system and with angular velocity $\boldsymbol{\omega}^i_i$ and angular acceleration $\dot{\boldsymbol{\omega}}^i_i$, a resultant torque or moment of force vector $\mathbf{M}^i_{o_i}$ is developed with respect to the center of rotation which, as shown in Chapter 4, satisfies equation (4.3.36). This equation, expressed in link coordinate system orientation, is written here as

$$\mathbf{M}^i_{o_i} = \mathbf{M}^i_{c_i} + \mathbf{r}^i_{i,i} \times \mathbf{F}^i_{c_i} \tag{5.3.14}$$

where

$$\mathbf{M}^i_{c_i} = \mathbf{I}^i_{c_i} \dot{\boldsymbol{\omega}}^i_i + \tilde{\boldsymbol{\omega}}^i_i \mathbf{I}^i_{c_i} \boldsymbol{\omega}^i_i \tag{5.3.15}$$

is the resultant torque with respect to the center of mass and

$$\mathbf{F}^{i}_{C_i} = m_i \ddot{\mathbf{r}}^{i}_{i,i} \tag{5.3.16}$$

is the total force caused at the center of mass of the i-th link due to its acceleration. Using equation (5.3.16), we can write equation (5.3.14) in the form

$$\mathbf{M}^{i}_{O_i} = \mathbf{M}^{i}_{C_i} + m_i \tilde{\mathbf{r}}^{i}_{i,i} \ddot{\mathbf{r}}^{i}_{i,i}. \tag{5.3.17}$$

Equation (5.3.17) is the basic equation which describes the rotational motion of the i-th link. Now, if instead of the i-th link, the i-th augmented link actually experiences this rotational motion, then equation (5.3.17) needs to be changed to

$$\boldsymbol{\mu}^{i}_{i} = \mathbf{M}^{i}_{C_i} + m_i \tilde{\mathbf{r}}^{i}_{i,i} \ddot{\mathbf{r}}^{i}_{i,i} + \bar{m}_{i+1} \tilde{\mathbf{s}}^{i}_{i,i+1} \ddot{\mathbf{s}}^{i}_{i,i+1} \tag{5.3.18}$$

where the term $\bar{m}_{i+1} \tilde{\mathbf{s}}^{i}_{i,i+1} \ddot{\mathbf{s}}^{i}_{i,i+1}$ has been added to account for the torque which will be caused due to the presence of a mass equal to $\bar{m}_{i+1}$ at a point which has position vector $\mathbf{s}^{i}_{i,i+1}$ relative to the center of rotation. Obviously, the resultant torque vector is now denoted by a different vector, the vector $\boldsymbol{\mu}^{i}_{i}$. Moreover, as we have shown in Chapter 4, equation (5.3.17), which is actually another formulation for the generalized Euler equation, can be written in terms of the link inertia tensor as follows

$$\mathbf{M}^{i}_{O_i} = \mathbf{I}^{i}_{O_i} \dot{\boldsymbol{\omega}}^{i}_{i} + \tilde{\boldsymbol{\omega}}^{i}_{i} \mathbf{I}^{i}_{O_i} \boldsymbol{\omega}^{i}_{i} \tag{5.3.19}$$

where $\mathbf{I}^{i}_{O_i} = \mathbf{I}^{i}_{C_i} - m_i \tilde{\mathbf{r}}^{i}_{i,i} \tilde{\mathbf{r}}^{i}_{i,i}$ is the inertia tensor of the i-th link with respect to the origin of the i-th link coordinate system. Therefore, by analogy, we can say that equation (5.3.18) describes the generalized Euler equation for the i-th augmented link, when it is written with respect to the origin of the i-th link coordinate system. Obviously, as is the case with the i-th link, the generalized Euler equation can be written in terms of the inertia tensor $\mathbf{K}^{i}_{O_i}$ of the i-th augmented link and the vectors of the angular velocity $\boldsymbol{\omega}^{i}_{i}$ and angular acceleration $\dot{\boldsymbol{\omega}}_i^i$, as follows

$$\boldsymbol{\mu}_i^i = \mathbf{K}_{o_i}^i \dot{\boldsymbol{\omega}}_i^i + \tilde{\boldsymbol{\omega}}_i^i \mathbf{K}_{o_i}^i \boldsymbol{\omega}_i^i \tag{5.3.20}$$

where the inertia tensor $\mathbf{K}_{o_i}^i$ of the i-th augmented link is defined by equation (5.3.11). To see that equation (5.3.18) is indeed equivalent to equation (5.3.20), we proceed as follows. First we need to prove the following relations

$$\tilde{\mathbf{r}}_{i,i}^i \ddot{\mathbf{r}}_{i,i}^i = -[\tilde{\mathbf{r}}_{i,i}^i \tilde{\mathbf{r}}_{i,i}^i \dot{\boldsymbol{\omega}}_i^i + \tilde{\boldsymbol{\omega}}_i^i \tilde{\mathbf{r}}_{i,i}^i \tilde{\mathbf{r}}_{i,i}^i \boldsymbol{\omega}_i^i] \tag{5.3.21}$$

and

$$\tilde{\mathbf{s}}_{i,i+1}^i \ddot{\mathbf{s}}_{i,i+1}^i = -[\tilde{\mathbf{s}}_{i,i+1}^i \tilde{\mathbf{s}}_{i,i+1}^i \dot{\boldsymbol{\omega}}_i^i + \tilde{\boldsymbol{\omega}}_i^i \tilde{\mathbf{s}}_{i,i+1}^i \tilde{\mathbf{s}}_{i,i+1}^i \boldsymbol{\omega}_i^i]. \tag{5.3.22}$$

To prove (5.3.21), we note that since $\mathbf{r}_{i,i}^i$ is a constant vector, its absolute acceleration satisfies the equation

$$\ddot{\mathbf{r}}_{i,i}^i = \boldsymbol{\Omega}_i^i \mathbf{r}_{i,i}^i \tag{5.3.23}$$

or

$$\begin{aligned}\ddot{\mathbf{r}}_{i,i}^i &= \left[\dot{\tilde{\boldsymbol{\omega}}}_i^i + \tilde{\boldsymbol{\omega}}_i^i \tilde{\boldsymbol{\omega}}_i^i\right] \mathbf{r}_{i,i}^i \\ &= \dot{\tilde{\boldsymbol{\omega}}}_i^i \mathbf{r}_{i,i}^i + \tilde{\boldsymbol{\omega}}_i^i \tilde{\boldsymbol{\omega}}_i^i \mathbf{r}_{i,i}^i \\ &= -\tilde{\mathbf{r}}_{i,i}^i \dot{\boldsymbol{\omega}}_i^i - \tilde{\boldsymbol{\omega}}_i^i \tilde{\mathbf{r}}_{i,i}^i \boldsymbol{\omega}_i^i\end{aligned} \tag{5.3.24}$$

where equation (3.4.6) has been used in the last step. Therefore, we can write

$$\tilde{\mathbf{r}}_{i,i}^i \ddot{\mathbf{r}}_{i,i}^i = -\left[\tilde{\mathbf{r}}_{i,i}^i \tilde{\mathbf{r}}_{i,i}^i \dot{\boldsymbol{\omega}}_i^i + \tilde{\mathbf{r}}_{i,i}^i \tilde{\boldsymbol{\omega}}_i^i \tilde{\mathbf{r}}_{i,i}^i \boldsymbol{\omega}_i^i\right]. \tag{5.3.25}$$

Moreover, using equation (3.4.19b) we can write

$$\tilde{\mathbf{r}}_{i,i}^i \tilde{\boldsymbol{\omega}}_i^i \tilde{\mathbf{r}}_{i,i}^i = \tilde{\boldsymbol{\omega}}_i^i \tilde{\mathbf{r}}_{i,i}^i \tilde{\mathbf{r}}_{i,i}^i + \tilde{\mathbf{r}}_{i,i}^i \tilde{\mathbf{r}}_{i,i}^i \tilde{\boldsymbol{\omega}}_i^i - \frac{1}{2} tr[\tilde{\mathbf{r}}_{i,i}^i \tilde{\mathbf{r}}_{i,i}^i] \boldsymbol{\omega}_i^i. \tag{5.3.26}$$

Then, by substituting equation (5.3.26) into equation (5.3.25) and noting that $\tilde{\boldsymbol{\omega}}_i^i \boldsymbol{\omega}_i^i = 0$, equation (5.3.25) becomes equation (5.3.21). Also, since the

$(i+1)$-th joint has been assumed to be revolute, the vector $\mathbf{s}^i_{i,i+1}$ is constant, and so we have

$$\ddot{\mathbf{s}}^i_{i,i+1} = \mathbf{\Omega}^i_i \mathbf{s}^i_{i,i+1}. \tag{5.3.27}$$

Equation (5.3.22) can then be proved following the same arguments as in the proof of equation (5.3.21). Now, substituting equations (5.3.15), (5.3.21) and (5.3.22) into equation (5.3.18) and using the definition of $\mathbf{K}^i_{o_i}$ from equation (5.3.11), we get equation (5.3.20).

In this analysis, the generalized Euler equation for both the i-th link and the i-th augmented link has been stated in its vector form. However, as we have seen in Chapter 4, the generalized Euler equation can also be stated in a tensor formulation. Moreover, the tensor formulation of the generalized Euler equation assumes a simpler formulation when it is stated in terms of the rigid body Euler tensor. Therefore, for a simple tensor formulation of equation (5.3.20), we need to define the Euler tensor of the i-th generalized link. This can be easily done if one uses equation (4.3.6) which transforms the rigid body inertia tensor into the Euler tensor. Thus, we shall use the symbol $\hat{\mathbf{K}}^i_{o_i}$ (note: this is not a spatial quantity) to denote the Euler tensor of the i-th augmented link and we define it as follows

$$\hat{\mathbf{K}}^i_{o_i} = \frac{1}{2} tr[\mathbf{K}^i_{o_i}]\mathbf{1} - \mathbf{K}^i_{o_i} \tag{5.3.28}$$

where $\mathbf{K}^i_{o_i}$ is the inertia tensor of the i-th augmented link. Using this Euler tensor, we can state the generalized Euler equation of the i-th augmented link in a tensor form as follows,

$$\tilde{\boldsymbol{\mu}}^i_i = \mathbf{\Omega}^i_i \hat{\mathbf{K}}^i_{o_i} - [\mathbf{\Omega}^i_i \hat{\mathbf{K}}^i_{o_i}]^T. \tag{5.3.29}$$

where $\mathbf{\Omega}^i_i$ is the angular acceleration of the i-th augmented link. Obviously, from the skew-symmetric tensor $\tilde{\boldsymbol{\mu}}^i_i$, we can recover the vector invariant $\boldsymbol{\mu}^i_i$ by using the dual operator, i.e., we can write

$$\overset{i}{\boldsymbol{\mu}}_i = dual(\overset{i}{\tilde{\boldsymbol{\mu}}}_i).$$

From the foregoing, when the $(i+1)$-th joint is revolute, the dynamic analysis for the i-th augmented link is very simple. However, this analysis has to be modified when the $(i+1)$-th joint is prismatic because, in this case the i-th augmented link is not a rigid body. As we have mentioned above, the vector $\overset{i}{\mathbf{s}}_{i,i+1}$ is not constant in this case and therefore, the vector $\overset{\cdot\cdot i}{\mathbf{s}}_{i,i+1}$ is not equal to $\overset{i}{\boldsymbol{\Omega}}_i\overset{i}{\mathbf{s}}_{i,i+1}$. The correct expression for the vector $\overset{\cdot\cdot i}{\mathbf{s}}_{i,i+1}$ follows from equation (5.3.7d) and is as follows.

$$\overset{\cdot\cdot i}{\mathbf{s}}_{i,i+1} = \overset{i}{\boldsymbol{\Omega}}_i\overset{i}{\mathbf{s}}_{i,i+1} + 2\overset{i}{\tilde{\boldsymbol{\omega}}}_i\overset{i}{\mathbf{z}}_{i+1}\dot{q}_{i+1} + \overset{i}{\mathbf{z}}_{i+1}\ddot{q}_{i+1} \tag{5.3.30}$$

or,

$$\overset{\cdot\cdot i}{\mathbf{s}}_{i,i+1} = \overset{i}{\boldsymbol{\Omega}}_i\overset{i}{\mathbf{s}}_{i,i+1} + (1-\sigma_{i+1})\overset{i}{\boldsymbol{\zeta}}_{i,i+1} \tag{5.3.31}$$

where

$$\overset{i}{\boldsymbol{\zeta}}_{i,i+1} = 2\overset{i}{\tilde{\boldsymbol{\omega}}}_i\overset{i}{\mathbf{z}}_{i+1}\dot{q}_{i+1} + \overset{i}{\mathbf{z}}_{i+1}\ddot{q}_{i+1}. \tag{5.3.32}$$

and

$$\sigma_{i+1} = \begin{cases} 1 \text{ if the } (i+1)\text{–th joint is revolute} \\ 0 \text{ if the } (i+1)\text{–th joint is prismatic} \end{cases} \tag{5.3.33}$$

Therefore, when the $(i+1)$-th joint is prismatic we have to modify equation (5.3.20). In this case, we denote the resultant torque by the vector $\overset{i}{\mathbf{v}}_i$, i.e., we write equation (5.3.18) as

$$\overset{i}{\mathbf{v}}_i = \overset{i}{\mathbf{M}}_{\mathrm{C}_i} + m_i\overset{i}{\tilde{\mathbf{r}}}_{i,i}\overset{\cdot\cdot i}{\mathbf{r}}_{i,i} + \bar{m}_{i+1}\overset{i}{\tilde{\mathbf{s}}}_{i,i+1}\overset{\cdot\cdot i}{\mathbf{s}}_{i,i+1}. \tag{5.3.34}$$

Now, following an analysis similar to that applied to equation (5.3.18) and using equation (5.3.31) instead of equation (5.3.28), we can show that equation (5.3.34) can be written as

$$\overset{i}{\mathbf{v}}_i = \overset{i}{\mathbf{K}}_{\mathrm{O}_i}\overset{\cdot i}{\boldsymbol{\omega}}_i + \overset{i}{\tilde{\boldsymbol{\omega}}}_i\overset{i}{\mathbf{K}}_{\mathrm{O}_i}\overset{i}{\boldsymbol{\omega}}_i + (1-\sigma_{i+1})\bar{m}_{i+1}\overset{i}{\tilde{\mathbf{s}}}_{i,i+1}\overset{i}{\boldsymbol{\zeta}}_{i,i+1}.$$

$$= \mu_i^i + (1-\sigma_{i+1})\bar{m}_{i+1}\bar{s}_{i,i+1}^i\zeta_{i,i+1}^i. \tag{5.3.35}$$

From the foregoing, the resultant torque vector at the origin of the i-th coordinate system, due to the rotational motion of the i-th augmented link, can be described by a single equation as

$$\mu_i^i + (1-\sigma_{i+1})\bar{m}_{i+1}\bar{s}_{i,i+1}^i\zeta_{i,i+1}^i = \mathbf{M}_{C_i}^i + m_i\bar{r}_{i,i}^i\ddot{r}_{i,i}^i + \bar{m}_{i+1}\bar{s}_{i,i+1}^i\ddot{s}_{i,i+1}^i \tag{5.3.36}$$

where σ_{i+1} is defined by equation (5.3.33).

With this preliminary result, we now assume that the links are augmented links and we proceed to modify Algorithm 5.6 so as to make it applicable to manipulators whose modeling scheme utilizes the ideas of augmented and generalized links. Since the kinematic analysis of an augmented link is the same as that of the corresponding actual link, only the dynamic recurrence relations need to be modified. Thus we proceed by reformulating the recurrence relation for the moment vector $\dot{\eta}_i^i$, which is exerted on link i by link i-1. To do this, we first write (5.3.8a) in its expanded form *i.e.*, we write

$$\mathbf{f}_i^i = \sum_{j=i}^{n} \mathbf{F}_{C_j}^i \tag{5.3.37}$$

where, from (5.3.7e), $\mathbf{F}_{C_j}^i = m_j\ddot{\mathbf{r}}_{0,j}^i$. Now, since $\ddot{\mathbf{r}}_{0,j}^i = \ddot{\mathbf{s}}_{0,i}^i + \ddot{\mathbf{r}}_{i,j}^i$ for $j \geq i$, we have

$$\mathbf{f}_i^i = \sum_{j=i}^{n} m_j(\ddot{\mathbf{s}}_{0,i}^i + \ddot{\mathbf{r}}_{i,j}^i)$$

$$= \sum_{j=i}^{n} m_j\ddot{\mathbf{s}}_{0,i}^i + \sum_{j=i}^{n} m_j\ddot{\mathbf{r}}_{i,j}^i = \bar{m}_i\ddot{\mathbf{s}}_{0,i}^i + \ddot{\mathbf{U}}_{O_i}^i \tag{5.3.38}$$

where $\bar{m}_i$ is defined by (5.3.9) and

$$\ddot{\mathbf{U}}_{O_i}^i = \sum_{j=i}^{n} m_j\ddot{\mathbf{r}}_{i,j}^i. \tag{5.3.39}$$

Equation (5.3.39) is a consequence of (5.3.12). However, we do not need to use equation (5.3.39) for computing the vector $\ddot{\mathbf{U}}^i_{\mathrm{o}_i}$. This vector can be computed recursively as the following lemma shows.

Lemma 5.2 : The absolute derivative of the first moment of the i-th generalized link satisfies the following recursive relation

$$\ddot{\mathbf{U}}^i_{\mathrm{o}_i} = \ddot{\mathbf{u}}^i_{\mathrm{o}_i} + \mathbf{A}_{i+1}[\ddot{\mathbf{U}}^{i+1}_{\mathrm{o}_{i+1}} + (1-\sigma_{i+1})\bar{m}_{i+1}\boldsymbol{\zeta}^{i+1}_{i,i+1}] \tag{5.3.40}$$

where σ_{i+1} is defined by equation (5.3.33), $\bar{m}_{i+1}$ is the 0-th moment of the $(i+1)$-th generalized link and $\boldsymbol{\zeta}^i_{i,i+1}$ is defined by equation (5.3.32)

Proof: From equation (5.3.12) we have

$$\ddot{\mathbf{U}}^i_{\mathrm{o}_i} = m_i\ddot{\mathbf{r}}^i_{i,i} + \sum_{j=i+1}^{n} m_j\ddot{\mathbf{r}}^i_{i,j}$$

Now, since for $j>i$, $\ddot{\mathbf{r}}^i_{i,j} = \ddot{\mathbf{s}}^i_{i,i+1} + \mathbf{A}_{i+1}\ddot{\mathbf{r}}^{i+1}_{i+1,j}$, we have

$$\ddot{\mathbf{U}}^i_{\mathrm{o}_i} = m_i\ddot{\mathbf{r}}^i_{i,i} + \sum_{j=i+1}^{n} m_j\ddot{\mathbf{s}}^i_{i,i+1} + \mathbf{A}_{i+1}\sum_{j=i+1}^{n} m_j\ddot{\mathbf{r}}^{i+1}_{i+1,j}$$

$$= m_i\ddot{\mathbf{r}}^i_{i,i} + \bar{m}_{i+1}\ddot{\mathbf{s}}^i_{i,i+1} + \mathbf{A}_{i+1}\ddot{\mathbf{U}}^{i+1}_{\mathrm{o}_{i+1}} \tag{5.3.41}$$

Further, using equations (5.3.23) and (5.3.31) we can write

$$m_i\ddot{\mathbf{r}}^i_{i,i} + \bar{m}_{i+1}\ddot{\mathbf{s}}^i_{i,i+1} = \boldsymbol{\Omega}^i_i[m_i\mathbf{r}^i_{i,i} + \bar{m}_{i+1}\mathbf{s}^i_{i,i+1}] + (1-\sigma_{i+1})\bar{m}_{i+1}\boldsymbol{\zeta}^i_{i,i+1}$$

$$= \boldsymbol{\Omega}^i_i\mathbf{u}^i_{\mathrm{o}_i} + (1-\sigma_{i+1})\bar{m}_{i+1}\boldsymbol{\zeta}^i_{i,i+1}$$

$$= \ddot{\mathbf{u}}^i_{\mathrm{o}_i} + (1-\sigma_{i+1})\bar{m}_{i+1}\mathbf{A}_{i+1}\boldsymbol{\zeta}^{i+1}_{i,i+1} \tag{5.3.42}$$

where, in the last step, equation (5.3.10) has been used. Finally, equation (5.3.40) follows on substituting equation (5.3.42) into equation (5.3.41). □

Now, for the vector $\boldsymbol{\eta}^i_i$, we have from (5.3.8b)

$$\boldsymbol{\eta}^i_i = \mathbf{M}^i_{\mathrm{C}_i} + \tilde{\mathbf{r}}^i_{i,i}\mathbf{F}^i_{\mathrm{C}_i} + \tilde{\mathbf{s}}^i_{i,i+1}\mathbf{A}_{i+1}\mathbf{f}^{i+1}_{i+1} + \mathbf{A}_{i+1}\boldsymbol{\eta}^{i+1}_{i+1}$$

$$= \mathbf{M}_{\mathrm{C}_i}^{i} + m_i \tilde{\mathbf{r}}_{i,i}^{i} \ddot{\mathbf{r}}_{0,i}^{i} + \tilde{\mathbf{s}}_{i,i+1}^{i} [\bar{m}_{i+1} \ddot{\mathbf{s}}_{0,i+1}^{i} + \ddot{\mathbf{U}}_{\mathrm{o}_{i+1}}^{i}] + \mathbf{A}_{i+1} \boldsymbol{\eta}_{i+1}^{i+1}$$

$$= \mathbf{M}_{\mathrm{C}_i}^{i} + m_i \tilde{\mathbf{r}}_{i,i}^{i} [\ddot{\mathbf{s}}_{0,i}^{i} + \ddot{\mathbf{r}}_{i,i}^{i}] + \bar{m}_{i+1} \tilde{\mathbf{s}}_{i,i+1}^{i} [\ddot{\mathbf{s}}_{0,i}^{i} + \ddot{\mathbf{s}}_{i,i+1}^{i}] + \tilde{\mathbf{s}}_{i,i+1}^{i} \ddot{\mathbf{U}}_{\mathrm{o}_{i+1}}^{i} + \mathbf{A}_{i+1} \boldsymbol{\eta}_{i+1}^{i+1}$$

$$= \mathbf{M}_{\mathrm{C}_i}^{i} + m_i \tilde{\mathbf{r}}_{i,i}^{i} \ddot{\mathbf{r}}_{i,i}^{i} + \bar{m}_{i+1} \tilde{\mathbf{s}}_{i,i+1}^{i} \ddot{\mathbf{s}}_{i,i+1}^{i} + [m_i \tilde{\mathbf{r}}_{i,i}^{i} + \bar{m}_{i+1} \tilde{\mathbf{s}}_{i,i+1}^{i}] \ddot{\mathbf{s}}_{0,i}^{i} + \mathbf{s}_{i,i+1}^{i} \ddot{\mathbf{U}}_{\mathrm{o}_{i+1}}^{i}$$

$$+ \mathbf{A}_{i+1} \boldsymbol{\eta}_{i+1}^{i+1}.$$

Further, using (5.3.10) and (5.3.36), we can write

$$\boldsymbol{\eta}_i^{i} = \boldsymbol{\mu}_i^{i} + (1 - \sigma_{i+1}) \bar{m}_{i+1} \tilde{\mathbf{s}}_{i,i+1}^{i} \boldsymbol{\zeta}_{i,i+1}^{i} + \tilde{\mathbf{u}}_{\mathrm{o}_i}^{i} \ddot{\mathbf{s}}_{0,i}^{i} + \tilde{\mathbf{s}}_{i,i+1}^{i} \ddot{\mathbf{U}}_{\mathrm{o}_{i+1}}^{i} + \mathbf{A}_{i+1} \boldsymbol{\eta}_{i+1}^{i+1}$$

$$= \boldsymbol{\mu}_i^{i} + \tilde{\mathbf{u}}_{\mathrm{o}_i}^{i} \ddot{\mathbf{s}}_{0,i}^{i} + \tilde{\mathbf{s}}_{i,i+1}^{i} [\ddot{\mathbf{U}}_{\mathrm{o}_{i+1}}^{i} + (1 - \sigma_{i+1}) \bar{m}_{i+1} \boldsymbol{\zeta}_{i,i+1}^{i}] + \mathbf{A}_{i+1} \boldsymbol{\eta}_{i+1}^{i+1} \quad (5.3.43)$$

We can see from (5.3.43) that in computing the vector $\boldsymbol{\eta}_i^{i}$ we do not use the vectors $\mathbf{F}_{\mathrm{C}_i}^{i}$ and $\mathbf{f}_i^{i}$. Therefore, we do not need to compute equations (5.3.7e) and (5.3.8a). In their place, we use equation (5.3.40). Also, we do not need to compute the vector $\mathbf{M}_{\mathrm{C}_i}^{i}$, since in (5.3.43), we use the vector $\boldsymbol{\mu}_i^{i}$.

We are now in a position to formally outline an algorithm which efficiently computes the joint actuator torques for a rigid-link open-chain robot manipulator.

ALGORITHM 5.7

Step 0 : Initialization

$$\sigma_i = \begin{cases} 1 & \text{revolute } i\text{–th joint} \\ 0 & \text{prismatic } i\text{–th joint} \end{cases}, \qquad q_i = \begin{cases} \theta_i & \text{revolute } i\text{–th joint} \\ d_i & \text{prismatic } i\text{–th joint} \end{cases}$$

$$\boldsymbol{\omega}_0^0 = 0, \quad \dot{\boldsymbol{\omega}}_0^0 = 0, \quad \boldsymbol{\Omega}_0^0, \quad \ddot{\mathbf{s}}_{0,0}^{0} = -\mathbf{g}, \quad \mathbf{A}_{n+1} = 0, \quad \mathbf{z}_i^{i} = [\,0\,0\,1\,]^T$$

Step 1 : Backward recursion :- For $i = n, 1$ do

$$\bar{m}_i = m_i + \bar{m}_{i+1} \tag{5.3.44a}$$

$$\tilde{\mathbf{u}}^i_{\mathrm{O}_i} = m_i \tilde{\mathbf{r}}^i_{i,i} + \bar{m}_{i+1} \tilde{\mathbf{s}}^i_{i,i+1} \tag{5.3.44b}$$

$$\mathbf{K}^i_{\mathrm{O}_i} = \mathbf{I}^i_{\mathrm{C}_i} - m_i \tilde{\mathbf{r}}^i_{i,i} \tilde{\mathbf{r}}^i_{i,i} - \bar{m}_{i+1} \tilde{\mathbf{s}}^i_{i,i+1} \tilde{\mathbf{s}}^i_{i,i+1} \tag{5.3.44c}$$

$$\hat{\mathbf{K}}^i_{\mathrm{O}_i} = \frac{1}{2} tr[\mathbf{K}^i_{\mathrm{O}_i}]1 - \mathbf{K}^i_{\mathrm{O}_i} \tag{5.3.44d}$$

Step 2 : Forward recursion :- For $i = 0, n\text{-}1$ do

$$\omega^{i+1}_{i+1} = \mathbf{A}^T_{i+1}\, \omega^i_i + \sigma_{i+1} \mathbf{z}^{i+1}_{i+1} \dot{q}_{i+1} \tag{5.3.45a}$$

$$\dot{\omega}^{i+1}_{i+1} = \mathbf{A}^T_{i+1}\, \dot{\omega}^i_i + \sigma_{i+1}[\tilde{\omega}^{i+1}_i \mathbf{z}^{i+1}_{i+1} \dot{q}_{i+1} + \mathbf{z}^{i+1}_{i+1} \ddot{q}_{i+1}] \tag{5.3.45b}$$

$$\Omega^{i+1}_{i+1} = \dot{\tilde{\omega}}^{i+1}_{i+1} + \tilde{\omega}^{i+1}_{i+1} \tilde{\omega}^{i+1}_{i+1} \tag{5.3.45c}$$

$$\zeta^{i+1}_{i,i+1} = (1 - \sigma_{i+1})[2\tilde{\omega}^{i+1}_i \mathbf{z}^{i+1}_{i+1} \dot{q}_{i+1} + \mathbf{z}^{i+1}_{i+1} \ddot{q}_{i+1}] \tag{5.3.45d}$$

$$\ddot{\mathbf{s}}^{i+1}_{0,i+1} = \mathbf{A}^T_{i+1}[\ddot{\mathbf{s}}^i_{0,i} + \Omega^i_i \mathbf{s}^i_{i,i+1}] + (1 - \sigma_{i+1})\zeta^{i+1}_{i,i+1} \tag{5.3.45e}$$

$$\tilde{\mu}^{i+1}_{i+1} = \Omega^{i+1}_{i+1} \hat{\mathbf{K}}^{i+1}_{\mathrm{O}_{i+1}} - [\Omega^{i+1}_{i+1} \hat{\mathbf{K}}^{i+1}_{\mathrm{O}_{i+1}}]^T . \tag{5.3.45f}$$

$$\mu^{i+1}_{i+1} = dual(\tilde{\mu}^{i+1}_{i+1}) \tag{5.3.45g}$$

Step 3 : Backward recursion :- For $i = n, 1$ do

$$\ddot{\mathbf{U}}^i_{\mathrm{O}_i} = \Omega^i_i \mathbf{u}^i_{\mathrm{O}_i} + \mathbf{A}_{i+1}[\ddot{\mathbf{U}}^{i+1}_{\mathrm{O}_{i+1}} + (1 - \sigma_{i+1})\bar{m}_{i+1} \zeta^{i+1}_{i,i+1}] \tag{5.3.46a}$$

$$\eta^i_i = \mu^i_i + \tilde{\mathbf{u}}^i_{\mathrm{O}_i} \ddot{\mathbf{s}}^i_{0,i} + \tilde{\mathbf{s}}^i_{i,i+1}[\ddot{\mathbf{U}}^i_{\mathrm{O}_{i+1}} + (1 - \sigma_{i+1})\bar{m}_{i+1} \zeta^i_{i,i+1}] + \mathbf{A}_{i+1} \eta^{i+1}_{i+1} \tag{5.3.46b}$$

$$\tau_i = \sigma_i(\eta^i_i \cdot \mathbf{z}^i_i) + (1 - \sigma_i)\mathbf{z}^i_i \cdot [\bar{m}_i \ddot{\mathbf{s}}^i_{0,i} + \ddot{\mathbf{U}}^i_{\mathrm{O}_i}] \tag{5.3.46c}$$

end

As we can see, in Step 1 of Algorithm 5.7, we compute the dynamic parameters for the augmented links. These parameters are configuration independent when the augmented links have rigid body characteristics and in this case can be computed off-line. An augmented link, say the i-th, is not a rigid body when the $(i+1)$-th joint is a prismatic joint. Thus, for robot manipulators which have all joints of revolute type, e.g., a PUMA type robot, Step 1 of Algorithm 5.7 can be computed off-line. Even for robot manipulators with one prismatic joint, e.g., a Stanford-arm type robot, Step 1 of Algorithm 5.7 can almost all be computed off-line, because for this type of robot manipulators only minor modifications are needed and these can be easily incorporated in the on-line computations. Therefore, since almost all industrial robot manipulators are either of PUMA or of Stanford-arm type, we can say that Step 1 of Algorithm 5.7 can be computed off-line for almost all industrial robot manipulators in use today.

In the following subsection, we shall consider the numerical implementation of the algorithms derived in this section. In particular, first some observations will be made about the most computationally intensive operations appearing in these algorithms. Then the implementation of Algorithm 5.7 for robot manipulators which have all revolute joints will be examined in more detail. We examine this case in more detail since, as is well known, solving inverse dynamics for robot manipulators of this type is computationally more intensive than for robot manipulators which have some prismatic joints.

5.3.2 Implementation and Computational Considerations

In this section, we shall demonstrate how the algorithms developed earlier can be implemented efficiently. We consider two cases; robot manipulators with a general geometric structure and those for which the twist angle is, by design, either 0 or 90 degrees. We consider the latter case since most industrial robots manipulators have this characteristic.

In the following, we are concerned with the numerical implementation of Algorithms 5.6 and 5.7. Therefore, to be technically correct we have to rewrite these Algorithms in terms of the corresponding coordinate matrix equations. However, as we mentioned in Chapter 3, a tensor equation and its corresponding coordinate matrix equation (with respect to a Cartesian coordinate system) are formally the same. Therefore, in a coordinate matrix form these two algorithms have the same structure and appearance and therefore there is no need to actually rewrite these algorithms in a coordinate matrix form. Based on this observation, by a slight abuse of the notation, we shall refer to the computation of the coordinate matrix Ω_i^i of the tensor $\mathbf{\Omega}_i^i$ relative to the i-th link coordinate system as the computation of the matrix $\mathbf{\Omega}_i^i$.

It is clear from Algorithms 5.6 and 5.7 that the maximum number of operations required to implement them results from various matrix-vector or matrix-matrix multiplications. These matrix operations can be implemented in a straightforward manner by using general purpose standard subroutines. However, the structure of these matrices (e.g., symmetric or skew-symmetric) are standard and common to all robot manipulators that we are concerned with. Therefore, for efficient implementation, the structure of the matrices involved should be taken into account. This approach does not, of course, restrict the applicability of the algorithms to a general robot manipulator.

The matrix-vector multiplications which are involved in Algorithms 5.6 and 5.7 can be categorized as follows:

Class (a) : consists of those operations in which the matrix under consideration is a coordinate transformation matrix;

Class (b) : consists of those operations in which the matrix involved is a skew-symmetric matrix; and

Class (c) : consists of those operations in which the matrix involved is the matrix $\mathbf{\Omega}$.

For a general manipulator, in order to implement a matrix-vector multiplication of class (a), we need 8 scalar multiplications and 4 scalar additions. For the case of manipulators with twist angle α equal to 0 or 90 degrees, we need only 4 multiplications and 2 additions. A matrix-vector multiplication of class (b) can be implemented in 6 scalar multiplications and 3 scalar additions; and finally, for the multiplications in class (c), we need 9 scalar multiplications and 6 scalar additions. Also, the following observations are important for an efficient implementation. To compute the matrix Ω_{i+1}^{i+1}, we require a matrix-matrix multiplication. Moreover, since the product $\tilde{\omega}_{i+1}^{i+1}\tilde{\omega}_{i+1}^{i+1}$ is symmetric, and the matrices $\tilde{\omega}_{i+1}^{i+1}$ and $\dot{\tilde{\omega}}_{i+1}^{i+1}$ are skew-symmetric, we can do this with 6 scalar multiplications and 9 scalar additions. Similarly, since the matrix $\tilde{\mu}_{i+1}^{i+1}$ (or $\tilde{\mathbf{M}}_{\mathrm{C}_{i+1}}^{i+1}$) is skew-symmetric, only three of its elements need to be computed. Thus, by taking into account the symmetry of $\hat{\mathbf{K}}_{\mathrm{O}_{i+1}}^{i+1}$, we can compute the skew-symmetric matrix $\tilde{\mu}_{i+1}^{i+1}$ with only 15 scalar multiplications and 15 scalar additions. Moreover, for implementing the dual operator, we do not need any computations because of the one-to-one correspondence between a skew-symmetric matrix and its dual vector or vector invariant. Finally, since $\mathbf{z}_i^i = [0\ 0\ 1]^T$, evaluating the scalar torque (or force) τ_i does not require any operations in Algorithm 5.6. In Algorithm 5.7 we need only 1 multiplication and 1 addition if the joint is prismatic.

Besides these general observations, for an even more efficient implementation of these algorithms we note the following : For most of the equations, the initial conditions are zero. Therefore, the first cycle (iteration) in Steps 2 and 3 can be computed at almost no computational cost. For example, since $\Omega_0^0 = 0$, the functional expression for the vector $\ddot{\mathbf{s}}_{0,1}^1$ can be easily defined, especially when the gravity vector has only one non-zero component. Thus, with the proper initial conditions for the recursive equations, the computational cost of implementing these algorithms is reduced considerably. This is obviously also true for other algorithms for solving the inverse dynamics problem. However, for the algorithms proposed here, the effort for performing the first cycle (or even the second) by hand is tolerable. Also, as we have mentioned above, the general organization of the computations is important for an efficient implementation. Thus, for example in Step 3, when evaluating η_i^i for $i=1$, only the last component of that vector needs to be evaluated. Finally, knowledge of the geometry (zero components of various vectors or matrices) of a particular class of robot manipulators can considerably reduce the cost of computation.

Steps 2 & 3	General manipulator with revolute joints		General manipulator with revolute joints & $\alpha=0^o$ or 90°	
Equation	Multipl.	Additions	Multipl.	Additions
5.3.45a	$8(n-1)$	$5(n-1)$	$4(n-1)$	$3(n-1)$
5.3.45b	$10(n-1)$	$7(n-1)$	$6(n-1)$	$5(n-1)$
5.3.45c	$6(n-1)+1$	$9(n-1)$	$6(n-1)+1$	$9(n-1)$
5.3.45e	$17(n-1)+3$	$13(n-1)$	$13(n-1)+2$	$11(n-1)$
5.3.45f	$15(n-1)+5$	$15(n-1)+3$	$15(n-1)+5$	$15(n-1)+3$
5.3.45g	0	0	0	0
5.3.46a	$17(n-1)+9$	$13(n-1)+6$	$13(n-1)+9$	$11(n-1)+6$
5.3.46b	$20(n-1)+6$	$19(n-1)+6$	$16(n-1)+6$	$17(n-1)+6$
5.3.46c	0	0	0	0
Total	$93(n-1)+24$	$81(n-1)+15$	$73(n-1)+23$	$71(n-1)+15$
$n=6$	489	420	388	370

Table 5.1: Operations counts for implementing Algorithm 5.7.

A breakdown of the number of scalar multiplications and additions required by each equation of Algorithm 5.7, when this algorithm is applied to a robot manipulator which has all joints of revolute type, is given in Table 5.1. For this implementation of Algorithm 5.7 we have assumed, as is usually the practice, that the equations in the forward and backward recursions for $i = 0$ and $i = n$, respectively, are computed outside the main loops. Therefore, there are only $n - 1$ cycles to be performed in the actual implementation. For these iterations no effort has been made to reduce further the computations, since we wish to keep customization at a minimum. However,

Steps 2 & 3	General manipulator with revolute joints		General manipulator with revolute joints & $\alpha=0^o$ or 90°	
Equation	Multipl.	Additions	Multipl.	Additions
5.3.45a	$8n-12$	$5n-9$	$4n-7$	$3n-5$
5.3.45b	$10n-15$	$7n-11$	$6n-10$	$5n-8$
5.3.45c	$6n-5$	$9n-9$	$6n-5$	$9n-9$
5.3.45e	$17n-19$	$13n-16$	$13n-16$	$11n-14$
5.3.45f	$15n-13$	$15n-14$	$15n-13$	$15n-14$
5.3.45g	0	0	0	0
5.3.46a	$17n-17$	$13n-16$	$13n-13$	$11n-3$
5.3.46b	$20n-27$	$19n-25$	$16n-20$	$17n-22$
5.3.46c	0	0	0	0
Total	$93n-108$	$81n-100$	$73n-84$	$71n-75$
$n=6$	450	386	354	351

Table 5.2: Operations counts for implementing Algorithm 5.7 (Valid for $n \geq 2$).

some saving in computation is obvious (e.g., when $i = 2$ in Step 2 or when $i = 1$ in Step 3) and can be easily taken into account for a more efficient implementation as is shown in Table 5.2.

Algorithm	Multiplications	Additions
Hollerbach (4×4) [14]	$830n-592$ (4388)†	$675n-464$ (3586)
Hollerbach (3×3) [14]	$412n-277$ (2195)	$320n-201$ (1719)
Luh et al. [14]	$150n-48$ (852)	$131n-48$ (738)
Craig [34]	$126n-99$ (657)	$106n-92$ (544)
Khosla and Neuman [49]	$123n-60$ (678)	$96n-55$ (521)
Li [26]	$120n-104$ (616)	$98n-94$ (494)
Khalil et al. [50]	$105n-92$ (538)	$94n-86$ (478)
Angeles et al. [45]	$105n-109$ (521)††	$90n-105$ (435)
Alg. 5.6 in this monograph	$96n-77$ (499)	$84n-70$ (434)
Alg. 5.7 in this monograph	$93n-69$ (489)	$81n-66$ (420)
Alg. 5.7 in this monograph	$93n-108$ (450)††	$81n-100$ (386)

† Number of Operations for $n = 6$, †† Implementation Valid for $n \geq 2$

Table 5.3: Comparison of operations counts for solving the IDP.

For this implementation of Algorithm 5.7 we have assumed, as is usually the practice, that the equations in the forward and backward recursions for $i = 0$ and $i = n$, respectively, are computed outside the main loops. Therefore, there are only $n - 1$ cycles to be performed in the actual implementation. For these iterations no effort has been made to reduce further the computations, since we wish to keep customization at a minimum. However, some saving in computation is obvious (e.g., when $i = 2$ in Step 2 or when $i = 1$ in Step 3) and can be easily taken into account for an efficient implementation as is shown in Table 5.2.

The figures in Tables 5.1 and 5.2 represent the operations counts for steps 2 and 3 of Algorithm 5.7 for computing *all* the joint actuator torques for a particular point along a trajectory. For the sake of comparison, in Table 5.3, we have given the operations counts for a number of algorithms reported in the literature for computing the same vector of the joint actuator torques.

The computational effort required for a particular algorithm, shown in Table 5.3 depends on the degree of optimization in the operations involved in its implementation. Therefore, an accurate comparison of the relative performance of these algorithms requires that they be implemented fairly. Thus, for example, the algorithm by Khalil, Kleinfinger and Gautier which is included in Table 5.3, is implemented using a recursive symbolic procedure and is based on an analysis of the inertial parameters of the links. This helps to reduce the number of operations. However, despite such specialized features in some of the other algorithms presented in Table 5.3, the algorithms described in this monograph have a significantly higher computational efficiency. It may be noted that the computational efficiency of these algorithms results, mainly, from the use of the tensor Ω in evaluating linear accelerations and Euler's equation. For example, the implementation of Euler's equation in Algorithm 5.3, i.e., in its traditional vector formulation, requires $24(n-1) + 8$ scalar multiplications and $18(n-1) + 6$ scalar additions, whereas the corresponding computations in Algorithms 5.6 and 5.7

where Euler's equation is stated in its tensor formulation, require $15(n-1)+5$ scalar multiplications and $15(n-1)+3$ scalar additions. This clearly indicates that the tensor representation for the angular velocity leads to a description of rigid body angular motion (Euler's equation) which is computationally far more efficient than the classical vector description. Therefore, the question which was raised in the previous section, concerning the computational efficiency of a tensor description of rigid body angular motion has been answered here in the affirmative.

5.4 The Use of Euler-Lagrange and Kane's Formulations in Deriving Algorithm 5.7

In this section, we shall demonstrate that the computationally efficient algorithm (Algorithm 5.7) which was in the previous section derived using the Newton-Euler equations, can also be derived using Kane's or the Euler-Lagrange dynamic equations of motion. This demonstration will make clear that the computational efficiency of an algorithm for solving inverse dynamics is indeed completely independent of the particular procedure of classical mechanics which has been used to derive that algorithm. For the sake of simplicity in this demonstration we consider manipulators with revolute joints only. However, the analysis can be easily extended to also include prismatic joints.

5.4.1 The Euler-Lagrange Formulation

As we mentioned in Section 5.2, a number of algorithms for solving inverse dynamics have been derived based on the Euler-Lagrange formulation and among them is the recursive Algorithm 5.2. This algorithm has been devised by Hollerbach [14] in an attempt to implement the following equation efficiently.

$$\tau_i = \sum_{j=i}^{n} \left\{ tr \left[m_j \frac{\partial \mathbf{s}_{0j}}{\partial q_i} \ddot{\mathbf{s}}_{0j}^T + m_j \frac{\partial \mathbf{s}_{0j}}{\partial q_i} ({}^j\mathbf{r}_{j,j})^T \ddot{\mathbf{W}}_j^T + m_j \frac{\partial \mathbf{W}_j}{\partial q_i} {}^j\mathbf{r}_{j,j} \ddot{\mathbf{s}}_{0j}^T \right.\right.$$

$$\left.\left. + \frac{\partial \mathbf{W}_j}{\partial q_i} {}^j\mathbf{J}_{o_j} \ddot{\mathbf{W}}_j^T \right] - m_j \mathbf{g}^T \frac{\partial \mathbf{W}_i}{\partial q_i} {}^i\mathbf{r}_{i,j} \right\}, \quad i = 1, \cdots, n \qquad (5.4.1)$$

which has been derived from the Euler-Lagrange dynamical equations of motion. However, if we can show that the right-hand side of the equation (5.4.1) is exactly the same as the right-hand side of the equation

$$\tau_i = {}^i\mathbf{z}_i \cdot {}^i\eta_i, \qquad i = 1, \cdots, n, \qquad (5.4.2)$$

which is simply equation (5.3.46c) of Algorithm 5.7 (stated here for revolute joints only), then it is clear that Algorithm 5.7 can also be derived from the Euler-Lagrange equations. Therefore, our aim in this section is to show that equation (5.4.1) can assume the same formulation as equation (5.4.2). We proceed as follows:

Our first objective is to eliminate from equation (5.4.1) the term which contains the effects of gravity. To achieve this, we notice that from a simple comparison of equations (A.7) and (A.10) in Appendix A, we have

$$\frac{\partial \mathbf{W}_i}{\partial q_i} {}^i\mathbf{r}_{i,j} = \frac{\partial \mathbf{s}_{0j}}{\partial q_i} + \frac{\partial \mathbf{W}_j}{\partial q_i} {}^j\mathbf{r}_{j,j}.$$

Therefore, using the fact that the dot product of two vectors **a** and **b** satisfies the equation

$$\mathbf{a} \cdot \mathbf{b} = tr[\mathbf{a}\,\mathbf{b}^T], \qquad (5.4.3)$$

we can write the term which contains the gravitational effects in equation (5.4.1) as follows,

$$m_j \mathbf{g}^T \frac{\partial \mathbf{W}_i}{\partial q_i} {}^i\mathbf{r}_{i,j} = tr\left[m_j \frac{\partial \mathbf{s}_{0j}}{\partial q_i} \mathbf{g}^T \right] + tr\left[m_j \frac{\partial \mathbf{W}_j}{\partial q_i} {}^j\mathbf{r}_{j,j} \mathbf{g}^T \right]. \qquad (5.4.4)$$

Now, substitution of equation (5.4.4) into (5.4.1) yields

$$\tau_i = \sum_{j=i}^{n} tr\left[m_j \frac{\partial \mathbf{s}_{0,j}}{\partial q_i} (\ddot{\mathbf{s}}_{0,j} - \mathbf{g})^T + m_j \frac{\partial \mathbf{s}_{0,j}}{\partial q_i} (\mathbf{r}^j_{j,j})^T \ddot{\mathbf{W}}_j^T \right.$$

$$\left. + m_j \frac{\partial \mathbf{W}_j}{\partial q_i} \mathbf{r}^j_{j,j} (\ddot{\mathbf{s}}_{0,j} - \mathbf{g})^T + \frac{\partial \mathbf{W}_j}{\partial q_i} \mathbf{J}^j_{o_j} \ddot{\mathbf{W}}_j^T \right], \quad i = 1, \cdots, n$$

$$= \sum_{j=i}^{n} tr\left[m_j \frac{\partial \mathbf{s}_{0,j}}{\partial q_i} \ddot{\mathbf{s}}_{0,j}^T + m_j \frac{\partial \mathbf{s}_{0,j}}{\partial q_i} (\mathbf{r}^j_{j,j})^T \ddot{\mathbf{W}}_j^T \right.$$

$$\left. + m_j \frac{\partial \mathbf{W}_j}{\partial q_i} \mathbf{r}^j_{j,j} \ddot{\mathbf{s}}_{0,j}^T + \frac{\partial \mathbf{W}_j}{\partial q_i} \mathbf{J}^j_{o_j} \ddot{\mathbf{W}}_j^T \right] \quad i = 1, \cdots, n \tag{5.4.5}$$

where the initial condition $\ddot{\mathbf{s}}_{0,0}$ for the vector $\ddot{\mathbf{s}}_{0,j}$ is now equal to $-\mathbf{g}$, instead of being zero as is usually the case in the Lagrangian formulation. Furthermore, since

$$\ddot{\mathbf{r}}_{j,j} = \ddot{\mathbf{W}}_j \mathbf{r}^j_{j,j} \quad \text{and} \quad \ddot{\mathbf{r}}_{0,j} = \ddot{\mathbf{s}}_{0,j} + \ddot{\mathbf{r}}_{j,j},$$

equation (5.4.5) can be simplified to

$$\tau_i = \sum_{j=i}^{n} tr\left[m_j \frac{\partial \mathbf{s}_{0,j}}{\partial q_i} \ddot{\mathbf{r}}_{0,j}^T + m_j \frac{\partial \mathbf{W}_j}{\partial q_i} \mathbf{r}^j_{j,j} \ddot{\mathbf{s}}_{0,j}^T + \frac{\partial \mathbf{W}_j}{\partial q_i} \mathbf{J}^j_{o_j} \ddot{\mathbf{W}}_j^T \right], \quad i = 1, \cdots, n. \tag{5.4.6}$$

Moreover, as we have shown in Appendices B and C, we can write

$$\frac{\partial \mathbf{s}_{0,j}}{\partial q_i} = \tilde{\mathbf{z}}_i \mathbf{s}_{i,j} \tag{5.4.7}$$

$$\frac{\partial \mathbf{W}_j}{\partial q_i} = \tilde{\mathbf{z}}_i \mathbf{W}_j \tag{5.4.8}$$

$$tr\left[\frac{\partial \mathbf{W}_j}{\partial q_i} \mathbf{J}_{o_j} \ddot{\mathbf{W}}_j^T \right] = \mathbf{z}_i \cdot \mathbf{M}_{o_j} \tag{5.4.9}$$

Therefore, using equations (5.4.7)-(5.4.9) and equation (5.4.3), equation (5.4.6) can be simplified further to yield

$$\tau_i = \sum_{j=i}^{n} \left\{ m_j \tilde{\mathbf{z}}_i \mathbf{s}_{i,j} \cdot \ddot{\mathbf{r}}_{0,j} + m_j \tilde{\mathbf{z}}_i \mathbf{r}_{j,j} \cdot \ddot{\mathbf{s}}_{0,j} + \mathbf{z}_i \cdot \mathbf{M}_{o_j} \right\}, \quad i = 1,\cdots,n. \tag{5.4.10}$$

Finally, since for any vectors $\mathbf{a}$, $\mathbf{b}$ and $\mathbf{c}$ we have

$$\tilde{\mathbf{a}}\mathbf{c} \cdot \mathbf{b} = \mathbf{a} \cdot \tilde{\mathbf{c}}\mathbf{b},$$

equation (5.4.10) can be written as

$$\tau_i = \mathbf{z}_i \cdot \sum_{j=i}^{n} \left\{ m_j \tilde{\mathbf{s}}_{i,j} \ddot{\mathbf{r}}_{0,j} + m_j \tilde{\mathbf{r}}_{j,j} \ddot{\mathbf{s}}_{0,j} + \mathbf{M}_{o_j} \right\}, \qquad i = 1, \cdots ,n,$$

from which we get

$$\tau_i = \overset{i}{\mathbf{z}}_i \cdot \sum_{j=i}^{n} \left\{ m_j \overset{i}{\tilde{\mathbf{s}}}_{i,j} \overset{i}{\ddot{\mathbf{r}}}_{0,j} + m_j \overset{i}{\tilde{\mathbf{r}}}_{j,j} \overset{i}{\ddot{\mathbf{s}}}_{0,j} + \overset{i}{\mathbf{M}}_{o_j} \right\}, \quad i = 1, \cdots ,n \tag{5.4.11}$$

since the dot product is invariant under coordinate transformations.

Now, for $i = 1, 2, \cdots ,n$, let us define the vectors

$$\overset{i}{\upsilon}_i = \sum_{j=i}^{n} \left\{ m_j \overset{i}{\tilde{\mathbf{s}}}_{i,j} \overset{i}{\ddot{\mathbf{r}}}_{0,j} + m_j \overset{i}{\tilde{\mathbf{r}}}_{j,j} \overset{i}{\ddot{\mathbf{s}}}_{0,j} + \overset{i}{\mathbf{M}}_{o_j} \right\}. \tag{5.4.12}$$

We shall show that for $i = 1, \cdots ,n$, the vector $\overset{i}{\upsilon}_i$ is equal to the following vector

$$\overset{i}{\eta}_i = \overset{i}{\mu}_i + \overset{i}{\tilde{\mathbf{u}}}_{o_i} \overset{i}{\ddot{\mathbf{s}}}_{0,i} + \overset{i}{\tilde{\mathbf{s}}}_{i,i+1} \overset{i}{\ddot{\mathbf{U}}}_{o_{i+1}} + \overset{i}{\eta}_{i+1} \tag{5.4.13}$$

which results from equation (5.3.46b) of Algorithm 5.7 when the $(i+1)$-th joint is assumed to be of revolute type.

First, since for any i the vector $\overset{i}{\mathbf{s}}_{i,i}$ is equal to zero by definition, we notice that the vector $\overset{i}{\upsilon}_i$ can also be written as

$$\boldsymbol{\upsilon}_i^i = m_i \tilde{\mathbf{r}}_{i,i}^i \ddot{\mathbf{s}}_{0,i}^i + \mathbf{M}_{\mathrm{o}_i}^i + \sum_{j=i+1}^{n} \left[m_j [\tilde{\mathbf{s}}_{i,i+1}^i + \tilde{\mathbf{s}}_{i+1,j}^i] \ddot{\mathbf{r}}_{0,j}^i + m_j \tilde{\mathbf{r}}_{j,j}^i \ddot{\mathbf{s}}_{0,j}^i + \mathbf{M}_{\mathrm{o}_j}^i \right]$$

$$= m_i \tilde{\mathbf{r}}_{i,i}^i \ddot{\mathbf{s}}_{0,i}^i + \mathbf{M}_{\mathrm{o}_i}^i + \tilde{\mathbf{s}}_{i,i+1}^i \sum_{j=i+1}^{n} m_j \ddot{\mathbf{r}}_{0,j}^i + \boldsymbol{\upsilon}_{i+1}^i. \tag{5.4.14}$$

Moreover, since

$$\sum_{j=i+1}^{n} m_j \ddot{\mathbf{r}}_{0,j}^i = \sum_{j=i+1}^{n} m_{i+1} \left[\ddot{\mathbf{s}}_{0,i}^i + \ddot{\mathbf{s}}_{i,i+1}^i + \ddot{\mathbf{r}}_{i+1,j}^i \right]$$

$$= \bar{m}_{i+1} \ddot{\mathbf{s}}_{0,i}^i + \bar{m}_{i+1} \ddot{\mathbf{s}}_{i,i+1}^i + \sum_{j=i+1}^{n} m_j \ddot{\mathbf{r}}_{i+1,j}^i$$

$$= \bar{m}_{i+1} \ddot{\mathbf{s}}_{0,i}^i + \bar{m}_{i+1} \ddot{\mathbf{s}}_{i,i+1}^i + \ddot{\mathbf{U}}_{\mathrm{o}_{i+1}}^i, \tag{5.4.15}$$

equation (5.4.14) can be written as

$$\boldsymbol{\upsilon}_i^i = (m_i \tilde{\mathbf{r}}_{i,i}^i + \bar{m}_{i+1} \tilde{\mathbf{s}}_{i,i+1}^i) \ddot{\mathbf{s}}_{0,i}^i + \mathbf{M}_{\mathrm{o}_i}^i + \bar{m}_{i+1} \tilde{\mathbf{s}}_{i,i+1}^i \ddot{\mathbf{s}}_{i,i+1}^i + \tilde{\mathbf{s}}_{i,i+1}^i \ddot{\mathbf{U}}_{\mathrm{o}_i}^i + \boldsymbol{\upsilon}_{i+1}^i. \tag{5.4.16}$$

Now, using equations (5.3.10), (5.3.17) and (5.3.18), we have

$$\tilde{\mathbf{u}}_{\mathrm{o}_i}^i = m_i \tilde{\mathbf{r}}_{i,i}^i + \bar{m}_{i+1} \tilde{\mathbf{s}}_{i,i+1}^i$$

and

$$\boldsymbol{\mu}_i^i = \mathbf{M}_{\mathrm{o}_i}^i + \bar{m}_{i+1} \tilde{\mathbf{s}}_{i,i+1}^i \ddot{\mathbf{s}}_{i,i+1}^i,$$

Therefore, equation (5.4.16) can be written in the following form

$$\boldsymbol{\upsilon}_i^i = \tilde{\mathbf{u}}_{\mathrm{o}_i}^i \ddot{\mathbf{s}}_{0,i}^i + \boldsymbol{\mu}_i^i + \tilde{\mathbf{s}}_{i,i+1}^i \ddot{\mathbf{U}}_{\mathrm{o}_{i+1}}^i + \boldsymbol{\upsilon}_{i+1}^i, \tag{5.4.17}$$

which shows that indeed $\boldsymbol{\upsilon}_i^i = \boldsymbol{\eta}_i^i$ for $i = 1, \cdots, n$. From the foregoing, equations (5.4.1) and (5.4.2) are equivalent and this shows that Algorithm 5.7 can also be derived using the Lagrangian formulation.

5.4.2 Kane's Formulation

In this section, we shall show that, as with the Lagrangian formulation, Algorithm 5.7 can also be derived using Kane's equations. As we mentioned in Section 5.2, Angeles et al. have shown [45] that, based on Kane's equations, we can determine the actuator torques τ_i for a robot manipulator with all revolute joints using the following equation:

$$\tau_i = {}^i\mathbf{z}_i \cdot \sum_{j=i}^{n} \left\{ {}^i\mathbf{M}_{C_j} + m_j {}^i\tilde{\mathbf{r}}_{i,j} {}^i\ddot{\mathbf{r}}_{0,j} \right\} , \qquad i = 1, \cdots ,n. \tag{5.4.18}$$

To evaluate equation (5.4.18), Angeles et al. [45] proposed an algorithm which has the same structure as Algorithm 5.3. As with the Euler-Lagrange case, we shall show that equation (5.4.18) is equivalent to equation (5.4.2). To do this let us define the following vector for $i = 1, \cdots ,n$.

$$\mathbf{h}_i^i = \sum_{j=i}^{n} \left\{ {}^i\mathbf{M}_{C_j} + m_j {}^i\tilde{\mathbf{r}}_{i,j} {}^i\ddot{\mathbf{r}}_{0,j} \right\}. \tag{5.4.19}$$

As before, our aim is to show that $\mathbf{h}_i^i = \mathbf{\eta}_i^i$ for all i, $i = 1, \cdots ,n$, where $\mathbf{\eta}_i^i$ is defined by equation (5.4.13). By expanding the summation in equation (5.4.19), we obtain the following equation after a few manipulations:

$$\mathbf{h}_i^i = {}^i\mathbf{M}_{C_i} + m_i {}^i\tilde{\mathbf{r}}_{i,i} {}^i\ddot{\mathbf{r}}_{0,i} + {}^i\tilde{\mathbf{s}}_{i,i+1} \sum_{j=i+1}^{n} m_j {}^i\ddot{\mathbf{r}}_{0,j} + {}^i\mathbf{h}_{i+1}.$$

Furthermore, using equations (5.3.10), (5.3.18) and (5.4.15), we can simplify the above equation to

$$\mathbf{h}_i^i = \mathbf{\mu}_i^i + {}^i\tilde{\mathbf{u}}_{O_i} {}^i\ddot{\mathbf{s}}_{0,i} + {}^i\tilde{\mathbf{s}}_{i,i+1} {}^i\ddot{\mathbf{U}}_{O_{i+1}} + {}^i\mathbf{h}_{i+1} \tag{5.4.20}$$

which obviously shows that $\mathbf{h}_i^i = \mathbf{\eta}_i^i$ for all i. Thus, Algorithm 5.7 can also be derived from Kane's equations.

From the foregoing, starting from the Euler-Lagrange or Kane's equations and following an appropriate analysis, we can derive not only equivalent but exactly the same formulations for the vector of the generalized forces, namely, equation (5.4.2). This equation was derived earlier (in Section 5.3) using the Newton-Euler formulation. Therefore, independent of which approach from classical mechanics is used to derive the dynamical equations of motion, we can devise the same computational algorithm for their implementation. This result clearly indicates that apart from personal preference or experience there is nothing to be gained, in terms of computational efficiency, by choosing one approach over another for solving the manipulator IDP. However, it should be noted that the choice of a particular approach is important because it determines the nature of the analysis and the amount of effort needed to devise an algorithm for solving the IDP.

5.5 Concluding Remarks

In this chapter, the Cartesian tensor methodology, developed in Chapter 4, has been used to analyze the dynamic equations of motion of rigid-link, open-chain robot manipulators. Also, the ideas of augmented and generalized links have been used in the underlying modeling scheme for this class of robot manipulators. Based on this modeling scheme, we proposed an algorithm for computing manipulator inverse dynamics which allows us to compute off-line several configuration independent parameters of the manipulator. At the same time, the Cartesian tensor formulation for the quantities to be computed on-line enables us to propose implementations for this algorithm which are computationally very efficient. In fact, we have shown, by comparing the computational complexity of this algorithm with that of other existing ones, that the proposed algorithm is computationally the most efficient non-customized algorithm which is available today for solving the problem of manipulator inverse dynamics. The computational efficiency of this algorithm has been achieved mainly because a tensor representation, instead of a vector one, has been used for the angular velocity. Finally, in this chapter, we have shown that the Newton-Euler, Euler-Lagrange or Kane's formulations of robot dynamics, with appropriate analysis, can lead us to the same computational algorithms. Thus, we have established that from an algorithmic point of view, the solution of the inverse dynamics problem does not depend on which of these formulations is used for deriving the equations of motion. This result clearly indicates that apart from personal preference or experience there is nothing to be gained, in terms of computational efficiency, by choosing one approach over another for solving the problem of inverse dynamics for rigid-link open-chain robot manipulators.

5.6 References

[1] O. Khatib, "Dynamic Control of Manipulators in Operational Space", *6th IFTOMM Congress on Theory of Machines and Mechanisms*, New Delhi, Dec. 15-20, pp. 1-10, 1983.

[2] T. Yoshikawa, "Dynamic Hybrid Position/Force Control of Robot Manipulators, Description of Hand Constraints and Calculation of Joint Driving Force", *Proc. 1986 IEEE Int. Conf. Robotics and Automation*, San Francisco, CA, pp. 1393-1398, Apr. 1986.

[3] P. Misra, R. V. Patel and C. A. Balafoutis, "Robust Control of Robot Manipulators using Linearized Dynamic Models", *Recent Trends in Robotics : Modeling, Control and Education*, M. Jamshidi, J. Y. S. Luh and M. Shahinpoor, Eds., North-Holland, Elsevier Science Publishing Co., Inc., New York, 1986.

[4] P. Misra, R. V. Patel and C. A. Balafoutis, "Robust Control of Robot manipulators in Cartesian Space", *Proc. American Control Conference*, pp. 1351-1356, Atlanta, Georgia, June 15-17, 1988.

[5] M. W. Spong, J. S. Thorp and J. M. Kleinwaks, "The Control of Robot Manipulators with Bounded Input", *IEEE Trans. on Automatic Control,* Vol. AC-31, No. 6, pp. 483-490, 1986.

[6] K. G. Shin and N. D. Mckay, "A Dynamic Programming Approach to Trajectory Planning of Robotic Manipulators", *IEEE Trans. on Automatic Control,* Vol. AC-31, No. 6, pp. 491-500, 1986.

[7] H. H. Tan and R. B. Potts, "Minimum-Time Trajectory Planner for the Discrete Dynamic Robot Model With Dynamic Constraints", *IEEE J. of Robotics and Automation,* Vol. RA-4, No. 2, pp. 174-185, 1988.

[8] J. M. Hollerbach, "Dynamic Scaling of Manipulator Trajectories", *ASME J. of Dynamic Systems, Measurement, and Control,* Vol. 106, pp. 102-106, 1984.

[9] T. Yoshikawa, "Dynamic Manipulability of Robot Manipulators", *J. of Robotic Systems,* Vol. 2, No. 1, pp. 113-124, 1985.

[10] J. J. Murray and C. P. Neuman, "ARM : An Algebraic Robot Dynamic Modeling Program", *Proc. 1st Int. IEEE Conf. on Robotics,* pp. 103-114, Atlanta, GA, Mar. 13-15, 1984.

[11] A. P. Tzes, S. Yurkovich and F. D. Langer, "A Symbolic Manipulation Package for Modeling of Rigid or Flexible Manipulators", *Proc. 1986 IEEE Int. Conf. Robotics and Automation,* Philadelphia, PA, pp. 1526-1531, Apr. 1988.

[12] J. J. Uicker, *On the Dynamic Analysis of Spatial Linkages using* 4×4 *Matrices,* Ph.D. Dissertation, Northwestern University, August 1965.

[13] M. E. Kahn, *The Near Minimum-Time Control of Open Articulated Kinematic Chains,* Ph.D. Thesis, Stanford University, 1969.

[14] J. M. Hollerbach, "A Recursive Lagrangian Formulation of Manipulator Dynamics and a Comparative Study of Dynamics Formulation Complexity", *IEEE Trans. on Systems, Man, and Cybernetics,* Vol. SMC-10, no. 11, pp. 730-736, 1980.

[15] R. Paul, "Modeling, Trajectory Calculation, and Servoing of a Computer Controlled Arm", *A. I. Memo. 177*, Stanford Artificial Intelligence Lab., Sept. 1972.

[16] A. K. Bejczy, "Robot Arm Dynamics and Control", *Memo. 33-669*, Jet Propulsion Labs. Tech. Feb. 1974.

[17] M. Brady *et al.*, Eds., *Robot Motion : Planning and Control*, MIT Press, Cambridge, MA, 1982.

[18] J. S. Albus, "A New Approach to Manipulator Control : The Cerebellar Model Articulation Controller (CMAC)", *ASME J. Dynamics Systems, Measurement, Control*, Vol. 97, pp. 270-277, 1975.

[19] M. H. Raibert, "Analytical Equations vs. Table Look-up for Manipulation : A Unifying Concept", *Proc IEEE Conf. Decision and Control*, New Orleans, pp. 576-579, Dec. 1977.

[20] B. K. P. Horn and M. H. Raibert, "Configuration Space Control", *The Industrial Robot*, pp. 69-73, June, 1978.

[21] R. C. Waters, "Mechanical Arm Control", M.I.T. Artificial Intelligence Lab. *Memo. 549*, Oct. 1979.

[22] D. Fischer, *Theoretical Foundation for the Mechanics of Living Mechanisms* (in German), Teubner, Leipzig, 1906.

[23] I. J. Wittenburg, *Dynamics of Systems of Rigid Bodies*, B. G. Teubner, Stuttgart, 1977.

[24] M. Renaud, "An Efficient Iterative Analytical Procedure for Obtaining a Robot Manipulator Dynamic Model", *Proc. 1st International Symp. on Robotics Research*, Bretton Woods, New Hampshire, pp. 749-762, 1983.

[25] M. Vucobratovic, S. Li and N. Kircanski, "An Efficient Procedure for Generating Dynamic Manipulator Models", *Robotica*, Vol. 3, No. 3, pp. 147-152, 1985.

[26] C. J. Li, "A New Method of Dynamics for Robot Manipulators", *IEEE Trans. on Systems, Man and Cybernatics,* Vol. 18, No. 1, pp. 105-114, 1988.

[27] W. W. Hooker and G. Margulies, "The Dynamical Attitude Equations for an n-Body Satellite", *J. Astronautical Sciences,* Vol. 12, No. 4, pp. 123-128, 1965.

[28] Y. Stepanenko and M. Vucobratovic, "Dynamics of Articulated Open-Chain Active Mechanisms", *Mathematical Biosciences,* Vol. 28, pp. 137-170, 1976.

[29] M. Vucobratovic, "Dynamics of Active Articulated Mechanisms and Synthesis of Artificial Motion", *Mechanism and Machine Theory,* Vol. 13, pp. 1-56, 1978.

[30] J. Y. L. Ho, "Direct Path Method for Flexible Multibody Spacecraft Dynamics", *AIAA J. Spacecraft and Rockets* , Vol. 14, No. 2, pp. 102-110, 1977.

[31] P. C. Hughes, "Dynamics of a Chain of Flexible bodies", *J. Astronautical Sciences,* Vol. 27, No. 4, pp. 359-380, 1979.

[32] D. E. Orin, R. B. McGhee, M. Vucobratovic, and G. Hartoch, "Kinematic and Kinetic Analysis of Open-Chain Linkages Utilizing Newton-Euler Methods", *Mathematical Biosciences,* Vol. 43, No. 1/2, pp. 107-130, 1979.

[33] J. Y. S. Luh, M. W. Walker, and R. P. Paul, "On-Line Computational Scheme for Mechanical Manipulators", *ASME J. Dyn. Syst. Meas. and Contr.,* Vol. 102, pp. 69-79, 1980.

[34] J. J. Craig, *Introduction to Robotics : Mechanics & Control, 2nd* ed. Addison-Wesley, Reading, MA, 1989.

[35] W. M. Silver, "On the Equivalence of Lagrangian and Newton-Euler Dynamics for Manipulators", *Int. J. Robotics Research,* Vol. 1, pp. 60-70, 1982.

[36] R. Featherstone, *Robot Dynamics Algorithms*, Kluwer Academic Publishers, Boston, MA, 1987.

[37] G. Rodriguez, "Kalman Filtering, Smoothing and Recursive Robot Arm Forward and Inverse Dynamics", *IEEE J. of Robotics and Automation*, Vol. RA-3, No. 6, pp. 624-639, 1987.

[38] G. Rodriguez and K. Kreutz, "Recursive Mass Matrix Factorization and Inversion : An Operator Approach to Open- and Closed-Chain Multibody Dynamics", *JPL Publication 88-11*, March 15, 1988.

[39] T. R. Kane, "Dynamics of Holonomic Systems", *ASME J. Applied Mechanics*, Vol. 28, pp. 574-578, 1961.

[40] T. R. Kane and C. F. Wang, "On the Derivation of Equations of Motion", *J. Soc. for Ind. and Appl. Math.*, Vol. 13, pp. 487-492, 1965.

[41] R. L. Huston, C. E. Passerello and M. W. Harlow, "Dynamics of Multirigid-Body Systems", *ASME J. Applied Mechanics*, Vol. 45, pp. 889-894, 1978.

[42] R. L. Huston and F. A. Kelly, "The Development of Equations of Motion of Single-Arm Robots", *IEEE Trans. Systems, Man and Cybernetics*, Vol. SMC-12, No. 3. pp. 259-266, 1982.

[43] H. Faessler, "Computer-Assisted Generation of Dynamical Equations for Multi-Body systems", *Int. J. of Robotics Research*, Vol. 5, No. 3, pp. 129-141, 1986.

[44] T. R. Kane and D. A. Levinson, "The Use of Kane's Dynamical Equations in Robotics", *Int. J. of Robotics Research*, Vol. 2, No. 3, pp. 3-21, 1983.

[45] J. Angeles, O. Ma and A. Rojas, "An Algorithm for the Inverse Dynamics of n-Axis General Manipulators Using Kane's Equations", *Computers Math. Applic.*, Vol. 17, No. 12, pp. 1545-1561, 1989.

[46] C. P. Neuman and J. J. Murray, "The Complete Dynamic Model and Customized Algorithms of the Puma Robot", *IEEE Trans. on Systems,*

Man, and Cybernetics, Vol. SMC-17, No. 4, pp. 635-644, 1987.

[47] J. J. Murray and C. P. Neuman, "Organizing Customized Robot Dynamics Algorithms for Efficient Numerical Evaluation", *IEEE Trans. on Systems, Man, and Cybernetics,* Vol. SMC-18, No. 1, pp. 115-125, 1988.

[48] P. K. Khosla and C. P. Neuman, "Computational Requirements of Customized Newton-Euler Algorithms", *J. of Robotic Systems,* Vol. 2, No. 3, pp. 309-327, 1985.

[49] M. Renaud, "Quasi-Minimal Computation of the Dynamic Model of a Robot Manipulator Utilizing the Newton-Euler Formalism and the Notion of Augmented Body", *Proc. 1987 IEEE Int. Conf. on Robotics and Automation,* Raleigh, NC, pp. 1677-1682, Apr. 1987.

[50] W. Khalil and J. F. Kleinfinger, and M. Gautier, "Reducing the Computational Burden of the Dynamic Models of Robots", *Proc. 1986 IEEE Int. Conf. on Robotics and Automation,* San Francisco, CA, pp. 525-531, Apr. 1986.

[51] W. Khalil and J. F. Kleinfinger, "Minimum Operations and Minimum Parameters of the Dynamic Models of Tree Structure Robots", *IEEE J. Robotics and Automation,* Vol. RA-3, No. 6, pp. 517-526, 1987.

[52] J. W. Burdick, "An Algorithm for Generation of Efficient Manipulator Dynamic Equations, *Proc. 1986 IEEE Int. Conf. on Robotics and Automation,* San Francisco, CA, pp. 212-218, Apr. 1986.

[53] B. Armstrong, O. Khatib, and J. Burdick, "The Explicit Dynamic Model and Inertia Parameters of the PUMA 560 Arm", *Proc. 1986 IEEE Int. Conf. on Robotics and Automation,* San Francisco, CA, pp. 510-518, Apr. 1986.

[54] K. Youcef-Toumi and H. Asada, "The Design of Open-Loop Manipulator Arms With Decoupled and Configuration-Invariant Inertia Tensors", *Proc. 1986 IEEE Int. Conf. on Robotics and Automation,* San

Francisco, CA, pp. 2018-2026, Apr. 1986.

[55] D. C. H. Yang and S. W. Tzeng, "Simplification and Linearization of Manipulator Dynamics by the Design of Inertia Distribution", *Int. J. of Robotics Research,* Vol. 5, No. 3, pp. 120-128, 1986.

[56] S. Ramos and P. K. Khosla, "Scheduling Parallel Computation of Inverse Dynamics Formulation" *Robotics and Manufacturing : Recent Trends in Research, Education, and Applications,* M. Jamshidi, J. Y. S. Luh, H. Seraji and G. P. Starr, Eds., ASME Press, New York, 1988.

[57] H. Kasahara and S. Narita, "Parallel Processing of Robot-Arm Control Computation on a Multi-microprocessor System", *IEEE J. Robotics and Automation,* Vol. RA-1, No. 2, pp. 104-113, 1985.

[58] R. Nigam and C. S. G. Lee, "A Multiprocessor-Based Controller for the Control of Mechanical Manipulators", *IEEE J. Robotics and Automation,* Vol. RA-1, No. 4, pp. 173-182, 1985.

[59] C. S. G. Lee and P. R. Chang, "Efficient Parallel Algorithm for Robot Inverse Dynamics Computation", *IEEE Trans. on Systems, Man and Cybernetics,* Vol. SMC-16, No. 4, pp. 532-542, 1986.

[60] M. Vucobratovic, N. Kircanski and S. G. Li, "An Approach to Parallel Processing of Dynamic Robot Models", *Int. J. of Robotics Research,* Vol. 7, No. 2, pp. 64-71, 1988.

[61] K. Kazerounian and K. C. Gupta, "Manipulator Dynamics Using the Extended Zero Reference Position Description", *IEEE J. Robotics and Automation,* Vol. RA-2, No. 4, pp. 221-224, 1986.

[62] L. T. Wang and B. Ravani, "Recursive Computations of Kinematics and Dynamics Equations for Mechanical Manipulators", *IEEE J. Robotics and Automation,* Vol. RA-1, No. 3, pp. 124-131, 1985.

[63] C. A. Balafoutis, R. V. Patel and P. Misra, "Efficient Modeling and Computation of Manipulator Dynamics Using Orthogonal Cartesian Tensors", *IEEE J. of Robotics and Automation,* Vol. 4, No. 6, pp. 665-

676, 1988.

[64] C. A. Balafoutis, R. V. Patel and J. Angeles, "A Comparative Study of Lagrange, Newton-Euler and Kane's Formulations for Robot Manipulator Dynamics" *Robotics and Manufacturing : Recent Trends in Research, Education, and Applications*, M. Jamshidi, J. Y. S. Luh, H. Seraji and G. P. Starr, Eds., ASME Press, New York, 1988.

Chapter 6

Manipulator Forward Dynamics

6.1 Introduction

The problem of evaluating *forward* or *direct* dynamics involves the calculation of joint accelerations (and through integration, joint velocities and positions) given the actuator torques/forces and any external torques/forces exerted on the last link of the manipulator. Forward dynamics computation is used primarily in simulation, so that, it is not so important for this computation to meet the stringent speed requirements of inverse dynamics applications unless real-time simulation is required in which case computational efficiency is an important issue. Real-time simulation is often desirable since it provides more powerful, flexible, and economic ways of developing new robot designs and new control algorithms. Also, as fast 3-D computer graphics is becoming more easily available, robot kinematic motions have begun to be displayed on graphic work stations. However, to study dynamic robot motions completely, real-time dynamics of robot manipulators need to be included in computer simulation of robotic systems.

Mathematically, forward dynamics can be described by a vector differential equation of the form

$$\ddot{\mathbf{q}}(t) = \mathbf{h}(\mathbf{q}(t), \dot{\mathbf{q}}(t), \tau(t), \text{manipulator parameters}, \mathbf{f}(t)) \quad (6.1.1)$$

where, $\mathbf{q}(t)$ is the vector of generalized coordinates (joint variables), $\dot{\mathbf{q}}(t)$ and $\ddot{\mathbf{q}}(t)$ are its derivatives with respect to time, $\tau(t)$ is the (input) generalized force vector, i.e., the vector of joint torques and/or joint forces, "manipulator parameters" are all those parameters which characterize the particular geometry and dynamics of a robot manipulator, and $\mathbf{f}(t)$ is the vector of the external torques/forces. In general, equation (6.1.1) is not a simple equation for which an analytic solution can be provided easily. For a general robot manipulator equation (6.1.1) is very complex since it is highly nonlinear with strong coupling between the joint variables. Hence, to solve equation (6.1.1) for $\mathbf{q}$, requires complicated procedures for evaluating $\mathbf{h}$ and for performing numerical integration. These procedures, as in the case of inverse dynamics, are defined by structured algorithms which are evaluated in stages.

In this chapter, we shall review the basic approaches taken to solve the forward dynamics problem (FDP). By introducing a new algorithm, we shall improve upon the computational efficiency of one of these methods, namely, the *composite rigid body* method which is currently the most efficient one available for computing forward dynamics. The outline of this chapter is as follows: Section 6.2 contains a review of existing methods for solving the forward dynamics problem. In Section 6.3, a new algorithm is devised for computing efficiently the *generalized inertia tensor* of a robot manipulator, which is a basic ingredient of the composite rigid body method. In Section 6.4 the computational complexity of this algorithm is analyzed. The computational cost of the composite rigid body method is examined when algorithms derived in this and the previous chapter are used to solve basic subproblems associated with this method. Finally, Section 6.5 concludes this chapter.

6.2 Previous Results on Manipulator Forward Dynamics

In the past few years, two basic approaches have been taken for solving the FDP, which may be outlined as follow:

a) Obtain and solve a set of simultaneous equations in the unknown joint accelerations.

b) Calculate, recursively, the coefficients which propagate motion and force constraints along the mechanism allowing the problem to be solved directly.

Most of the published algorithms for solving the FDP adopt the first approach which involves the *composite rigid body* method. Algorithms which are derived based on this approach may have $o(n^3)$ computational complexity, as oppose to $o(n)$ which results from the second approach. This is so, since in the first approach, a set of n simultaneous equations has to be solved. However, these algorithms can be computationally very efficient for small values of n, since the coefficient of n^3 in the measure of complexity is very small.

In the framework of the first approach, a well defined scheme for deriving algorithms for computing forward dynamics has been proposed by Walker and Orin [1]. To define the algorithms, the dynamic equations of motion of a robot manipulator are written in a vector form as

$$\boldsymbol{\tau} = \mathbf{D}(\mathbf{q})\ddot{\mathbf{q}} + \mathbf{C}(\mathbf{q},\dot{\mathbf{q}}) + \mathbf{G}(\mathbf{q}) + \mathbf{J}(\mathbf{q})^T \mathbf{f} \tag{6.2.1}$$

where τ is the vector of the applied joint torques/forces, $\mathbf{D}(\mathbf{q})$ is the $n \times n$ positive definite generalized inertia tensor of the robot manipulator, $\mathbf{q}$ ($\dot{\mathbf{q}}, \ddot{\mathbf{q}}$) is the vector of the joint positions (velocities, accelerations), $\mathbf{C}(\mathbf{q},\dot{\mathbf{q}})$ is the vector of Coriolis and centrifugal forces, $\mathbf{G}(\mathbf{q})$ is the vector of the gravitational effects, $\mathbf{J}(\mathbf{q})$ is the $n \times n$ Jacobian tensor and $\mathbf{f}$ is a vector of external forces. Equation (6.2.1) can be written in a more compact form as

$$\tau = \mathbf{D}(\mathbf{q})\ddot{\mathbf{q}} + \mathbf{b}(\mathbf{q},\dot{\mathbf{q}},\mathbf{g},\mathbf{f}) \qquad (6.2.2)$$

where $\mathbf{b}$ is a *bias* vector containing the gravity, centrifugal, Coriolis and external forces, i.e.,

$$\mathbf{b} = \mathbf{C}(\mathbf{q},\dot{\mathbf{q}}) + \mathbf{G}(\mathbf{q}) + \mathbf{J}(\mathbf{q})^T \mathbf{f}. \qquad (6.2.3)$$

Now, based on equation (6.2.2), the solution of the FDP can be derived by solving the following subproblems:

(i) Computation of the generalized inertia tensor: $\mathbf{D}(\mathbf{q})$.

(ii) Computation of the bias vector: $\mathbf{b}(\mathbf{q}, \dot{\mathbf{q}}, \mathbf{g}, \mathbf{f})$

(iii) Solution of the linear system of equations: $\mathbf{D}(\mathbf{q})\ddot{\mathbf{q}} = (\tau - \mathbf{b})$

(iv) Solution of a set of ordinary differential equations.

From the foregoing, for the dynamic simulation of a robot manipulator one has to solve problems directly related to manipulator dynamics (steps (i) and (ii)) and problems of numerical analysis (steps (iii) and (iv)). Therefore, since from a dynamic analysis point of view, one is concerned with the problems in steps (i) and (ii), we shall review methods for solving these two problems only. The problems of the type (iii) and (iv) have been extensively studied in the numerical analysis literature and efficient methods of solving them exist, e.g. see [18-21].

To solve part (ii), one can use non-recursive inverse dynamics algorithms, which explicitly calculate the terms of the bias vector $\mathbf{b}$ as is shown in equation (6.2.3). However, as it turns out, this approach is not computationally efficient. Walker and Orin [1], have proposed another more efficient method for computing the bias vector $\mathbf{b}$. In this method, a recursive inverse dynamics algorithm is used to solve for the actuator torques/forces, assuming that the accelerations are zero, i.e., $\ddot{\mathbf{q}} = 0$. As is obvious from equation (6.2.2), since the vector of the generalized forces τ is equal to the bias vector $\mathbf{b}$, an inverse dynamics algorithm suffices for solving problem (*ii*). Moreover, with significant improvements in the computational efficiency of algorithms for solving the inverse dynamics problem (see Chapter 5), the solution for the bias vector **b** by this method, can be computed very efficiently. Furthermore, to improve the computational efficiency of this method, we can formulate a specialized version of an inverse dynamics algorithm, with the assumption $\ddot{\mathbf{q}} = 0$ built in. Therefore, using inverse dynamics algorithms, this subproblem of forward dynamics can be solved in an efficient manner.

Solving the first problem, i.e., computing the manipulator inertia tensor $\mathbf{D}(\mathbf{q})$ is the point where most of the existing algorithms for solving the FDP using the first approach really differ. Walker and Orin [1] considered three methods for computing the inertia tensor $\mathbf{D}(\mathbf{q})$. The first two methods are based on an algorithm which solves the inverse dynamics problem from which all the velocity terms, the gravitational effects and the effects due to the external forces and torques have been eliminated. In this approach, as we can see from equation (6.2.2), the columns of the inertia matrix D, which represent the generalized inertia tensor **D** in joint space coordinates, are computed by applying a unit vector acceleration to the joints. That is, for the i-th column of D we have

$$d_i = (\tau - \mathbf{b})|_{\ddot{\mathbf{q}} = [0, \cdots, 1, \cdots, 0]^T} \tag{6.2.4}$$

where the 1 is the i-th component of $\ddot{\mathbf{q}}$. By repeating the above process n times (not necessarily recursively), all the components of D may be computed. The first two methods are basically the same, with the exception that in the second method, since D is symmetric, only the diagonal and the bottom half of the off-diagonal elements of D are computed. However, as Walker and Orin have shown in their computational complexity analysis, this approach for computing the manipulator joint space inertia matrix is computational expensive. The third method by Walker and Orin is known as the *composite rigid body* method and is computationally more efficient than their previous two methods. The basic idea in the composite rigid body method is as follows:

As in the first two methods, we assume that unit acceleration is applied to a joint (for instance at joint i, i.e., $\ddot{q}_i = 1$ at joint i) with all joint velocities and other joint accelerations equal to zero. Under this action the manipulator chain is divided into two sets of composite rigid bodies with one degree of freedom between them. The lower composite body, i.e., links 1 to $i-1$, is stationary.

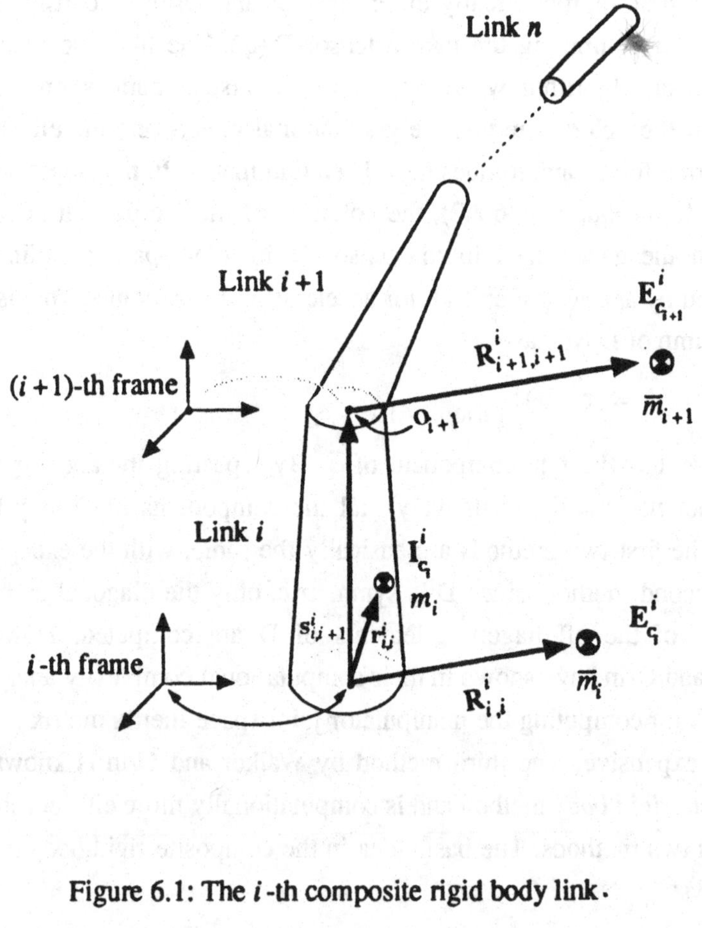

Figure 6.1: The *i* -th composite rigid body link.

The upper composite body, which is shown in Figure 6.1, is composed of links i through n and moves as a single rigid body with a *composite mass* ($\bar{m}_i$), *composite center of mass* ($\mathbf{R}^i_{i,i}$), relative to the origin of the i-th coordinate system, and *composite moment of inertia* or *inertia tensor* ($\mathbf{E}^i_{c_i}$), with respect to the composite center of mass.

Due to the motion of the i-th composite rigid body, forces and moments will be developed at the joints $1,\cdots,i$, which can be computed as follows: The force and moment at the i-th joint can be computed by applying Newton's and Euler's equations to the i-th composite rigid body. Moreover, since the acceleration at the joints $1,\cdots,i-1$, is assumed to be zero, the forces and moments at these joints result only from the propagation down the chain of the forces and moments of the i-th joint. Now, having compute these forces or moments, to define the elements of the joint space inertia matrix D, we simply need to consider their projections onto the appropriate joint axes of the manipulator.

From this computational scheme, it is obvious that the computation of $\bar{m}_i$, $\mathbf{R}^i_{i,i}$ and $\mathbf{E}^i_{c_i}$ is important (as far as computational efficiency is concerned) for the determination of the generalized inertia tensor $\mathbf{D}$. To achieve computational efficiency for these quantities, Walker and Orin proposed linear recurrence relations. These relations, formulated in the notation of this monograph, can be stated as follows:

$$\bar{m}_i = \bar{m}_{i+1} + m_i \tag{6.2.5}$$

$$\mathbf{R}^i_{i,i} = \frac{1}{\bar{m}_i}[m_i \mathbf{r}^i_{i,i} + \bar{m}_{i+1}(\mathbf{s}^i_{i,i+1} + \mathbf{A}_{i+1}\mathbf{R}^{i+1}_{i+1,i+1})] \tag{6.2.6}$$

$$\begin{aligned}\mathbf{E}^i_{c_i} = \mathbf{I}^i_{c_i} &+ m_i \left[(\mathbf{r}^i_{i,i} - \mathbf{R}^i_{i,i}) \cdot (\mathbf{r}^i_{i,i} - \mathbf{R}^i_{i,i})\mathbf{1} - (\mathbf{r}^i_{i,i} - \mathbf{R}^i_{i,i})(\mathbf{r}^i_{i,i} - \mathbf{R}^i_{i,i})^T\right] \\ &+ \bar{m}_{i+1}\Big[(\mathbf{s}^i_{i,i+1} + \mathbf{R}^i_{i+1,i+1} - \mathbf{R}^i_{i,i}) \cdot (\mathbf{s}^i_{i,i+1} + \mathbf{R}^i_{i+1,i+1} - \mathbf{R}^i_{i,i})\mathbf{1} \\ &- (\mathbf{s}^i_{i,i+1} + \mathbf{R}^i_{i+1,i+1} - \mathbf{R}^i_{i,i})(\mathbf{s}^i_{i,i+1} + \mathbf{R}^i_{i+1,i+1} - \mathbf{R}^i_{i,i})^T\Big] + \mathbf{A}_{i+1}\mathbf{E}^{i+1}_{c_{i+1}}\mathbf{A}^T_{i+1}\end{aligned} \tag{6.2.7}$$

Based on these equations and using the Newton-Euler equations to analyze the composite rigid body dynamics, Walker and Orin proposed an algorithm for computing the upper triangular part of the symmetric joint space inertia matrix D which, in the notation of this monograph, can be stated as follows:

ALGORITHM 6.1

Step 0: Initialization

$$\bar{m}_{n+1} = 0, \quad \mathbf{R}^{n}_{n+1,n+1} = 0, \quad \mathbf{E}^{n+1}_{\mathbf{C}_{n+1}} = 0, \quad \mathbf{A}_{n+1} = 0, \quad z^{i}_{i} = [\,0\,0\,1\,]^{T}$$

$$\sigma_i = \begin{cases} 1 & \text{revolute } i\text{–}th \text{ joint} \\ 0 & \text{prismatic } i\text{–}th \text{ joint} \end{cases}$$

Step 1: For $i = n, 1$ **do**

$$\bar{m}_i = \bar{m}_{i+1} + m_i \tag{6.2.8a}$$

$$\mathbf{R}^{i}_{i,i} = \frac{1}{\bar{m}_i}[m_i \mathbf{r}^{i}_{i,i} + \bar{m}_{i+1}(\mathbf{s}^{i}_{i,i+1} + \mathbf{A}_{i+1}\mathbf{R}^{i+1}_{i+1,i+1})] \tag{6.2.8b}$$

$$\begin{aligned}\mathbf{E}^{i}_{\mathbf{C}_i} = \mathbf{I}^{i}_{\mathbf{C}_i} &+ m_i \left[(\mathbf{r}^{i}_{i,i} - \mathbf{R}^{i}_{i,i}) \cdot (\mathbf{r}^{i}_{i,i} - \mathbf{R}^{i}_{i,i})\mathbf{1} - (\mathbf{r}^{i}_{i,i} - \mathbf{R}^{i}_{i,i})(\mathbf{r}^{i}_{i,i} - \mathbf{R}^{i}_{i,i})^{T}\right] \\ &+ \bar{m}_{i+1}\left[(\mathbf{s}^{i}_{i,i+1} + \mathbf{R}^{i}_{i+1,i+1} - \mathbf{R}^{i}_{i,i}) \cdot (\mathbf{s}^{i}_{i,i+1} + \mathbf{R}^{i}_{i+1,i+1} - \mathbf{R}^{i}_{i,i})\mathbf{1}\right. \\ &\left. -(\mathbf{s}^{i}_{i,i+1}+\mathbf{R}^{i}_{i+1,i+1}-\mathbf{R}^{i}_{i,i})(\mathbf{s}^{i}_{i,i+1}+\mathbf{R}^{i}_{i+1,i+1}-\mathbf{R}^{i}_{i,i})^{T}\right] + \mathbf{A}_{i+1}\mathbf{E}^{i+1}_{\mathbf{C}_{i+1}}\mathbf{A}^{T}_{i+1}\end{aligned} \tag{6.2.8c}$$

$$\mathbf{F}^{i}_{\mathbf{C}_i} = \sigma_i(\mathbf{z}^{i}_{i} \times \bar{m}_i \mathbf{R}^{i}_{i,i}) + (1 - \sigma_i)(\bar{m}_i \mathbf{z}^{i}_{i}) \tag{6.2.8d}$$

$$\mathbf{M}^{i}_{\mathbf{C}_i} = \sigma_i(\mathbf{E}^{i}_{\mathbf{C}_i} \cdot \mathbf{z}^{i}_{i}) \tag{6.2.8e}$$

$$\mathbf{f}^{i}_{i,i} = \mathbf{F}^{i}_{\mathbf{C}_i} \tag{6.2.8f}$$

$$\boldsymbol{\eta}^{i}_{i,i} = \mathbf{M}^{i}_{\mathbf{c}_i} + \mathbf{R}^{i}_{i,i} \times \mathbf{F}^{i}_{\mathbf{c}_i} \tag{6.2.8g}$$

$$d_{i,i} = \sigma_i(\boldsymbol{\eta}^{i}_{i,i} \cdot \mathbf{z}^{i}_{i}) + (1 - \sigma_i)(\mathbf{f}^{i}_{i,i} \cdot \mathbf{z}^{i}_{i}) \tag{6.2.8h}$$

For $j = i-1, 1$ **do**

$$\mathbf{f}^{j}_{j,i} = \mathbf{A}_{j+1}\mathbf{f}^{j+1}_{j+1,i} \tag{6.2.9a}$$

$$\boldsymbol{\eta}^{j}_{j,i} = \mathbf{A}_{j+1}\boldsymbol{\eta}^{j+1}_{j+1,i} + \mathbf{s}^{j}_{j,j+1} \times \mathbf{f}^{j}_{j+1,i} \tag{6.2.9b}$$

$$d_{j,i} = \sigma_i(\boldsymbol{\eta}^{j}_{j,i} \cdot \mathbf{z}^{i}_{i}) + (1 - \sigma_i)(\mathbf{f}^{j}_{j,i} \cdot \mathbf{z}^{i}_{i}) \tag{6.2.9c}$$

end

For its implementation, Algorithm 6.1 requires $12n^2 + 26n + 27$ scalar multiplications and $7n^2 + 67n - 56$ scalar additions (see Table 6.2), which for $n = 6$ amounts to 741 scalar multiplications and 601 scalar additions, respectively. Featherstone has shown [2] that based on the composite rigid body method and using a spatial notation to combine the representation of rotational and translational quantities, and spatial algebra to manipulate these quantities efficiently, another more efficient algorithm can be devised which requires $10n^2 + 31n - 41$ scalar multiplications and $6n^2 + 40n - 46$ scalar additions. For $n = 6$, this involves 505 scalar multiplications and 410 scalar additions. However, the computational efficiency of Featherstone's algorithm results from the special purpose spatial arithmetic package which he developed to handle spatial operations efficiently.

Walker and Orin also describe another method for calculating $\ddot{\mathbf{q}}$ which by-passes the need to calculate $\mathbf{D}$ explicitly (Method 4 in [1]). This method uses an iterative technique, namely, the conjugate gradient technique for solving the linear system of equations in step (*iii*). Based on an initial estimate for the joint acceleration the method uses successive adjustments to these variables until they converge to the correct solution. If there are no round-off errors, the solution for $\ddot{\mathbf{q}}$ can be achieved in a maximum of n

iterations. The complexity of this method is $o(n^2)$, but the coefficient of n^2 is large enough to make it less efficient for all but very large values of n ($n \geq 12$) than the composite rigid body method, which has $o(n^3)$ computational complexity.

Following a similar decomposition of forward dynamics into subproblems, as suggested by Walker and Orin, Angeles and Ma [3] have proposed a method for computing the generalized inertia tensor **D** which in terms of computational efficiency is comparable to the composite rigid body method. The basic idea in Angeles and Ma's approach is as follows:

First the $6n$-dimensional vector of *generalized twist*

$$\mathbf{t} \equiv [\mathbf{t}_1^T, \cdots \mathbf{t}_n^T]^T \tag{6.2.10}$$

and the $6n \times 6n$ block diagonal tensor of *generalized extended mass*

$$\mathbf{M} \equiv \mathrm{diag}\,(\mathbf{M}_1, \cdots \mathbf{M}_n), \tag{6.2.11}$$

are defined, where $\mathbf{t}_i$ is a 6-dimensional vector representing the twist of the i-th link, namely,

$$\mathbf{t}_i \equiv \begin{bmatrix} \boldsymbol{\omega}_i \\ \dot{\mathbf{r}}_{0,i} \end{bmatrix} \tag{6.2.12}$$

and $\mathbf{M}_i$ is a 6×6 tensor defined as

$$\mathbf{M}_i \equiv \begin{bmatrix} \mathbf{I}_{C_i} & 0 \\ 0 & m_i 1 \end{bmatrix} \tag{6.2.13}$$

in which 1 is the 3×3 identity tensor and 0 is the 3×3 zero tensor. Then, from the linear transformation

$$\mathbf{t} = \mathbf{T}\dot{\mathbf{q}} \tag{6.2.14}$$

the $6n \times n$ tensor **T** is defined. (The tensor **T** is referred to as *the natural orthogonal complement* since, as shown in [4], it is an orthogonal complement of the tensor which defines the kinematic constraints of the manipulator). Moreover, from kinetic energy considerations the generalized inertia tensor **D** may be defined as

$$\mathbf{D}(\mathbf{q}) = \mathbf{T}^T \mathbf{M} \mathbf{T}.$$

Furthermore, the tensor **M** can be factored as $\mathbf{M} = \mathbf{N}^T \mathbf{N}$ since it is symmetric and positive definite. From the foregoing, the generalized inertia tensor **D** can be decomposed as

$$\mathbf{D} = \mathbf{P}^T \mathbf{P} \tag{6.2.15}$$

where $\mathbf{P} \equiv \mathbf{N}\mathbf{T}$ is a lower block triangular tensor. Based on this analysis, Angeles and Ma [3] proposed an algorithm for computing the generalized inertia tensor **D** which, as we mentioned above, (see also Table 6.2) has almost the same computational complexity as Algorithm 6.1. Angeles and Ma also proposed another method which avoids the determination of the generalized inertia tensor **D**. In this method, based on (6.2.15), the linear system of equations

$$\mathbf{D}\ddot{\mathbf{q}} = \boldsymbol{\tau} - \mathbf{b} \tag{6.2.16}$$

is decomposed as follows,

$$\mathbf{P}^T \mathbf{x} = \boldsymbol{\tau} - \mathbf{b} \tag{6.2.17a}$$

$$\mathbf{P}\ddot{\mathbf{q}} = \mathbf{x} \tag{6.2.17b}$$

where **x** is a $6n$-dimensional vector. Equation (6.2.17a) represents an underdetermined system (n equations with $6n$ unknowns) and equation (6.2.17b) represents an overdetermined system ($6n$ equations with n unknowns). Based on these equations, Angeles and Ma have shown that the vector $\ddot{\mathbf{q}}$ can be computed as the least squares approximation to equation (6.2.17b), if **x** is first computed as the minimum norm solution of equation (6.2.17a). The computational efficiency of this second method of Angeles and Ma is comparable (see [3]) to that of their first method.

As we have mentioned above, an approach for solving the forward dynamics problem is to calculate recursively the coefficients which propagate motion and force constraints along the mechanism allowing the problem to be solved directly. However, although it is theoretically more sound, currently few methods adopt this approach. This is because first, it requires an extensive analysis and second, and more important, algorithms derived from this approach are computationally expensive despite the fact that one can usually achieve $o(n)$ computational complexity. This is so, since the coefficient of n in the measure of complexity is quite large.

Probably, the best known method in this approach is the *articulated-body* method proposed by Featherstone [2]. The basic idea in this method is to regard the robot as consisting of a base member (whose motion is known), a single joint, and a single moving link which is in fact an articulated body (i.e., a collection of rigid bodies connected by joints) representing the rest of the robot. The forward dynamics problem for this one-joint robot is easily solved once the apparent inertia of the moving link is known. Having found the acceleration of the first joint, the articulated body itself can be treated as a robot and the same process applied to obtain the acceleration of the next joint, and so on. So the articulated-body method consists of the calculation of a series of articulated-body inertias which are used to solve the forward dynamics problem one joint at a time. Thus, this approach leads to algorithms which have $o(n)$ computational complexity. To facilitate the analysis of his method, Featherstone introduced a spatial notation which provides a uniform combined representation of rotational and translational quantities, and he developed a spatial algebra for manipulating these spatial quantities. Also, to implement his algorithm efficiently (see Table 6.3), Featherstone developed a spatial arithmetic package with special-purpose arithmetic functions to operate on these compact spatial representations.

Other examples of methods which solve the FDP by the constraint propagation approach are described by Armstrong [5], Rodriguez [7] and

Rodriguez and Kreutz [8]. In particular, Armstrong's method also achieves $o(n)$ complexity, and uses recursion coefficients playing a similar role to articulated-body inertias. This method, in its basic form, is applicable to robots with spherical joints but a modification applicable for revolute joints is outlined in one of the appendixes in Armstrong's paper. However, this modification increases the computational requirement significantly, although the complexity remains $o(n)$. Rodriguez and Kreutz [8] have developed a two-step algorithm for computing forward dynamics which has $o(n)$ computational complexity. Based on a linear operator approach for formulating and analyzing the manipulator dynamics developed by Rodriguez [6,7], the two-step algorithm by Rodriguez and Kreutz first computes and subtracts out the Coriolis, centrifugal, gravity and contact force bias terms, exactly as in Walker and Orin's approach, to obtain a "bias-free" robot dynamic equation. Then, in the second step, using techniques for solving linear operator equations by operator factorization, the joint space accelerations are obtained in $o(n)$ iterations. Also, using certain operator identities, they propose alternative algorithms for which the need for a preliminary bias vector computation and subtraction is avoided. The dynamic analysis of these algorithms, as in Featherstone's approach, is based on spatial notation and spatial algebra. The approach by Rodriguez and Kreutz is important because it provides a method to formulate, analyze and understand spatial recursions in multibody dynamics. This analysis leads them to a simple factorization of the generalized inertia tensor $\mathbf{D}$ from which an immediate inversion of $\mathbf{D}$ is readily available. In particular, they established the following factorization for the generalized inertia tensor $\mathbf{D}$ and its inverse:

$$\mathbf{D} = \left[1 + \mathbf{H\Psi L}\right]\hat{\mathbf{D}}\left[1 + \mathbf{H\Psi L}\right]^T \tag{6.2.18}$$

and

$$\mathbf{D}^{-1} = \left[1 - \mathbf{H\Phi L}\right]^T\hat{\mathbf{D}}^{-1}\left[1 - \mathbf{H\Phi L}\right] \tag{6.2.19}$$

where $\mathbf{H}$ and $\boldsymbol{\Psi}$ are given by known geometric link parameters, and $\mathbf{L}$, $\boldsymbol{\Phi}$

and $\hat{\mathbf{D}}$ are obtained recursively by a spatial discrete-step Kalman filter and by the corresponding Riccati equation associated with this filter. The factors $(1 + \mathbf{H\Psi L})$ and $(1 - \mathbf{H\Phi L})$ are lower triangular tensors which are inverses of each other, and $\hat{\mathbf{D}}$ is a diagonal tensor. This analytic factorization and inversion is obviously important because it avoids numerical triangular decomposition and inversion, and consequently the problems associated with build-up to numerical round-off errors in such computations. However, a computational complexity analysis of these algorithms has not been included in [8] and this makes a fair comparison of their method with others difficult.

Finally, an approach for solving the FDP which is quite different from those presented above has been proposed by Chou, Baciu and Kesavan [9,10]. Their formulation uses graph-theoretic models for the joints and the open-loop kinematic chains of rigid bodies. Euler parameters are used instead of the conventional direction cosines to describe relative orientations. The final mathematical model derived by this formulation is a large system ($20n$ scalar equations with $20n$ unknowns) of differential and algebraic equations. A complete computational complexity analysis has not been provided for the method. However, because of the large system of equations that has to be solved, the approach is almost certainly very expensive computationally.

Concluding this review on forward dynamics computation, it is worth mentioning that, as with inverse dynamics computation, to improve computational efficiency parallel algorithms and special architectures have been proposed. For example, parallel processing techniques have been proposed by Lee and Chang [11] and systolic architectures have been used by Amin-Javaheri and Orin [12]. Also in [13], Han has examined possible applications of parallel and pipeline processing as well as VLSI systolic array processors for computing forward dynamics in real-time.

6.3 The Generalized Manipulator Inertia Tensor

As we mentioned in section 6.2, one of the methods which may be used for computation of the generalized inertia tensor **D** is the composite rigid body method. This method leads to algorithms (e.g., Algorithm 6.1) which compute efficiently the manipulator inertia tensor **D** by utilizing recurrence relations for some of its basic equations. However, a drawback of this method is that it leads to algorithms which require all the quantities to be computed online. Moreover, as we shall see in this section, the recurrence relations on which these algorithms are based can be stated in computationally more efficient formulations.

In order to reduce the computational complexity of the above mentioned algorithm, two other methods are proposed in this section. The first method is similar to the third method proposed by Walker and Orin [1]. In particular, a similar decomposition, i.e., a set of stationary and moving links, is used as the underlying modeling scheme and the dynamic analysis is based on the Newton-Euler equations. However, the dynamic analysis of the moving set of links in this method uses the concepts of generalized and augmented links instead of that of the composite rigid body alone. The concepts of augmented and generalized links have been introduced in Chapter 5 to facilitate more efficient solutions of the inverse dynamics problem. Here, as in the case of inverse dynamics, these concepts will allow us to devise an algorithm which is applicable to general robot manipulators and, in almost all cases, its computational burden may be split into computations which can be performed *off-line* and computations which have to be performed *online*. Moreover, based on these concepts and using Cartesian tensor analysis, computationally more efficient recurrence relations will be devised to facilitate the online computations. The second method also uses the concepts of augmented and generalized links and Cartesian tensor analysis, but the dynamic analysis is based now on the Euler-Lagrange equations instead of the Newton-Euler ones.

As we shall see, both methods lead to the same algorithm for computing the generalized inertia tensor **D** of a robot manipulator. Thus, from an algorithmic point of view, it may seem that this duplication in the analysis is unnecessary, and this is definitely true. However, the main reason for the duplication is to show, as we did with inverse dynamics, that apart from personal preference or experience there is nothing to be gained, in terms of computational efficiency, by choosing one or the other set of dynamic equations in our analysis.

6.3.1 Generalized Links and their Inertia Tensor

A generalized link has been defined in Chapter 5 (see Definition 5.2). It is obvious from this definition that a generalized link is simply a composite rigid body as defined by Walker and Orin. However, since the analysis to follow is different from that presented by Walker and Orin, we shall continue to refer to the set of moving links as a generalized link. Basically, the analysis here is different from that of Walker and Orin in that all moments concerning a generalized link are considered about the origin of one of the link coordinate systems instead of its center of mass. This modification allows us to use the inertia tensor of an augmented link (which, can be computed off-line for most industrial robots) for a computationally more efficient formulation of the inertia tensor of a generalized link.

The definition of an augmented link has been given in Chapter 5 (see Definition 5.1). Also, in Chapter 5, the definitions of the first and second moments of an augmented link, as well as, the definition of the first moment of a generalized link have been given. Here these definitions are repeated for quick reference and the list of these definitions is completed with the definition of the second moment (inertia tensor) of a generalized link. These definitions for the i-th augmented and the i-th generalized link may be stated as follows:

(1) First moment of the i-th augmented link about the origin o_i of the i-th link coordinate system:

$$\mathbf{u}^{i}_{o_i} = m_i \mathbf{r}^{i}_{i,i} + \bar{m}_{i+1} \mathbf{s}^{i}_{i,i+1} \tag{6.3.1}$$

(2) Second moment (inertia matrix) of the i-th augmented link about o_i:

$$\mathbf{K}^{i}_{o_i} = \mathbf{I}^{i}_{c_i} - m_i \tilde{\mathbf{r}}^{i}_{i,i} \tilde{\mathbf{r}}^{i}_{i,i} - \bar{m}_{i+1} \tilde{\mathbf{s}}^{i}_{i,i+1} \tilde{\mathbf{s}}^{i}_{i,i+1} \tag{6.3.2}$$

(3) First moment of the i-th generalized link about o_i:

$$\begin{aligned} \mathbf{U}^{i}_{o_i} &= \sum_{j=i}^{n} m_j \mathbf{r}^{i}_{i,j}. \\ &= \mathbf{u}^{i}_{o_i} + A_{i+1} \mathbf{U}^{i+1}_{o_{i+1}}. \end{aligned} \tag{6.3.3}$$

where the last step follows from Lemma 5.1, and

(4) Second moment (inertia matrix) about o_i of the i-th generalized link:

$$\mathbf{E}^{i}_{o_i} = \sum_{k=i}^{n} [\mathbf{I}^{i}_{c_k} - m_k \tilde{\mathbf{r}}^{i}_{i,k} \tilde{\mathbf{r}}^{i}_{i,k}]. \tag{6.3.4}$$

Also, as we may recall from Chapter 5, the zero-th moment or mass of the i-th augmented link which is equal to the zero-th moment of the i-th generalized link is defined as

$$\bar{m}_i = m_i + \bar{m}_{i+1} \tag{6.3.5}$$

The formulation of equation (6.3.4), which defines the inertia tensor of the i-th generalized link, is important because it provides physical insight into the structure of this inertia tensor. However, this formulation is obviously computationally very expensive for any use in practical applications. Therefore, to be able to use generalized links effectively, we have to define their inertia tensor by using a computationally more efficient equation. This equation is provided by the following lemma [17].

Lemma 6.1: The inertia tensor of the i-th generalized link with respect to the origin o_i of the i-th link coordinate system may be defined by the following recurrence relation:

$$\mathbf{E}^i_{o_i} = \mathbf{K}^i_{o_i} - \left[\tilde{\mathbf{s}}^i_{i,i+1}\tilde{\mathbf{U}}^i_{o_{i+1}} + \tilde{\mathbf{U}}^i_{o_{i+1}}\tilde{\mathbf{s}}^i_{i,i+1}\right] + \mathbf{A}_{i+1}\mathbf{E}^{i+1}_{o_{i+1}}\mathbf{A}^T_{i+1} \tag{6.3.6}$$

where $\mathbf{K}^i_{o_i}$ is the inertia tensor of the i-th augmented link with respect to the origin o_i, $\tilde{\mathbf{s}}^i_{i,i+1}$ is the dual tensor of the position vector of o_{i+1} relative to o_i and $\tilde{\mathbf{U}}^i_{o_{i+1}}$ is the dual tensor of the first moment about o_{i+1} of the $(i+1)$-th generalized link.

Proof: Since, for $k>i$, $\tilde{\mathbf{r}}^i_{i,k} = \tilde{\mathbf{s}}^i_{i,i+1} + \tilde{\mathbf{r}}^i_{i+1,k}$, we have

$$\tilde{\mathbf{r}}^i_{i,k}\tilde{\mathbf{r}}^i_{i,k} = \tilde{\mathbf{s}}^i_{i,i+1}\tilde{\mathbf{s}}^i_{i,i+1} + \tilde{\mathbf{s}}^i_{i,i+1}\tilde{\mathbf{r}}^i_{i+1,k} + \tilde{\mathbf{r}}^i_{i+1,k}\tilde{\mathbf{s}}^i_{i,i+1} + \tilde{\mathbf{r}}^i_{i+1,k}\tilde{\mathbf{r}}^i_{i+1,k}.$$

Equation (6.3.4) can be written as

$$\begin{aligned}
\mathbf{E}^i_{o_i} &= \mathbf{I}^i_{c_i} - m_i\tilde{\mathbf{r}}^i_{i,i}\tilde{\mathbf{r}}^i_{i,i} - \sum_{k=i+1}^{n} m_k\tilde{\mathbf{s}}^i_{i,i+1}\tilde{\mathbf{s}}^i_{i,i+1} \\
&\quad - \sum_{k=i+1}^{n} [m_k(\tilde{\mathbf{s}}^i_{i,i+1}\tilde{\mathbf{r}}^i_{i+1,k} + \tilde{\mathbf{r}}^i_{i+1,k}\tilde{\mathbf{s}}^i_{i,i+1})] + \sum_{k=i+1}^{n} [\mathbf{I}^i_{c_k} - m_k\tilde{\mathbf{r}}^i_{i+1,k}\tilde{\mathbf{r}}^i_{i+1,k}] \\
&= \mathbf{I}^i_{c_i} - m_i\tilde{\mathbf{r}}^i_{i,i}\tilde{\mathbf{r}}^i_{i,i} - \bar{m}_{i+1}\tilde{\mathbf{s}}^i_{i,i+1}\tilde{\mathbf{s}}^i_{i,i+1} - [\tilde{\mathbf{s}}^i_{i,i+1}\tilde{\mathbf{U}}^i_{o_{i+1}} + \tilde{\mathbf{U}}^i_{o_{i+1}}\tilde{\mathbf{s}}^i_{i,i+1}] \\
&\quad + \mathbf{A}_{i+1}\sum_{k=i+1}^{n} [\mathbf{I}^{i+1}_{c_k} - m_k\tilde{\mathbf{r}}^{i+1}_{i+1,k}\tilde{\mathbf{r}}^{i+1}_{i+1,k}]\mathbf{A}^T_{i+1} \\
&= \mathbf{K}^i_{o_i} - \left[\tilde{\mathbf{s}}^i_{i,i+1}\tilde{\mathbf{U}}^i_{o_{i+1}} + \tilde{\mathbf{U}}^i_{o_{i+1}}\tilde{\mathbf{s}}^i_{i,i+1}\right] + \mathbf{A}_{i+1}\mathbf{E}^{i+1}_{o_{i+1}}\mathbf{A}^T_{i+1}
\end{aligned}$$

where the definitions of the inertia tensors for the i-th augmented and i-th generalized links have been used in the last step. Thus, equation (6.3.6) is valid and this completes the proof. □

As we mentioned above, a generalized link is another name for the concept of a composite rigid body. Therefore, it is obvious that equations (6.2.6) and (6.2.7) which define the composite center of mass and the composite inertia tensor, respectively, of the i-th composite rigid body are closely related to equations (6.3.3) and (6.3.6) which define the first and second moments, respectively, of the i-th generalized link. To see this, by substituting $\mathbf{u}_{o_i}^{i}$ from equation (6.3.1) into equation (6.3.3), and using equation (6.2.6), we can write the first moment of the i-th generalized link as follows:

$$\mathbf{U}_{o_i}^{i} = \bar{m}_i \mathbf{R}_{i,i}^{i}. \tag{6.3.7}$$

where $\mathbf{R}_{i,i}^{i}$ is the position vector of the center of mass of the i-th composite rigid body (or the i-th generalized link) with respect to the origin o_i of the i-th link coordinate system. Note that, based on physical considerations, we could use equation (6.3.7) instead of equation (6.3.3) as the definition of the first moment of the i-th generalized link with respect to the origin o_i of the i-th link coordinate system. However, the chosen definition provides physical insight which facilitates the analysis of the dynamic equations of motion. Similarly, using the parallel axis theorem, the inertia tensor of the i-th generalized link $\mathbf{E}_{o_i}^{i}$ with respect to the origin o_i can be defined by the equation

$$\mathbf{E}_{o_i}^{i} = \mathbf{E}_{c_i}^{i} - \bar{m}_i \tilde{\mathbf{R}}_{i,i}^{i} \tilde{\mathbf{R}}_{i,i}^{i} \tag{6.3.8}$$

where $\mathbf{E}_{c_i}^{i}$ (see equation (6.2.7)) is the inertia tensor of the i-th generalized link with respect to its center of mass, and $\tilde{\mathbf{R}}_{i,i}^{i}$ is the dual tensor of the position vector $\mathbf{R}_{i,i}^{i}$.

Now, using these moments of the augmented and generalized links, we can proceed to derive an algorithm for computing efficiently the joint-space matrix D of the generalized inertia tensor $\mathbf{D}$ of a rigid-link open-chain robot manipulator.

6.3.2 The Use of Newton-Euler Equations in Computing the Manipulator Inertia Tensor

To simplify the analysis, we shall assume that all joints are of revolute type. However, the final equations in Algorithm 6.2 have been modified to make the algorithm applicable to both revolute and prismatic joints. These modifications are simple and self explanatory.

Following the approach by Walker and Orin, let us assume that unit acceleration is applied at the i-th joint of a robot manipulator with all joint velocities and other joint accelerations equal to zero. Under these assumptions only the i-th generalized link moves. To describe the motion of the i-th generalized link, we shall use Newton's and Euler's equations.

As is well known, Newton's and Euler's equations describe the motion of a rigid body relative to an inertia frame and are given [14] by

$$\mathbf{F_c} = m\ddot{\mathbf{r}}_\mathbf{c} \tag{6.3.9}$$

$$\mathbf{M_c} = \mathbf{I_c} \cdot \dot{\boldsymbol{\omega}} + \tilde{\boldsymbol{\omega}}\, \mathbf{I_c} \cdot \boldsymbol{\omega} \tag{6.3.10}$$

respectively, where m is the mass, $\ddot{\mathbf{r}}_\mathbf{c}$ is the absolute acceleration of the center of mass, $\mathbf{I_c}$ is the inertia tensor of the rigid body about its center of mass and $\boldsymbol{\omega}$ ($\dot{\boldsymbol{\omega}}$) is the absolute angular velocity (acceleration) of the motion.

In the case of the i-th composite link , equations (6.3.9) and (6.3.10) take the form

$$\mathbf{F}_{\mathbf{c}_i} = \bar{m}_i \ddot{\mathbf{R}}_{0,i} \tag{6.3.11}$$

and

$$\mathbf{M}_{\mathbf{c}_i} = \mathbf{E}_{\mathbf{c}_i} \cdot \dot{\boldsymbol{\omega}}_i + \tilde{\boldsymbol{\omega}}_i \mathbf{E}_{\mathbf{c}_i} \cdot \boldsymbol{\omega}_i \,. \tag{6.3.12}$$

Note that, (6.3.11) and (6.3.12) are expressed in the base frame orientation. However, since by assumption links 1 to $i-1$ are stationary, we can assume that the origin of the inertia frame is at the same position as the origin o_i of the i-th link coordinate frame. This implies that $\ddot{\mathbf{R}}_{0,i} = \ddot{\mathbf{R}}_{i,i}$. Therefore, using equations (6.3.7) and (6.3.8) we can rewrite equations (6.3.11) and (6.3.12) in the following form

$$\mathbf{F}_{\mathbf{c}_i} = \ddot{\mathbf{U}}_{\mathbf{o}_i} \tag{6.3.13}$$

and

$$\mathbf{M}_{\mathbf{c}_i} = \mathbf{E}_{\mathbf{o}_i} \cdot \dot{\boldsymbol{\omega}}_i + \tilde{\boldsymbol{\omega}}_i \mathbf{E}_{\mathbf{o}_i} \cdot \boldsymbol{\omega}_i + \frac{1}{\bar{m}_i}\left[\tilde{\mathbf{U}}_{\mathbf{o}_i} \tilde{\mathbf{U}}_{\mathbf{o}_i} \cdot \dot{\boldsymbol{\omega}}_i + \tilde{\boldsymbol{\omega}}_i \tilde{\mathbf{U}}_{\mathbf{o}_i} \tilde{\mathbf{U}}_{\mathbf{o}_i} \cdot \boldsymbol{\omega}_i\right] \tag{6.3.14}$$

Moreover, under the assumptions made above, we have

$$\boldsymbol{\omega}_i = 0 \quad \text{and} \quad \dot{\boldsymbol{\omega}}_i = \mathbf{z}_i . \tag{6.3.15}$$

which implies that $\boldsymbol{\Omega}_j = \tilde{\mathbf{z}}_i$ for all $j \geq i$. Therefore, using Lemma 5.2, we can show that

$$\ddot{\mathbf{U}}_{\mathbf{o}_i} = \tilde{\mathbf{z}}_i \mathbf{U}_{\mathbf{o}_i} . \tag{6.3.16}$$

From the foregoing, equations (6.3.13) and (6.3.14) can be written as

$$\mathbf{F}_{\mathbf{c}_i} = \tilde{\mathbf{z}}_i \mathbf{U}_{\mathbf{o}_i} \tag{6.3.17}$$

and

$$\mathbf{M}_{\mathbf{c}_i} = \mathbf{E}_{\mathbf{o}_i} \cdot \mathbf{z}_i + \frac{1}{\bar{m}_i} \tilde{\mathbf{U}}_{\mathbf{o}_i} \tilde{\mathbf{U}}_{\mathbf{o}_i} \mathbf{z}_i \tag{6.3.18}$$

respectively. Moreover, equations (6.3.17) and (6.3.18) as tensor equations are invariant under coordinate transformations. Therefore, following the usual approach in the Newton-Euler formulation of robot dynamics, we express them in the i-th frame orientation as

$$\mathbf{F}^i_{\mathbf{c}_i} = \tilde{\mathbf{z}}^i_i \mathbf{U}^i_{\mathbf{o}_i} \tag{6.3.19}$$

and

$$\mathbf{M}^i_{\mathbf{c}_i} = \mathbf{E}^i_{\mathbf{o}_i} \cdot \mathbf{z}^i_i + \frac{1}{\bar{m}_i} \tilde{\mathbf{U}}^i_{\mathbf{o}_i} \tilde{\mathbf{U}}^i_{\mathbf{o}_i} \mathbf{z}^i_i \tag{6.3.20}$$

respectively.

Now, due to the motion of the i-th joint, forces ($\mathbf{f}^j_{j,i}$) and moments ($\boldsymbol{\eta}^j_{j,i}$) will be developed at all joints j for $j \le i$. These forces and moments are the inter-link constraints which keep the links in the lower part of the manipulator fixed with respect to each other. Projections of these forces or moments onto the joint axes are just the elements of the desired joint-space inertia matrix D of the generalized inertia tensor $\mathbf{D}$. To determine these constraint forces and moments we use a backward recursion from joint i through to joint 1. At the initial step of this recursion, i.e., for $j=i$, $\mathbf{f}^i_{i,i}$ and $\boldsymbol{\eta}^i_{i,i}$ can be determined by simply resolving $\mathbf{F}^i_{\mathrm{c}_i}$ and $\mathbf{M}^i_{\mathrm{c}_i}$ to the origin of the i-th coordinates. Thus, we have

$$\mathbf{f}^i_{i,i} = \mathbf{F}^i_{\mathrm{c}_i} \tag{6.3.21}$$

and

$$\boldsymbol{\eta}^i_{i,i} = \mathbf{M}^i_{\mathrm{c}_i} + \tilde{\mathbf{R}}^i_{i,i}\mathbf{F}^i_{\mathrm{c}_i} \tag{6.3.22}$$

Further, by using (6.3.19) and (6.3.20), we can write equation (6.3.22) as

$$\boldsymbol{\eta}^i_{i,i} = \mathbf{E}^i_{\mathrm{o}_i}\cdot\mathbf{z}^i_i + \frac{1}{\bar{m}_i}\tilde{\mathbf{U}}^i_{\mathrm{o}_i}\tilde{\mathbf{U}}^i_{\mathrm{o}_i}\mathbf{z}^i_i + \tilde{\mathbf{R}}^i_{i,i}\tilde{\mathbf{z}}^i_i\mathbf{U}^i_{\mathrm{o}_i} \tag{6.3.23}$$

and finally, since $\tilde{\mathbf{z}}^i_i\mathbf{U}^i_{\mathrm{o}_i} = -\tilde{\mathbf{U}}^i_{\mathrm{o}_i}\mathbf{z}^i_i$ and $\tilde{\mathbf{R}}^i_{i,i} = \dfrac{1}{\bar{m}_i}\tilde{\mathbf{U}}^i_{\mathrm{o}_i}$, we have

$$\boldsymbol{\eta}^i_{i,i} = \mathbf{E}^i_{\mathrm{o}_i}\cdot\mathbf{z}^i_i. \tag{6.3.24}$$

In the rest of the recursion, i.e., for $j < i$, the constraint forces and moments are computed using the equations

$$\mathbf{f}^j_{j,i} = \mathbf{A}_{j+1}\mathbf{f}^{j+1}_{j+1,i} \tag{6.3.25}$$

and

$$\boldsymbol{\eta}^j_{j,i} = \mathbf{A}_{j+1}\left[\boldsymbol{\eta}^{j+1}_{j+1,i} + \tilde{\mathbf{s}}^{j+1}_{j,j+1}\mathbf{f}^{j+1}_{j+1,i}\right]. \tag{6.3.26}$$

Note that these equations can also be derived from a compatible inverse dynamics algorithm, say Algorithm 5.4, when the accelerations at all joints j, for $j \neq i$, are assumed to be zero. Finally, the elements of the symmetric joint-space inertia matrix D of the generalized inertia tensor **D** are determined by using the following equation

$$d_{j,i} = {}^{j}\boldsymbol{\eta}_{j,i} \cdot {}^{j}\mathbf{z}_{j}. \tag{6.3.27}$$

Based on the above equations, we can state the following algorithm for computing the elements of the joint-space inertia matrix D.

ALGORITHM 6.2

Step 0: Initialization:

$$\bar{m}_{n+1} = 0, \quad \mathbf{A}_{n+1} = 0, \quad \mathbf{U}_{o_{n+1}}^{n+1} = 0, \quad {}^{i}z_i = [\,0\,0\,1\,]^T$$

$$\sigma_i = \begin{cases} 1 & \text{revolute } i\text{-th joint} \\ 0 & \text{prismatic } i\text{-th joint} \end{cases}$$

Step 1: For $i = n, 1$, **do**

$$\bar{m}_i = m_i + \bar{m}_{i+1} \tag{6.3.28a}$$

$$\mathbf{u}_{o_i}^{i} = m_i {}^{i}\mathbf{r}_{i,i} + \bar{m}_{i+1} {}^{i}\mathbf{s}_{i,i+1} \tag{6.3.28b}$$

$$\mathbf{K}_{o_i}^{i} = \mathbf{I}_{c_i}^{i} - m_i {}^{i}\tilde{\mathbf{r}}_{i,i} {}^{i}\tilde{\mathbf{r}}_{i,i} - \bar{m}_{i+1} {}^{i}\tilde{\mathbf{s}}_{i,i+1} {}^{i}\tilde{\mathbf{s}}_{i,i+1} \tag{6.3.28c}$$

Step 2: For $i = n, 1$, **do**

$$\mathbf{U}_{o_i}^{i} = \mathbf{u}_{o_i}^{i} + \mathbf{A}_{i+1} \mathbf{U}_{o_{i+1}}^{i+1} \tag{6.3.29a}$$

$${}^{i+1}\mathbf{s}_{i,i+1} = \mathbf{A}_{i+1}^{T} {}^{i}\mathbf{s}_{i,i+1} \tag{6.3.29b}$$

$$\mathbf{E}_{o_i}^{i} = \mathbf{K}_{o_i}^{i} + \mathbf{A}_{i+1} \left\{ \mathbf{E}_{o_{i+1}}^{i+1} - \left[{}^{i+1}\tilde{\mathbf{s}}_{i,i+1} \tilde{\mathbf{U}}_{o_{i+1}}^{i+1} + ({}^{i+1}\tilde{\mathbf{s}}_{i,i+1} \tilde{\mathbf{U}}_{o_{i+1}}^{i+1})^T \right] \right\} \mathbf{A}_{i+1}^{T} \tag{6.3.29c}$$

$$\mathbf{f}^i_{i,i} = \sigma_i \tilde{\mathbf{z}}^i_i \mathbf{U}^i_{\mathrm{o}_i} + (1 - \sigma_i)\overline{m}_i \mathbf{z}^i_i \tag{6.3.29d}$$

$$\eta^i_{i,i} = \sigma_i \mathbf{E}^i_{\mathrm{o}_i} \cdot \mathbf{z}^i_i \tag{6.3.29e}$$

$$d_{i,i} = \sigma_i(\eta^i_{i,i} \cdot \mathbf{z}^i_i) + (1 - \sigma_i)(\mathbf{f}^i_{i,i} \cdot \mathbf{z}^i_i) \tag{6.3.29f}$$

Step 3: For $j = i\text{-}1, 1$**, do**

$$\mathbf{f}^j_{j,i} = \mathbf{A}_{j+1} \mathbf{f}^{j+1}_{j+1,i} \tag{6.3.30a}$$

$$\eta^j_{j,i} = \mathbf{A}_{j+1} \left[\tilde{\mathbf{s}}^{j+1}_{j,j+1} \mathbf{f}^{j+1}_{j+1,i} + \eta^{j+1}_{j+1,i} \right] \tag{6.3.30b}$$

$$d_{j,i} = \sigma_j(\eta^j_{j,i} \cdot \mathbf{z}^j_j) + (1 - \sigma_j)(\mathbf{f}^j_{j,i} \cdot \mathbf{z}^j_j) \tag{6.3.30c}$$

end

Note that since the joint-space inertia matrix D is symmetric, Algorithm 6.2 computes only the upper triangular part of it. Also, as we have mentioned above, certain equations of this algorithm have been modified to allow for both revolute and prismatic joints.

6.3.3 The Use of Euler-Lagrange Equations in Computing the Manipulator Inertia Tensor

In this section we demonstrate how the Euler-Lagrange equations can be used to derive Algorithm 6.2. To simplify the derivation, we consider revolute joints only. However, with slight modifications the analysis can be extended to include prismatic joints as well.

In the Euler-Lagrange approach, we first compute the Lagrangian of the manipulator, which is defined by

$$\mathbf{L} = \mathbf{\Phi} - \mathbf{P} \tag{6.3.31}$$

where $\mathbf{\Phi}$ is the kinetic energy and $\mathbf{P}$ is the potential energy of the manipulator. Then to derive the generalized torques (i.e., forces and torques acting at the joints), we use the Euler-Lagrange equations

$$\tau_i = \frac{d}{dt}\frac{\partial L}{\partial \dot{q}_i} - \frac{\partial L}{\partial q_i} \qquad i = 1,2, ..., n \tag{6.3.32}$$

where, q_i ($\dot{q}_i$) are the generalized coordinates (velocities) of the manipulator. Performing the differentiation involved in (6.3.32), we can write the equation for the generalized torques (when there are no external forces acting on the manipulator) in vector form as

$$\boldsymbol{\tau} = \mathbf{D}(\mathbf{q})\ddot{\mathbf{q}} + \mathbf{C}(\mathbf{q},\dot{\mathbf{q}}) + \mathbf{G}(\mathbf{q}) \tag{6.3.33}$$

where $\mathbf{D}(\mathbf{q})$ is the generalized inertia tensor of the manipulator, $\mathbf{C}(\mathbf{q},\dot{\mathbf{q}})$ is a vector which contains Coriolis and centrifugal forces and $\mathbf{G}(\mathbf{q})$ is the vector of gravitational forces.

Since potential energy is independent of the joint velocities, it is obvious that the generalized inertia tensor $\mathbf{D}(\mathbf{q})$ results from kinetic energy only. Actually, if we write the kinetic energy of the manipulator in the form

$$\boldsymbol{\Phi} = \frac{1}{2}\dot{\mathbf{q}}^T \mathbf{H}(\mathbf{q})\dot{\mathbf{q}} \tag{6.3.34}$$

where $\mathbf{H}(\mathbf{q})$ is the kinetic energy tensor, we can compute the generalized inertia tensor $\mathbf{D}(\mathbf{q})$ directly by setting

$$\mathbf{D}(\mathbf{q}) = \mathbf{H}(\mathbf{q}) \tag{6.3.35}$$

i.e., the generalized inertia tensor of a manipulator is simply the kinetic energy tensor. Therefore, to compute the tensor $\mathbf{D}(\mathbf{q})$, we have to derive the kinetic energy of the manipulator in the form given by equation (6.3.34).

Kinetic energy is one of the most important physical quantities in rigid body dynamics and is defined by a number of equivalent equations [14]. Here, to define the kinetic energy of the k-th link of a robot manipulator, we use the following equation [14]

$$\boldsymbol{\Phi}_k = \frac{1}{2} m_k \dot{\mathbf{r}}_{0,k} \cdot \dot{\mathbf{r}}_{0,k} + \frac{1}{2}\boldsymbol{\omega}_k \cdot \mathbf{I}_{C_k} \cdot \boldsymbol{\omega}_k \tag{6.3.36}$$

where the first term defines the translational kinetic energy and the second term defines the rotational kinetic energy of the k-th link. Now, using the superposition theorem, we get the total kinetic energy of a robot manipulator as

$$\mathbf{\Phi} = \frac{1}{2}\sum_{k=1}^{n} [m_k \dot{\mathbf{r}}_{0,k} \cdot \dot{\mathbf{r}}_{0,k} + \boldsymbol{\omega}_k \cdot \mathbf{I}_{\mathbf{C}_k} \cdot \boldsymbol{\omega}_k]. \qquad (6.3.37)$$

The absolute linear and angular velocities of the k-th link can be defined explicitly by the following equations

$$\dot{\mathbf{r}}_{0,k} = \sum_{i=1}^{k} (\mathbf{z}_i \times \mathbf{r}_{i,k})\dot{q}_i \qquad (6.3.38)$$

$$\boldsymbol{\omega}_k = \sum_{i=1}^{k} \mathbf{z}_i \dot{q}_i . \qquad (6.3.39)$$

Now, by substituting (6.3.38) and (6.3.39) into (6.3.37) and noticing that from equation (3.4.9) we have

$$(\mathbf{z}_j \times \mathbf{r}_{j,k}) \cdot (\mathbf{z}_i \times \mathbf{r}_{i,k}) = - \mathbf{z}_j \cdot \tilde{\mathbf{r}}_{j,k} \tilde{\mathbf{r}}_{i,k} \cdot \mathbf{z}_i ,$$

we can write (6.3.39) as

$$\mathbf{\Phi} = \frac{1}{2}\sum_{k=1}^{n} \sum_{i,j=1}^{k} \left[\mathbf{z}_j \cdot (\mathbf{I}_{\mathbf{C}_k} - m_k \tilde{\mathbf{r}}_{j,k} \tilde{\mathbf{r}}_{i,k}) \cdot \mathbf{z}_i \right] \dot{q}_i \dot{q}_j . \qquad (6.3.40)$$

Finally, the permutation of the two summation symbols in (6.3.40) gives

$$\mathbf{\Phi} = \frac{1}{2}\sum_{i,j=1}^{n} \sum_{k=max(i,j)}^{n} \left[\mathbf{z}_j \cdot (\mathbf{I}_{\mathbf{C}_k} - m_k \tilde{\mathbf{r}}_{j,k} \tilde{\mathbf{r}}_{i,k}) \cdot \mathbf{z}_i \right] \dot{q}_i \dot{q}_j . \qquad (6.3.41)$$

Therefore, from (6.3.34), (6.3.35) and (6.3.41), we have that the elements of the joint-space inertia matrix D of the generalized inertia tensor $\mathbf{D}(\mathbf{q})$ satisfy the following equation,

$$d_{j,i} = \mathbf{z}_j \cdot \left[\sum_{k=max(i,j)}^{n} (\mathbf{I}_{\mathbf{C}_k} - m_k \tilde{\mathbf{r}}_{j,k} \tilde{\mathbf{r}}_{i,k}) \right] \cdot \mathbf{z}_i \qquad (6.3.42)$$

Equation (6.3.42) can be simplified if one uses equations (6.3.3) and (6.3.4) and Lemma 6.1. To see this, let us assume that $j \leq i$, then since by definition $\mathbf{r}_{j,k} = \mathbf{s}_{j,i} + \mathbf{r}_{i,k}$, we have

$$\sum_{k=i}^{n} [\mathbf{I}_{c_k} - m_k \tilde{\mathbf{r}}_{j,k} \tilde{\mathbf{r}}_{i,k}] = \sum_{k=i}^{n} [\mathbf{I}_{c_k} - m_k \tilde{\mathbf{r}}_{i,k} \tilde{\mathbf{r}}_{i,k}] - \sum_{k=i}^{n} m_k \tilde{\mathbf{s}}_{j,i} \tilde{\mathbf{r}}_{i,k}$$

$$= \mathbf{E}_{o_i} - \tilde{\mathbf{s}}_{j,i} \tilde{\mathbf{U}}_{o_i} \tag{6.3.43}$$

Therefore, we can write equation (6.3.42) as

$$d_{j,i} = \mathbf{z}_j \cdot \left[\mathbf{E}_{o_i} - \tilde{\mathbf{s}}_{j,i} \tilde{\mathbf{U}}_{o_i} \right] \cdot \mathbf{z}_i \tag{6.3.44}$$

Moreover, equation (6.3.44) as a tensor equation is invariant under coordinate transformations. Therefore, it can be written in the j-th frame orientation as

$$d_{j,i} = \mathbf{z}_j^{j} \cdot \left[\mathbf{E}_{o_i}^{j} - \tilde{\mathbf{s}}_{j,i}^{j} \tilde{\mathbf{U}}_{o_i}^{j} \right] \cdot \mathbf{z}_i^{j} \tag{6.3.45}$$

Now, we shall show that (6.3.45) is equivalent to (6.3.27). First we note that the vector $\boldsymbol{\eta}_{j,i}^{j}$, defined by (6.3.26), can also be written as

$$\boldsymbol{\eta}_{j,i}^{j} = \left[\mathbf{E}_{o_i}^{j} - \tilde{\mathbf{s}}_{j,i}^{j} \tilde{\mathbf{U}}_{o_i}^{j} \right] \cdot \mathbf{z}_i^{j} \tag{6.3.46}$$

To see this, we write equation (6.3.26) in its expanded form as

$$\begin{aligned} \boldsymbol{\eta}_{j,i}^{j} &= A_{j+1} \boldsymbol{\eta}_{j+1,i}^{j+1} + \tilde{\mathbf{s}}_{j,j+1}^{j} \mathbf{f}_{j,i}^{j} \\ &= \tilde{\mathbf{s}}_{j,j+1}^{j} \mathbf{f}_{j,i}^{j} + \tilde{\mathbf{s}}_{j+1,j+2}^{j} \mathbf{f}_{j,i}^{j} + \cdots + \tilde{\mathbf{s}}_{i-1,i}^{j} \mathbf{f}_{j,i}^{j} + \mathbf{E}_{o_i}^{j} \mathbf{z}_i^{j} \\ &= \tilde{\mathbf{s}}_{j,i}^{j} \mathbf{f}_{j,i}^{j} + \mathbf{E}_{o_i}^{j} \mathbf{z}_i^{j} \end{aligned}$$

and since, $\mathbf{f}_{j,i}^{j} = \tilde{\mathbf{z}}_i^{j} \mathbf{U}_{o_i}^{j} = - \tilde{\mathbf{U}}_{o_i}^{j} \mathbf{z}_i^{j}$, we get equation (6.3.46).

From the foregoing, equations (6.3.26) and (6.3.46) are equivalent. Moreover, since $\mathbf{s}_{i,i}^{i} = 0$ for all i, it is obvious that equation (6.3.46) contains equation (6.3.24). Thus the joint-space inertia matrix of a robot manipulator can be computed using either (6.3.27) or (6.3.45). Therefore, Newton-Euler and Euler-Lagrange formulations both lead to the same equations for obtaining the generalized inertia tensor of a robot manipulator.

6.4 Implementation and Computational Considerations

In this section, we shall demonstrate how Algorithm 6.2, can be implemented numerically in an efficient manner. Similar observations to those made in the numerical implementation of Algorithms 5.6 and 5.7 in Chapter 5, can be made here. Thus, for example, when we talk about the computation of the matrix $\mathbf{E}^{i}_{\mathbf{c}_i}$ we actually mean the computation of the coordinate matrix $E^{i}_{\mathbf{c}_i}$ of the tensor $\mathbf{E}^{i}_{\mathbf{c}_i}$. Moreover, since most robot manipulators have usually one prismatic joint, we shall assume that the moments of an augmented link, i.e., Step 1 of Algorithm 6.2, can be computed off-line. Therefore, in the following, we shall be concerned with the numerical implementation of the second and third steps of Algorithm 6.2, and we shall consider two cases: robot manipulators with a general geometric structure and robot manipulators for which the twist angle is, by design, either 0 or 90 degrees.

From the structure of the equations in these two steps, it is clear that the maximum number of operations required for implementing Algorithm 6.2 results from various matrix-vector and matrix-matrix multiplications. As we mentioned in Section 5.3.2, for a matrix-vector multiplication where the matrix under consideration is a coordinate transformation matrix, we need 8 scalar multiplications and 4 scalar additions. When the twist angle α in the transformation matrix, is either 0 or 90 degrees, we need only 4 scalar multiplications and 2 scalar additions. For a matrix-vector multiplication where the matrix under consideration is a skew-symmetric matrix, we need 6 scalar multiplications and 3 scalar additions. Finally, the implementation of the product of two skew-symmetric matrices requires 9 scalar multiplications and 3 scalar additions.

Now, as we can see from Algorithm 6.2, the most computationally intensive equation is equation (6.3.29c) which computes the inertia tensor of a generalized link. For a computationally efficient implementation of this equation we notice the following: The symmetric matrix $\left[(\tilde{s}_{i,i+1}^{i+1}\tilde{U}_{o_{i+1}}^{i+1} + (\tilde{s}_{i,i+1}^{i+1}\tilde{U}_{o_{i+1}}^{i+1})^T\right]$ can be implemented with only 9 scalar multiplications and 9 scalar additions. Moreover, it can be shown that by using the following trigonometric identities

$$a)\quad \sin(2\theta) = 2\sin(\theta)\cos(\theta)$$

$$b)\quad \cos^2(\theta) = \frac{1 + \cos(2\theta)}{2}$$

$$c)\quad \sin^2(\theta) = \frac{1 - \cos(2\theta)}{2}$$

the transformation of a symmetric matrix from one coordinate system to another can be implemented very efficiently. Thus, it can be shown that when the twist angle of the transformation matrix $\mathbf{A}_i$ is different from 0 or 90 degrees, for the transformation of a symmetric matrix we require 21 scalar multiplications and 18 scalar additions. When the twist angle of $\mathbf{A}_i$ is equal to 0 or 90 degrees, we need 11 scalar multiplications and 11 scalar additions. Thus, to implement equation (6.3.29c), in the general case, we need 30 scalar multiplications and 39 scalar additions.

Based on these general remarks, we shall now analyze the implementation of Algorithm 6.2, when it is applied to a robot with all revolute joints. We first notice that since $z_i^i = [\,0\,0\,1\,]^T$, equations (6.3.29d)-(6.3.29f) and (6.3.30c) do not require any computations for their implementation. To implement equation (6.3.29b) we can avoid the matrix-vector multiplication, since, as we can see from equations (2.3.3) and (2.3.4), the vector $\mathbf{s}_{i,i+1}^{i+1}$ can be defined as

$$\mathbf{s}_{i,i+1}^{i+1} = \begin{bmatrix} a_i cos(q_{i+1}) \\ -a_i sin(q_{i+1}) \\ d_i \end{bmatrix}$$

where a_i and d_i are known link parameters. Thus, we can implement equation (6.3.29b) with only 2 scalar multiplications. Moreover, for $i = n$, since $\mathbf{A}_{n+1} = 0$, we have $\mathbf{U}_{o_n}^{n} = \mathbf{u}_{o_n}^{n}$ and $\mathbf{E}_{o_n}^{n} = \mathbf{K}_{o_n}^{n}$.

Steps 2 & 3	General manipulator with revolute joints		General manipulator with revolute joints & $\alpha=0^o$ or 90°	
Equation	Multipl.	Additions	Multipl.	Additions
6.3.29a	$8n-16$	$7n-10$	$4n-8$	$5n-10$
6.3.29b	$2n-2$	0	$2n-2$	0
6.3.29c	$30n-42$	$39n-52$	$20n-25$	$32n-42$
6.3.29d	0	0	0	0
6.3.29e	0	0	0	0
6.3.29f	0	0	0	0
6.3.30a	$4n^2-12n+8$	$2n^2-6n+4$	$2n^2-6n+4$	n^2-3n+2
6.3.30b	$7n^2-17n+10$	$3.5n^2-7.5n+4$	$5n^2-13n+8$	$2.5n^2-5.5n+3$
6.3.30c	0	0	0	0
Total	$11n^2+11n-42$	$5.5n^2+32.5n-54$	$7n^2+7n-23$	$3.5n^2+28.5n-47$
$n=6$	420	339	271	250

Table 6.1: Operations counts for implementing Algorithm 6.2.

From the foregoing, no computations are involved in Step 2, when $i = n$. Also, since $d_{1,1}$ is the (3,3) element of the matrix $\mathbf{E}_{o_1}^{1}$, we need to compute only the (3,3) element of $\mathbf{E}_{o_1}^{1}$ and not the complete matrix when $i = 1$. This also implies that $\mathbf{U}_{o_1}^{1}$ need not be computed. In Step 3, equations (6.3.30a) and (6.3.30b) each need to be evaluated $n(n-1)/2$ times, since there

Authors	Remarks	Multiplications	Additions
Walker and Orin	Composite rigid bodies	$12n^2+56n-27$ (741)†	$7n^2+67n-56$ (601)
Angeles and Ma	Natural Orthogonal Complement	$n^3+17n^2-15n+8$ (746)	$n^3+14n^2-16n+5$ (629)
Featherstone	Composite rigid bodies	$10n^2+31n-41^*$ (505)	$6n^2+40n-46^*$ (410)
Algorithm 6.2 in this monograph	Generalized and augmented links	$11n^2+11n-42$ (420)	$5.5n^2+32.5n-54$ (339)

† Number of operations for $n = 6$.

* A spatial arithmetic package has been used for this implementation.

Table 6.2: Comparison of computational complexities of several algorithms for computing the joint-space inertia matrix.

are $n(n-1)/2$ off-diagonal elements in the upper triangular part of the joint space inertia matrix D. However, when $j = 1$, there is no need to compute the vector $\mathbf{f}_{1,i}^{1}$ (since it is not used anywhere) and from the vector $\boldsymbol{\eta}_{1,i}^{1}$ we need only compute its last entry. Thus, some saving can be made in computing the $n-1$ elements $d_{1,i}$ of D in Step 3 of the algorithm if these considerations are taken into account.

Following the observations made above, a breakdown of the number of scalar multiplications and additions required for the online implementation of each equation of Algorithm 6.2 is given in Table 6.1. The total figure represents the operations count for computing the inertia matrix (upper triangular part) of a robot manipulator with all joints of revolute type, and is valid for $n \geq 2$. Also, for the sake of comparison, the operations counts for a number of algorithms reported in the literature for computing the inertia matrix of a robot manipulator are given in Table 6.2. From Table 6.2, it is obvious that the proposed algorithm is computationally more efficient than other well known algorithms reported in the literature. The significantly higher computational efficiency of Algorithm 6.2 is obtained primarily through appropriate modeling and use of tensor analysis in the formulation of its basic equations.

As we mentioned in Section 6.2, one of the basic approaches for solving the forward dynamics problem is to obtain and solve a set of simultaneous equations in the unknown joint accelerations, i.e., steps (*i*)–(*iii*) in Walker and Orin's approach. Following this approach, one can use Algorithm 6.2 for computing the joint space inertia matrix D in step (*i*), Algorithm 5.7 for computing the bias vector **b** in step (*ii*) and any standard method [18-21] for solving the system of linear equations in step (*iii*). The total computational cost for solving these three subproblems of forward dynamics is given in Table 6.3. Also, for the sake of comparison, Table 6.3 contains the computational cost of computing the joint accelerations by other similar methods reported in the literature.

Authors	Remarks	Multipl.	Additions
Kazerounian and Gupta	Zero reference positions	2468	1879
Walker and Orin	Composite bodies	1627	1261
Wang and Ravani	Modified Walker and Orin method	1659	1252
Featherstone	Articulated bodies	1533	1415
Featherstone	Composite bodies	1303	1019
Angeles and Ma	Natural orthogonal complement	1353	1165
(i) Algorithm 6.2 (ii) Algorithm 5.7 (iii) Soln. of linear eq. [19]	Generalized and augmented links	956	790

Table 6.3: Computational cost for solving steps (i)-(iii) of the forward dynamics problem for $n = 6$.

6.5 Concluding Remarks

In this chapter, we have presented a new algorithm for computing the joint space inertia matrix D of the generalized inertia tensor **D** of a robot manipulator. We have shown that this algorithm can be derived using either Newton-Euler or Euler-Lagrange formulations of robot dynamics. Thus, we have established that from an algorithmic point of view, the solution of the forward dynamics problem does not depend on which of these formulations is used. A comparison of the computational complexity of this algorithm with that of other existing ones shows that the proposed algorithm is significantly more efficient. This efficiency is achieved mainly because the underlying modeling scheme used here for the dynamic analysis allows us to compute several quantities off-line. Moreover, the computational efficiency is improved since the tensor formulation for the equations to be computed online is computationally more efficient than the traditional vector formulation. Finally, we have shown that by using inverse dynamics algorithms from Chapter 5 to evaluate the bias vector **b**, and the proposed algorithm to evaluate the generalized inertia tensor **D**, the computational cost for solving the forward dynamics problem can be reduced considerably.

6.6 References

[1] M. W. Walker and D. E. Orin, "Efficient Dynamic Computer Simulation of Robotic Mechanisms", *ASME J. Dynamic Systems, Measurement and Control*, Vol. 104, pp. 205-211, 1982.

[2] R. Featherstone, *Robot Dynamics Algorithms*, Kluwer Academic Publishers, Boston, MA, 1987.

[3] J. Angeles and O. Ma, "Dynamic Simulation of n-Axis Serial Robotic Manipulators Using a Natural Orthogonal Complement", *Int. J. of Robotics Research*, Vol. 7, No. 5, pp. 32-45, 1988.

[4] J. Angeles and S. K. Lee, "The Formulation of Dynamical Equations of Holonomic Mechanical Systems Using a Natural Orthogonal Complement", *ASME J. of Applied Mechanics*, Vol. 55, pp. 243-244, 1988.

[5] W. W. Armstrong, "Recursive Solution to the Equations of Motion of an n-Link Manipulator", *Proc. 5th World Congress on the Theory of Machines and Mechanisms*, Vol. 2, pp. 1343-1346, Montreal, July, 1979.

[6] G. Rodriguez, "Kalman Filtering, Smoothing and Recursive Robot Arm Forward and Inverse Dynamics", *IEEE J. of Robotics and Automation*, Vol. RA-3, No. 6, pp. 624-639, 1987.

[7] G. Rodriguez, "Recursive Forward Dynamics for Multiple Robot Arms Moving a Common Task Object" *Robotics and Manufacturing : Recent Trends in Research, Education, and Applications*, M. Jamshidi, J. Y. S. Luh, H. Seraji and G. P. Starr, Eds., ASME Press, New York, 1988.

[8] G. Rodriguez and K. Kreutz, "Recursive and Mass Matrix Factorization and Inversion : An Operator Approach to Manipulator Forward Dynamics", *JPL Publication 88-11*, March 1988.

[9] J. C. K. Chou, G. Baciu and H. K. Kesavan, "Graph-Theoretic Models for Simulating Robot Manipulators", *Proc. IEEE Int. Conf. on Robotics and Automation*, Raleigh, NC, pp 953-959, 1987.

[10] J. C. K. Chou, G. Baciu and H. K. Kesavan, "Computational Scheme for Simulating Robot Manipulators", *Proc. IEEE Int. Conf. on Robotics and Automation*, Raleigh, NC, pp 961-966, 1987.

[11] C. S. G. Lee and P. R. Chang, "Efficient Parallel Algorithms For Robot Forward Dynamics Computation", *Proc. IEEE Int. Conf. on Robotics and Automation*, Raleigh, NC, pp 654-659, 1987.

[12] M. Amin-Javaheri and D. E. Orin, "A Systolic Architecture for Computation of the Manipulator Inertia Matrix", *Proc. IEEE Int. Conf. on*

Robotics and Automation, Raleigh, NC, pp 647-653, 1987.

[13] J. Y. Han, "Computational Aspects of Real-Time Simulation of Robotic Systems", *Proc. IEEE Int. Conf. on Robotics and Automation*, Raleigh, NC, pp 967-972, 1987.

[14] J. Wittenburg, *Dynamics of Systems of Rigid Bodies*, Stuttgart : B. G. Teubner, 1977.

[15] K. Kazerounian and K. C. Gupta, "Manipulator Dynamics Using the Extended Zero Reference Position Description", *IEEE J. of Robotics and Automation*, Vol. RA-2, pp 221-224, 1986.

[16] L. T. Wang and B. Ravani, "Recursive Computations of Kinematic and Dynamic Equations for Mechanical Manipulators", *IEEE J. of Robotics and Automation*, Vol. RA-1, pp 124-131, 1985.

[17] C. A. Balafoutis and R. V. Patel, "Efficient Computation of Manipulator Inertia Matrices and the Direct Dynamics Problem", *IEEE Trans. on Systems, Man, and Cybernetics*, Vol. SMC-19, No. 5, pp. 1313-1321, 1989.

[18] G. W. Stewart, *Introduction to Matrix Computations*, Academic Press, N.Y., 1973.

[19] J. J. Dongarra et al., *LINPACK : user's guide*, SIAM, Philadelphia, PA, 1973.

[20] G. Golub and C. Van Loan, *Matrix Computations*, 2nd Edition, Johns Hopkins Univ. Press, Baltimore, MD, 1989.

[21] R. L. Burden, J. D. Faires and A. C. Reynolds, *Numerical Analysis*, Prindle, Weder & Schmidt, Boston, MA, 1978.

Chapter 7

Linearized Dynamic Robot Models

7.1 Introduction

As is well known, our modeling approach to the real world is based on idealizations, and usually the working conditions are not the predicted ones. Therefore, in practice, one has to take into account the effects of perturbations on the applications being considered. For example, in the trajectory tracking problem, an important objective is to ensure that the end-effector of a manipulator tracks a desired "nominal" trajectory as closely as possible. Under ideal conditions a dynamic robot model, such as that presented in Chapter 5, will provide the generalized force $\boldsymbol{\tau}$ which will drive the end-effector of a robot manipulator along the desired trajectory. However, in practice, if only *feedforward* control signals (calculated from a nonlinear dynamic model) are applied, the end-effector will not necessarily track the nominal trajectory. This is due to factors such as modeling uncertainties, gear backlash and friction, actuator and sensor errors, and payload variations which are not taken into account in the dynamic model. Therefore, a feedback/feedforward control system is needed to remedy this situation.

In general, manipulator control is a challenging problem since, as we showed in Chapter 5, a dynamic robot model is described by equations which are highly nonlinear and dynamically coupled. Obviously, for these inherently nonlinear dynamical systems, much of the well established linear control theory is not directly applicable. However, many proposed solutions to the manipulator control problem [1-6] involve the use of methods from linear control theory. For example, in the aforementioned trajectory tracking problem, a control strategy which allows linear control methods to be used can be outlined as follows: First, feedforward control signals are applied to drive the manipulator along a desired nominal trajectory. Then, a correction term for the feedforward signals is generated by *feedback* of the perturbations (errors) in the physical state-variables (positions and velocities) from their nominal values. This is done using a control algorithm designed for the linearized dynamic equations of the manipulator -henceforth called the *linearized dynamic robot model.* Thus, linearized dynamic robot models which can be obtained in a computationally efficient manner are important in manipulator control. Furthermore, linearized dynamic robot models may be used in other aspects of robotics as well. For example, linearized robot models lead naturally to *trajectory sensitivity functions* [7-9] which characterize the sensitivity of the manipulator motion (along a nominal path) to fixed kinematic or dynamic parameters. Among other applications, these functions can be used [9,10] to describe the variations in a manipulator's trajectory introduced by an unknown payload.

The outline of this chapter is as follows: In Section 7.2 we briefly discuss some basic approaches taken to linearize the dynamic equations of robot manipulators. In Section 7.3 we present a procedure for deriving joint space linearized robot models in a computationally efficient manner. In Section 7.4, we introduce the Cartesian configuration space description for the dynamic equations of robot manipulators and propose a method for their linearization.

7.2 Linearization Techniques

In this section, we briefly discuss various linearization techniques that have been employed to linearize the dynamic equations of a robot manipulator. Broadly, these techniques may be categorized as *global* linearization techniques and *local* linearization techniques. As the name suggests, in global linearization the resulting linearized dynamic robot model is valid over the whole domain where the nonlinear model itself is valid. In local linearization the resulting linearized dynamic robot model is valid only at a particular point of a given trajectory which is known as the *nominal trajectory.* In local linearization, if the manipulator does not operate over a small range of its variables, then as the manipulator moves, the operating point changes. Therefore, at each new operating point a new (local) linearization has to be performed. Briefly, global and local linearization can be achieved by using the following methods.

7.2.1 Global Linearization

Global linearization resulted from efforts to control nonlinear systems, and can be achieved by using either feedback action alone or feedback action combined with coordinate transformations.

In the first approach, i.e., when feedback action alone is used, the main idea is to "cancel" the nonlinearities in the nonlinear system by using an appropriate feedback law which makes the overall closed-loop system behave as a linear system. The resolved acceleration technique [5] is probably the simplest and most common technique used for achieving "feedback linearization" of a robotic system. Also, another interesting feedback linearization technique is described in [6]. In these methods, under the feedback action, the closed-loop response of the system (which may be in terms of error signals or end-effector position variables) is described by a linear second order differential equation which is viewed as the linearized dynamic robot model. The computational complexity in this "linearizing control" approach varies with the method and the particular feedback law. However, in general a large amount of computation is required which in some cases includes matrix inversion in real-time [5]. Moreover, besides the complexity, a practical difficulty with this approach is that inexact cancellation of the nonlinearities can result in a 'linearized' model which gives very poor stability performance for the closed-loop system.

The feedback and coordinate transformation approach for linearizing robot manipulators resulted from advances made in the past few decades in differential geometric system theory. The main idea in this approach, is to use a diffeomorphic† feedback-transformation (which includes a state-space change of coordinates, additive feedback and input space change of coordinates) and transfer the nonlinear dynamic robot model to an equivalent linear and output decoupled system. In general, a diffeomorphic feedback transformation of this type exists under certain necessary and sufficient conditions for a class of nonlinear systems [11-12]. Necessary and sufficient conditions for the existence of a diffeomorphic feedback-transformation applicable to dynamic robot models and a method for constructing it have been given in [13]. At first, this method of linearization is attractive because it guarantees exact linearization. However, it requires an extensive analysis and the feedback transformation law is quite complex. Thus, the algorithms which compute this control law are computationally expensive and, in practical applications, this offsets the advantages gained from the exact linearization.

7.2.2 Local Linearization

In this approach, fundamental to linearizing a nonlinear system is the concept of a nominal trajectory which will be denoted here by

† A diffeomorphic transformation is a one-to-one onto C^{∞} transformation between two manifolds such that its inverse exists and is also C^{∞}.

$(\mathbf{q}^o(t), \dot{\mathbf{q}}^o(t), \ddot{\mathbf{q}}^o(t))$. Also, corresponding to the nominal trajectory, there is a *nominal generalized force* vector $\boldsymbol{\tau}^o(t)$ which can be computed by using any of the nonlinear dynamic robot models of Chapter 5. In the following, for the sake of notational simplicity, we shall drop the time parameter t from all the time dependent terms.

Local linearization is based on the Taylor series expansion of the nonlinear dynamic equations about a nominal trajectory. The Taylor series expansion is applicable to these equations since they are analytic functions of their arguments. The approach is conceptually simple. We assume that the perturbations are small and we consider a first order approximation to this expansion. The nominal portion of this expansion is cancelled algebraically and the terms which are *linear* in the perturbation quantities are retained and define the linearized dynamic equations. Following this procedure, we can express the linearized dynamic equations of a robot manipulator by a closed-form linear vector equation or by a recursive algorithm which has the same structure as the recursive algorithms which solve the inverse dynamics problem.

A closed-form linearized dynamic robot model, which can be derived by applying the Taylor series expansion to one of the recursive dynamic robot models of Chapter 5, can be written in the form

$$\delta\boldsymbol{\tau} = \mathbf{D}^o\,\delta\ddot{\mathbf{q}} + \mathbf{V}^o\,\delta\dot{\mathbf{q}} + \mathbf{P}^o\,\delta\mathbf{q} \tag{7.2.1}$$

where $\delta\boldsymbol{\tau}, \delta\mathbf{q}, \delta\dot{\mathbf{q}}, \delta\ddot{\mathbf{q}} \in \mathbb{R}^n$ are small deviations about some nominal torque and nominal joint space trajectory. The coefficients in equation (7.2.1) are known as the *(coefficient) sensitivity matrices* of the linearized model. These are functions of the nominal trajectory $(\mathbf{q}^o, \dot{\mathbf{q}}^o, \ddot{\mathbf{q}}^o)$ and are independent of the perturbations. Here, following established terminology, we call the coefficients in equation (7.2.1) "matrices", although actually they are *2nd* order tensors. The tensor character of these quantities is obvious from their definitions which are presented next:

$\mathbf{D}^o$ The *inertial force-acceleration* sensitivity matrix is the Jacobian of the generalized force vector $\boldsymbol{\tau}$ with respect to $\ddot{\mathbf{q}}$ evaluated about the nominal trajectory i.e., $\mathbf{D}^o \equiv \nabla_{\ddot{\mathbf{q}}} \boldsymbol{\tau} \mid_{(\mathbf{q}^o, \dot{\mathbf{q}}^o, \ddot{\mathbf{q}}^o)}$.

$\mathbf{V}^o$ The *centrifugal and Coriolis force-velocity* sensitivity matrix is the Jacobian of $\boldsymbol{\tau}$ with respect to $\dot{\mathbf{q}}$ evaluated about the nominal trajectory i.e., $\mathbf{V}^o \equiv \nabla_{\dot{\mathbf{q}}} \boldsymbol{\tau} \mid_{(\mathbf{q}^o, \dot{\mathbf{q}}^o, \ddot{\mathbf{q}}^o)}$.

$\mathbf{P}^o$ The *force-position* sensitivity matrix is the Jacobian of $\boldsymbol{\tau}$ with respect to $\mathbf{q}$ evaluated about the nominal trajectory i.e., $\mathbf{P}^o \equiv \nabla_{\mathbf{q}} \boldsymbol{\tau} \mid_{(\mathbf{q}^o, \dot{\mathbf{q}}^o, \ddot{\mathbf{q}}^o)}$.

To derive a recursive linearized dynamic robot model we again apply the Taylor series expansion to a nonlinear recursive dynamic robot model. But in this case, the joint space perturbations $\delta\boldsymbol{\tau}$ of the generalized force vector $\boldsymbol{\tau}$ are computed component-wise by a recursive algorithm which has the same structure as the recursive algorithm which computes the nonlinear manipulator dynamics. The only difference is that now instead of propagating the actual velocities, accelerations, forces, etc. from link to link we propagate the perturbations of these quantities about the operating point of a nominal trajectory. Thus it can be shown [9] that recursive linearized dynamic robot models retain the $o(n)$ computational complexity of a recursive nonlinear dynamic robot model. However, recursive linearization has limited applications in manipulator control, since it is difficult to derive a *state-space representation* (which is used for time domain analysis and control) of the linearized dynamics directly from a recursive model. On the other hand, the state-space representation of the linearized dynamics can be derived easily from a closed-form linearized dynamic robot model.

Thus, for example, since the force acceleration matrix $\mathbf{D}^o \in \mathbb{R}^{n \times n}$ is positive definite [14], its inverse always exists and this allows us to write equation (7.2.1) as

$$\begin{bmatrix} \delta\dot{\mathbf{q}} \\ \delta\ddot{\mathbf{q}} \end{bmatrix} = \begin{bmatrix} 0_n & 1_n \\ -(\mathbf{D}^o)^{-1}\mathbf{P}^o & -(\mathbf{D}^o)^{-1}\mathbf{V}^o \end{bmatrix} \begin{bmatrix} \delta\mathbf{q} \\ \delta\dot{\mathbf{q}} \end{bmatrix} + \begin{bmatrix} 0 \\ (\mathbf{D}^o)^{-1} \end{bmatrix} \delta\tau \qquad (7.2.2)$$

where 0_n and 1_n are the zero and unity $n \times n$ matrices, respectively. Equation (7.2.2) is in the standard state-space form

$$\dot{\mathbf{x}}(\mathbf{t}) = \mathbf{A}\,\mathbf{x}(\mathbf{t}) + \mathbf{B}\,\mathbf{u}(\mathbf{t}) \qquad (7.2.3)$$

where $\mathbf{x}(\mathbf{t})\ (= [\delta\mathbf{q}^T(t)\ \delta\dot{\mathbf{q}}^T(t)]^T) \in \mathbb{R}^{2n}$, and $\mathbf{u}(\mathbf{t})\ (= \delta\tau(t)) \in \mathbb{R}^n$.

To compute the coefficient sensitivity matrices of a closed-form linearized dynamic robot model, we can use one of the following methods.

a) **Parameter identification techniques:** In this method, a discrete-time version of the linearized model is considered first and then an iterative scheme, such as the least-squares parameter identification algorithm, is used to evaluate the unknown parameters in the sensitivity matrices. This approach has been used in [4] to calculate directly the coefficient matrices **A** and **B** of the state space linearized robot model (7.2.3). However, it should be noted that identification schemes can be used effectively only when the parameters of the system are *slowly time-varying*. Also, even in slowly time-varying systems, the convergence of the iterative algorithm may present a problem.

b) **Analytic or recursive formulations:** In this approach we first obtain *analytic* or *recursive* expressions for the elements of the Jacobians, and then devise algorithms which evaluate them numerically or symbolically along points on the nominal trajectory. In this approach, we do not face the aforementioned restrictions of the parameter identification techniques, but this method can lead to computationally expensive algorithms. Linearized robot models based on this approach have been proposed in [8,15]. In both cases, the 4x4 Lagrangian formulation of robot dynamics has been used, and this results in linearized robot models which are computationally inefficient since they inherit the computational complexity associated with the 4x4 formulation of the Lagrangian dynamic robot model. Also, in [14] the linearization of the Newton-Euler formulation of robot dynamic models has been derived in parallel with the nonlinear dynamic equations. This approach is also computationally inefficient since, as has been estimated in [14], one has to evaluate $n^3 + 2n^2$ terms in order to obtain the coefficient sensitivity matrices.

In the following, using recursive formulations, we propose a new method for deriving the Jacobians which define the coefficient sensitivity matrices of a linearized dynamic robot model, and devise algorithms for evaluating these Jacobians in a computationally efficient manner.

7.3 Joint Space Linearized Dynamic Robot Models

Application of the Taylor series expansion to the joint space nonlinear dynamic equations of a robot manipulator, at least in principle, does not present any problems. However, the complexity of these nonlinear equations makes the task challenging, especially if one attempts to derive efficient computational algorithms for determining the coefficient sensitivity matrices of the closed-form linearized dynamic robot model given by equation (7.2.1). In this section we shall address this problem and demonstrate how the various coefficient sensitivity matrices in equation (7.2.1) can be evaluated in a computationally efficient manner.

7.3.1 Joint Space Coefficient Sensitivity Matrices

By definition, the coefficient sensitivity matrices $\mathbf{D}^o$, $\mathbf{V}^o$ and $\mathbf{P}^o$ of a *joint space linearized dynamic robot model* are the Jacobians of the generalized force vector $\boldsymbol{\tau}$ with respect to the generalized coordinates $\ddot{\mathbf{q}}$, $\dot{\mathbf{q}}$ and $\mathbf{q}$ respectively, evaluated at a point of a nominal trajectory $(\mathbf{q}^o, \dot{\mathbf{q}}^o, \ddot{\mathbf{q}}^o)$. Thus, for example, if the generalized force vector $\boldsymbol{\tau}$ is defined component-wise by equation (5.3.46c) of Algorithm 5.7 in Chapter 5, we can compute the elements of the sensitivity matrix $\mathbf{D}^o$ by using the following equation,

$$d_{ij}^o = \frac{\partial \tau_i}{\partial \ddot{q}_j}\Big|_{(\mathbf{q}^o,\dot{\mathbf{q}}^o,\ddot{\mathbf{q}}^o)}$$

$$= \sigma_i \left[\frac{\partial \boldsymbol{\eta}_i^i}{\partial \ddot{q}_j}\cdot \mathbf{z}_i^i\right]\Big|_{(\mathbf{q}^o,\dot{\mathbf{q}}^o,\ddot{\mathbf{q}}^o)} + (1-\sigma_i)\left[\frac{\partial \mathbf{f}_i^i}{\partial \ddot{q}_j}\cdot \mathbf{z}_i^i\right]\Big|_{(\mathbf{q}^o,\dot{\mathbf{q}}^o,\ddot{\mathbf{q}}^o)} \tag{7.3.1}$$

where σ_i is equal to one when the i-th joint is revolute and zero when the i-th joint is prismatic. The elements of the sensitivity matrices $\mathbf{V}^o$ and $\mathbf{P}^o$ are defined in an analogous manner. Therefore, it is obvious that if we have available the various partial derivatives of the vector functions $\boldsymbol{\eta}_i^i$ and $\mathbf{f}_i^i$, we can easily compute the joint space sensitivity matrices $\mathbf{D}^o$, $\mathbf{V}^o$ and $\mathbf{P}^o$ for robot manipulators with revolute and/or prismatic joints. In this section, in order to derive a procedure for computing the coefficient sensitivity matrices, we shall consider manipulators with revolute joints only. In the case of manipulators with some prismatic joints, the equations are valid with minor modifications. Moreover, for the sake of continuity, only the main results will be presented in this section. The intermediate steps (such as partial differentiation of various vector functions) for arriving at these results are outlined in Appendix C.

In the following, we shall assume that the generalized force vector $\boldsymbol{\tau}$ is defined by equation (5.3.46c) of Algorithm 5.7, and that this algorithm is applied to manipulators with revolute joints only. As we have shown in Chapter 5, the tensor formulation of Algorithm 5.7 makes its implementation computationally very efficient. Therefore, Cartesian tensor analysis will be used here as well and the basic equations will be stated in a tensor formulation.

Based on the definition of the appropriate Jacobians, the expressions for the elements of the sensitivity matrices $\mathbf{D}^o$, $\mathbf{V}^o$ and $\mathbf{P}^o$ may be determined as given below:

i) Inertial force-acceleration coefficient sensitivity matrix $\mathbf{D}^o$: By definition, we can write the (i,j)-th element of $\mathbf{D}^o$ as

$$d_{ij}^o = \frac{\partial \tau_i}{\partial \ddot{q}_j}\Big|_{(q^o,\dot{q}^o,\ddot{q}^o)}$$

$$= \left[{}^i\mathbf{z}_i \cdot \frac{\partial\, {}^i\boldsymbol{\eta}_i}{\partial \ddot{q}_j} \right]\Big|_{(q^o,\dot{q}^o,\ddot{q}^o)}$$

which may be simplified (see Appendix C) to

$$d_{ij} = \begin{cases} {}^i\mathbf{z}_i \cdot \left[{}^i\mathbf{E}_{o_i} - {}^i\tilde{\mathbf{U}}_{o_i}\, {}^i\tilde{\mathbf{s}}_{j,i} \right] \cdot {}^i\mathbf{z}_j & j \le i \\ {}^i\mathbf{z}_i \cdot {}^i\mathbf{W}_j \left[[{}^j\mathbf{E}_{o_j} - {}^j\tilde{\mathbf{s}}_{i,j}\, {}^j\tilde{\mathbf{U}}_{o_j}] \cdot {}^j\mathbf{z}_j \right] & j > i \end{cases}$$

$$= \begin{cases} {}^j\mathbf{z}_i \cdot \left[{}^j\mathbf{E}_{o_i} - {}^j\tilde{\mathbf{U}}_{o_i}\, {}^j\tilde{\mathbf{s}}_{j,i} \right] \cdot {}^j\mathbf{z}_j & j \le i \\ {}^i\mathbf{z}_i \cdot \left[{}^i\mathbf{E}_{o_j} - {}^i\tilde{\mathbf{s}}_{i,j}\, {}^i\tilde{\mathbf{U}}_{o_j} \right] \cdot {}^i\mathbf{z}_j & j > i \end{cases} \tag{7.3.2}$$

where the last step follows from the fact that the dot product is invariant under coordinate transformations.

ii) Centrifugal and Coriolis force-velocity coefficient sensitivity matrix $\mathbf{V}^o$: The elements of the sensitivity matrix $\mathbf{V}^o$ are defined by considering the following Jacobian:

$$v_{ij}^o = \frac{\partial \tau_i}{\partial \dot{q}_j}\Big|_{(q^o,\dot{q}^o,\ddot{q}^o)}$$

$$= \left[\mathbf{z}_i^i \cdot \frac{\partial \eta_i^i}{\partial \dot{q}_j} \right] \Big|_{(\mathbf{q}^o, \dot{\mathbf{q}}^o, \ddot{\mathbf{q}}^o)}$$

which on simplification (see Appendix C) gives

$$v_{ij} = 2 \begin{cases} \mathbf{z}_i^i \cdot \left[\mathbf{E}_{o_i}^i - \tilde{\mathbf{U}}_{o_i}^i \, \tilde{\mathbf{s}}_{j,i}^i \right] \cdot \dot{\mathbf{z}}_j^i - \mathbf{z}_i^i \cdot \left[\mathbf{L}_i^i + \tilde{\mathbf{U}}_{o_i}^i \, \tilde{\dot{\mathbf{s}}}_{j,i}^i \right] \cdot \mathbf{z}_j^i & j \le i \\ \mathbf{z}_i^i \cdot {}^i\mathbf{W}_j \left([\mathbf{E}_{o_j}^j - \tilde{\mathbf{s}}_{i,j}^j \tilde{\mathbf{U}}_{o_j}^j] \cdot \dot{\mathbf{z}}_j^j - \mathbf{z}_i^j \cdot [\mathbf{L}_j^j + \tilde{\mathbf{s}}_{i,j}^j \dot{\tilde{\mathbf{U}}}_{o_j}^j] \cdot \mathbf{z}_j^j \right) & j > i \end{cases}$$

$$= 2 \begin{cases} \mathbf{z}_i^j \cdot \left[\mathbf{E}_{o_i}^j - \tilde{\mathbf{U}}_{o_i}^j \, \tilde{\mathbf{s}}_{j,i}^j \right] \cdot \dot{\mathbf{z}}_j^j - \mathbf{z}_i^j \cdot \left[\mathbf{L}_i^j + \tilde{\mathbf{U}}_{o_i}^j \, \tilde{\dot{\mathbf{s}}}_{j,i}^j \right] \cdot \mathbf{z}_j^i & j \le i \\ \mathbf{z}_i^i \cdot \left[\mathbf{E}_{o_j}^i - \tilde{\mathbf{s}}_{i,j}^i \tilde{\mathbf{U}}_{o_j}^i \right] \cdot \dot{\mathbf{z}}_j^i - \mathbf{z}_i^i \cdot \left[\mathbf{L}_j^i + \tilde{\mathbf{s}}_{i,j}^i \dot{\tilde{\mathbf{U}}}_{o_j}^i \right] \cdot \mathbf{z}_j^i & j > i \end{cases} \tag{7.3.3}$$

where

$$\mathbf{L}_i^i = \tilde{\omega}_i^i \, \hat{\mathbf{K}}_{o_i}^i + [\tilde{\mathbf{s}}_{i,i+1}^i \dot{\tilde{\mathbf{U}}}_{o_{i+1}}^i + \tilde{\mathbf{U}}_{o_{i+1}}^i \tilde{\dot{\mathbf{s}}}_{i,i+1}^i] + \mathbf{A}_{i+1} \mathbf{L}_{i+1}^{i+1} \mathbf{A}_{i+1}^T,$$

and $\hat{\mathbf{K}}_{o_i}^i$ is the Euler tensor of the i-th augmented link which is defined by equation (5.3.44d).

iii) Force-position coefficient sensitivity matrix $\mathbf{P}^o$: As in the case of the elements of $\mathbf{D}^o$ and $\mathbf{V}^o$, we can write

$$p_{ij}^o = \frac{\partial \tau_i}{\partial q_j} \Big|_{(\mathbf{q}^o, \dot{\mathbf{q}}^o, \ddot{\mathbf{q}}^o)}$$

$$= \left[\mathbf{z}_i^i \cdot \frac{\partial \eta_i^i}{\partial q_j} \right] \Big|_{(\mathbf{q}^o, \dot{\mathbf{q}}^o, \ddot{\mathbf{q}}^o)}$$

which may be simplified (see Appendix C) to

$$p_{ij} = \mathbf{z}_i^i \cdot \left[\mathbf{E}_{o_i}^i - \tilde{\mathbf{U}}_{o_i}^i \tilde{\mathbf{s}}_{j,i}^i\right] \cdot \ddot{\mathbf{z}}_j^i - 2\mathbf{z}_i^i \cdot \left[\mathbf{L}_i^i + \tilde{\mathbf{U}}_{o_i}^i \dot{\tilde{\mathbf{s}}}_{j,i}^i\right] \cdot \dot{\mathbf{z}}_j^i + \mathbf{z}_i^i \cdot \left[\tilde{\mathbf{U}}_{o_i}^i \ddot{\tilde{\mathbf{s}}}_{0j}^i\right] \cdot \mathbf{z}_j^i$$

$$= \mathbf{z}_i^j \cdot \left[\mathbf{E}_{o_i}^j - \tilde{\mathbf{U}}_{o_i}^j \tilde{\mathbf{s}}_{j,i}^j\right] \cdot \ddot{\mathbf{z}}_j^j - 2\mathbf{z}_i^j \cdot \left[\mathbf{L}_i^j + \tilde{\mathbf{U}}_{o_i}^j \dot{\tilde{\mathbf{s}}}_{j,i}^j\right] \cdot \dot{\mathbf{z}}_j^j + \mathbf{z}_i^j \cdot \left[\tilde{\mathbf{U}}_{o_i}^j \ddot{\tilde{\mathbf{s}}}_{0j}^j\right] \cdot \mathbf{z}_j^j \qquad (7.3.4a)$$

when $j \le i$, and

$$p_{ij} = \mathbf{z}_i^j \cdot \left[\mathbf{E}_{o_j}^j - \tilde{\mathbf{s}}_{i,j}^j \tilde{\mathbf{U}}_{o_j}^j\right] \cdot \ddot{\mathbf{z}}_j^j - 2\mathbf{z}_i^j \cdot \left[\mathbf{L}_j^j + \tilde{\mathbf{s}}_{i,j}^j \dot{\tilde{\mathbf{U}}}_{o_j}^j\right] \cdot \dot{\mathbf{z}}_j^j$$

$$+ \mathbf{z}_i^j \cdot \left[\tilde{\mathbf{U}}_{o_j}^j \ddot{\tilde{\mathbf{s}}}_{0j}^j - \tilde{\mathbf{s}}_{i,j}^j \ddot{\tilde{\mathbf{U}}}_{o_j}^j - \tilde{\boldsymbol{\eta}}_j^j\right] \cdot \mathbf{z}_j^j$$

$$= \mathbf{z}_i^i \cdot \left[\mathbf{E}_{o_j}^i - \tilde{\mathbf{s}}_{i,j}^i \tilde{\mathbf{U}}_{o_j}^i\right] \cdot \ddot{\mathbf{z}}_j^i - 2\mathbf{z}_i^i \cdot \left[\mathbf{L}_j^i + \tilde{\mathbf{s}}_{i,j}^i \dot{\tilde{\mathbf{U}}}_{o_j}^i\right] \cdot \dot{\mathbf{z}}_j^i$$

$$+ \mathbf{z}_i^i \cdot \left[\tilde{\mathbf{U}}_{o_j}^i \ddot{\tilde{\mathbf{s}}}_{0j}^i - \tilde{\mathbf{s}}_{i,j}^i \ddot{\tilde{\mathbf{U}}}_{o_j}^i - \tilde{\boldsymbol{\eta}}_j^i\right] \cdot \mathbf{z}_j^i \qquad (7.3.4b)$$

when $j > i$. Note that for notational convenience, the explicit dependence on the nominal variables has not been shown in equations (7.3.2), (7.3.3) and (7.3.4).

At a first glance, these equations look very complex. However, as we shall show later in this section, all basic terms in equations (7.3.2)-(7.3.4) can be computed recursively. Before we derive the recursive formulations, it is worth examining the above formulations because they provide valuable insight into the structural characteristics and properties of the coefficient sensitivity matrices $\mathbf{D}^o$, $\mathbf{V}^o$ and $\mathbf{P}^o$.

We notice first that at a point on the nominal trajectory, equation (7.3.2) is equivalent to equation (6.3.45) which defines the joint space generalized inertia tensor of a robot manipulator. This is to be expected, since in a closed form representation of a nonlinear dynamic robot model (such as that in equation (5.2.8)), the generalized force vector $\boldsymbol{\tau}$ is linear in the joint acceleration $\ddot{\mathbf{q}}$, with the generalized inertia tensor $\mathbf{D}$ as the coefficient of linearity. Therefore, the force-acceleration sensitivity matrix $\mathbf{D}^o$ exhibits the properties of the generalized inertia tensor $\mathbf{D}$, i.e., it is symmetric (which can also be seen from equation (7.3.2)) and positive definite. The former property of $\mathbf{D}^o$ is obviously important in computational considerations. The latter is important in controller design since it implies that $\mathbf{D}^o$ is nonsingular and thus its inverse exists for all the points along a nominal trajectory.

Unfortunately, symmetry is not preserved in the coefficient sensitivity matrices $\mathbf{V}^o$ and $\mathbf{P}^o$. However, they have other important characteristics which may be used to simplify their evaluation at points along a nominal trajectory. For example, in [8] it has been shown (following a different analysis) that the element $v_{n,n}$ of $\mathbf{V}^o$ is zero for *any* configuration of the manipulator. This also follows from equation (7.3.3), if we note the following: When $i = j = n$, we have,

$$\mathbf{E}^{n}_{o_n} = \mathbf{K}^{n}_{o_n}$$

$$\dot{\mathbf{z}}^{n}_{n} = \tilde{\boldsymbol{\omega}}^{n}_{n}\mathbf{z}^{n}_{n}$$

$$\mathbf{s}^{n}_{n,n} = \dot{\mathbf{s}}^{n}_{n,n} = 0$$

and

$$\mathbf{L}^{n}_{n} = \tilde{\boldsymbol{\omega}}^{n}_{n}\hat{\mathbf{K}}^{n}_{o_n}$$

$$= \frac{1}{2} tr\,[\mathbf{K}^{n}_{o_n}]\tilde{\boldsymbol{\omega}}^{n}_{n} - \tilde{\boldsymbol{\omega}}^{n}_{n}\mathbf{K}^{n}_{o_n}.$$

Therefore, for $i = j = n$ equation (7.3.3) implies that

$$v_{n,n} = \mathbf{z}^{n}_{n}\cdot\left[\mathbf{K}^{n}_{o_n}\tilde{\boldsymbol{\omega}}^{n}_{n} + \tilde{\boldsymbol{\omega}}^{n}_{n}\mathbf{K}^{n}_{o_n} - \frac{1}{2} tr\,[\mathbf{K}^{n}_{o_n}]\tilde{\boldsymbol{\omega}}^{n}_{n}\right]\cdot\mathbf{z}^{n}_{n}$$

which, by using the tensor equation (3.4.26), can be simplified to

$$v_{n,n} = -\,\mathbf{z}^{n}_{n}\cdot dual\left[\mathbf{K}^{n}_{o_n}\,\boldsymbol{\omega}^{n}_{n}\right]\cdot\mathbf{z}^{n}_{n}$$

$$= \mathbf{z}_n^n \cdot \tilde{\mathbf{z}}_n^n \cdot \left[\mathbf{K}_{o_n}^n \boldsymbol{\omega}_n^n \right] = 0 \tag{7.3.5}$$

where the first equality follows from equation (3.4.4) and the last has been derived by applying equation (3.4.7).

Another important observation, concerning the formulation of the elements of the first columns of the sensitivity matrices $\mathbf{V}^o$ and $\mathbf{P}^o$, is the following: Since the angular velocity vector is always parallel to the first revolute joint of the manipulator, i.e., the vector $\boldsymbol{\omega}_1^1$ is parallel to $\mathbf{z}_1^1$ (as is usually the case in practice, we take the first joint to be revolute), then according to equation (3.4.5) we have

$$\tilde{\boldsymbol{\omega}}_1^1 \cdot \mathbf{z}_1^1 = 0 \tag{7.3.6}$$

which implies that

$$\dot{\mathbf{z}}_1^1 = 0$$

Furthermore, since equation (7.3.6) is invariant under coordinate transformations, we can write

$$\dot{\mathbf{z}}_1^i = \tilde{\boldsymbol{\omega}}_1^i \cdot \mathbf{z}_1^i = 0 \tag{7.3.7}$$

for all i. Also, using the same arguments, it can be shown that the vector $\ddot{\mathbf{z}}_1^i$ is zero for all i, i.e., we have

$$\ddot{\mathbf{z}}_1^i = \boldsymbol{\Omega}_1^i \cdot \mathbf{z}_1^i \equiv \left[\dot{\tilde{\boldsymbol{\omega}}}_1^i + \tilde{\boldsymbol{\omega}}_1^i \tilde{\boldsymbol{\omega}}_1^i \right] \cdot \mathbf{z}_1^i = 0 \tag{7.3.8}$$

Therefore, by using equations (7.3.7) and (7.3.8) in equations (7.3.3) and (7.3.4), we can simplify the formulations for the elements of the first columns of $\mathbf{V}^o$ and $\mathbf{P}^o$. Moreover, in the case where the first joint of the manipulator rotates about an axis which is parallel to the gravity field, the first column of the sensitivity matrix $\mathbf{P}^o$ becomes zero. This is true since, by assuming that the vector $\mathbf{z}_1^1$ is parallel to the gravity vector $\mathbf{g} = \ddot{\mathbf{s}}_{0,1}^1$, it follows from equation (3.4.5) that

$$\ddot{\mathbf{s}}_{0,1}^{1} \cdot \mathbf{z}_1^1 = 0 \tag{7.3.9}$$

Now, as with equation (7.3.6), in the i-th coordinate system we have

$$\ddot{\mathbf{s}}_{0,1}^{i} \cdot \mathbf{z}_1^i = 0 \tag{7.3.10}$$

and this, together with equation (7.3.6), implies that the first column of the coefficient sensitivity matrix $\mathbf{P}^o$ is zero in the case where the first revolute joint of the manipulator is parallel to the gravity field. An alternative proof of this result can be found in [8].

Based on these observations on the structure of the coefficient sensitivity matrices $\mathbf{D}^o$, $\mathbf{V}^o$ and $\mathbf{P}^o$, an implementation of equations (7.3.2)-(7.3.4) has been proposed in [16] which requires 2186 scalar multiplications and 2040 scalar additions. For more efficient computation of the coefficient sensitivity matrices $\mathbf{D}^o$, $\mathbf{V}^o$ and $\mathbf{P}^o$, we can modify equations (7.3.2)-(7.3.4) so that the basic terms of these equations take on recursive formulations. These recursive formulations are derived as follows:

Let

$$\mathbf{B}_{j,i}^{j} = \mathbf{z}_i^j \cdot [\mathbf{E}_{o_i}^j - \tilde{\mathbf{U}}_{o_i}^j \tilde{\mathbf{s}}_{j,i}^j] \qquad j \le i \tag{7.3.11a}$$

Then, using equations (3.4.6)-(3.4.8), we can write

$$\begin{aligned}\mathbf{B}_{j,i}^{j} &= \mathbf{z}_i^j \cdot \mathbf{E}_{o_i}^j - \tilde{\mathbf{z}}_i^j \mathbf{U}_{o_i}^j \cdot \tilde{\mathbf{s}}_{j,i}^j \\ &= \mathbf{z}_i^j \cdot \mathbf{E}_{o_i}^j - \mathbf{f}_i^j \cdot \tilde{\mathbf{s}}_{j,i}^j\end{aligned} \tag{7.3.11b}$$

where $\mathbf{f}_i^j = \tilde{\mathbf{z}}_i^j \mathbf{U}_{o_i}^j$. Now, since for $j \le i$ we have $\tilde{\mathbf{s}}_{j,i}^j = \tilde{\mathbf{s}}_{j,j+1}^j + \tilde{\mathbf{s}}_{j+1,i}^j$ (if $j = i$, $\mathbf{s}_{i,i}^i = 0$), we can write (7.3.11b) recursively as

$$\mathbf{B}_{j,i}^{j} = \mathbf{A}_{j+1} \mathbf{B}_{j+1,i}^{j+1} - \mathbf{f}_i^j \cdot \tilde{\mathbf{s}}_{j,j+1}^j \qquad j < i \tag{7.3.11c}$$

where $\mathbf{B}_{i,i}^{i} = \mathbf{z}_i^i \cdot \mathbf{E}_{o_i}^i$. Also, let

$$\mathbf{C}_{j,i}^{j} = \mathbf{z}_i^j \cdot [\mathbf{L}_i^j + \tilde{\mathbf{U}}_{o_i}^j \ddot{\tilde{\mathbf{s}}}_{j,i}^j] \qquad j \le i. \tag{7.3.12a}$$

Then, as equation (7.3.11a) was written in the recursive form (7.3.11c), this equation can be written as

$$\mathbf{C}^{j}_{j,i} = \mathbf{A}_{j+1}\mathbf{C}^{j+1}_{j+1,i} + \mathbf{f}^{j}_{i}\cdot\dot{\tilde{\mathbf{s}}}^{j}_{j\,j+1} \qquad j < i \tag{7.3.12b}$$

where, $\mathbf{C}^{i}_{i,i} = \mathbf{z}^{i}_{i}\cdot\mathbf{L}^{i}_{i}$. Moreover, if

$$\mathbf{H}^{i}_{i\,j} = [\mathbf{E}^{i}_{\mathrm{o}_j} - \tilde{\mathbf{s}}^{i}_{i\,j}\tilde{\mathbf{U}}^{i}_{\mathrm{o}_j}]\dot{\mathbf{z}}^{i}_{j} - [\mathbf{L}^{i}_{j} + \tilde{\mathbf{s}}^{i}_{i\,j}\dot{\tilde{\mathbf{U}}}^{i}_{\mathrm{o}_j}]\mathbf{z}^{i}_{j} \qquad j > i \tag{7.3.13a}$$

since $\dot{\mathbf{z}}^{i}_{j} = \tilde{\boldsymbol{\omega}}^{i}_{j}\mathbf{z}^{i}_{j}$, we can write

$$\mathbf{H}^{i}_{i\,j} = [\mathbf{E}^{i}_{\mathrm{o}_j}\tilde{\boldsymbol{\omega}}^{i}_{j} - \mathbf{L}^{i}_{j}]\mathbf{z}^{i}_{j} - \tilde{\mathbf{s}}^{i}_{i\,j}[\tilde{\mathbf{U}}^{i}_{\mathrm{o}_j}\tilde{\boldsymbol{\omega}}^{i}_{j} + \dot{\tilde{\mathbf{U}}}^{i}_{\mathrm{o}_j}]\mathbf{z}^{i}_{j}$$

$$= [\mathbf{E}^{i}_{\mathrm{o}_j}\tilde{\boldsymbol{\omega}}^{i}_{j} - \mathbf{L}^{i}_{j}]\mathbf{z}^{i}_{j} - \tilde{\mathbf{s}}^{i}_{i\,j}\hat{\mathbf{f}}^{i}_{j} \tag{7.3.13b}$$

where $\hat{\mathbf{f}}^{i}_{j} = [\tilde{\mathbf{U}}^{i}_{\mathrm{o}_j}\tilde{\boldsymbol{\omega}}^{i}_{j} + \dot{\tilde{\mathbf{U}}}^{i}_{\mathrm{o}_j}]\mathbf{z}^{i}_{j}$. Further, since for $j > i$ we have $\tilde{\mathbf{s}}^{i}_{i\,j} = \tilde{\mathbf{s}}^{i}_{i,i+1} + \tilde{\mathbf{s}}^{i}_{i+1\,j}$, this can be written in the recursive form

$$\mathbf{H}^{i}_{i\,j} = \mathbf{A}_{i+1}\mathbf{H}^{i+1}_{i+1\,j} - \tilde{\mathbf{s}}^{i}_{i,i+1}\hat{\mathbf{f}}^{i}_{j} \qquad i < j \tag{17.3.13c}$$

with initialization $\mathbf{H}^{j}_{j\,j} = [\mathbf{E}^{j}_{\mathrm{o}_j}\tilde{\boldsymbol{\omega}}^{j}_{j} - \mathbf{L}^{j}_{j}]\cdot\mathbf{z}^{j}_{j}$. Finally, for $j > i$, let

$$\mathbf{G}^{i}_{i\,j} = \left[\mathbf{E}^{i}_{\mathrm{o}_j} - \tilde{\mathbf{s}}^{i}_{i\,j}\tilde{\mathbf{U}}^{i}_{\mathrm{o}_j}\right]\ddot{\mathbf{z}}^{i}_{j} - 2\left[\mathbf{L}^{i}_{j} + \tilde{\mathbf{s}}^{i}_{i\,j}\dot{\tilde{\mathbf{U}}}^{i}_{\mathrm{o}_j}\right]\dot{\mathbf{z}}^{i}_{j}$$

$$+ \left[\tilde{\mathbf{U}}^{i}_{\mathrm{o}_j}\ddot{\tilde{\mathbf{s}}}^{i}_{0\,j} - \tilde{\mathbf{s}}^{i}_{i\,j}\ddot{\tilde{\mathbf{U}}}^{i}_{\mathrm{o}_j} - \tilde{\boldsymbol{\eta}}^{i}_{j}\right]\mathbf{z}^{i}_{j} \tag{7.3.14a}$$

or, since $\ddot{\mathbf{z}}^{i}_{j} = \boldsymbol{\Omega}^{i}_{j}\mathbf{z}^{i}_{j}$,

$$\mathbf{G}^{i}_{i\,j} = \left[\mathbf{E}^{i}_{\mathrm{o}_j}\boldsymbol{\Omega}^{i}_{j} - 2\mathbf{L}^{i}_{j}\tilde{\boldsymbol{\omega}}^{i}_{j} + \tilde{\mathbf{U}}^{i}_{\mathrm{o}_j}\ddot{\tilde{\mathbf{s}}}^{i}_{0\,j} - \tilde{\boldsymbol{\eta}}^{i}_{j}\right]\mathbf{z}^{i}_{j} - \tilde{\mathbf{s}}^{i}_{i\,j}\left[\tilde{\mathbf{U}}^{i}_{\mathrm{o}_j}\boldsymbol{\Omega}^{i}_{j} + 2\dot{\tilde{\mathbf{U}}}^{i}_{\mathrm{o}_j}\tilde{\boldsymbol{\omega}}^{i}_{j} + \ddot{\tilde{\mathbf{U}}}^{i}_{\mathrm{o}_j}\right]\mathbf{z}^{i}_{j}$$

$$= \left[\mathbf{E}^{i}_{\mathrm{o}_j}\boldsymbol{\Omega}^{i}_{j} - 2\mathbf{L}^{i}_{j}\tilde{\boldsymbol{\omega}}^{i}_{j} + \tilde{\mathbf{U}}^{i}_{\mathrm{o}_j}\ddot{\tilde{\mathbf{s}}}^{i}_{0\,j} - \tilde{\boldsymbol{\eta}}^{i}_{j}\right]\mathbf{z}^{i}_{j} - \tilde{\mathbf{s}}^{i}_{i\,j}\overline{\mathbf{f}}^{i}_{j} \tag{7.3.14b}$$

where $\overline{\mathbf{f}}^{i}_{j} = \left[\tilde{\mathbf{U}}^{i}_{\mathrm{o}_j}\boldsymbol{\Omega}^{i}_{j} + 2\dot{\tilde{\mathbf{U}}}^{i}_{\mathrm{o}_j}\tilde{\boldsymbol{\omega}}^{i}_{j} + \ddot{\tilde{\mathbf{U}}}^{i}_{\mathrm{o}_j}\right]\mathbf{z}^{i}_{j}$. Then, as before, we can write

$$\mathbf{G}^{i}_{i,j} = \mathbf{A}_{i+1}\mathbf{G}^{i+1}_{i+1,j} - \tilde{\mathbf{s}}^{i}_{i,i+1}\overline{\mathbf{f}}^{i}_{j} \qquad i < j \tag{7.3.14c}$$

where

$$\mathbf{G}^{j}_{j,j} = \left[\mathbf{E}^{j}_{o_j}\boldsymbol{\Omega}^{j}_{j} - 2\mathbf{L}^{j}_{j}\tilde{\boldsymbol{\omega}}^{j}_{j} + \tilde{\mathbf{U}}^{j}_{o_j}\ddot{\mathbf{s}}^{j}_{0,j} - \boldsymbol{\eta}^{j}_{j}\right]\cdot \mathbf{z}^{j}_{j}$$

From the foregoing, we can simplify equations (7.3.2), (7.3.3) and (7.3.4) to get

$$d_{ij} = \begin{cases} \mathbf{B}^{j}_{j,i}\cdot\mathbf{z}^{j}_{j} & j \le i \\ \mathbf{z}^{i}_{i}\cdot\mathbf{B}^{i}_{i,j} & j > i \end{cases} \tag{7.3.15}$$

$$v_{ij} = 2\begin{cases} \left[\mathbf{B}^{j}_{j,i}\cdot\tilde{\boldsymbol{\omega}}^{j}_{j} - \mathbf{C}^{j}_{j,i}\right]\cdot\mathbf{z}^{j}_{j} & j \le i \\ \mathbf{z}^{i}_{i}\cdot\mathbf{H}^{i}_{i,j} & j > i \end{cases} \tag{7.3.16}$$

and

$$p_{ij} = \begin{cases} \left[\mathbf{B}^{j}_{j,i}\cdot\boldsymbol{\Omega}^{j}_{j} - 2\mathbf{C}^{j}_{j,i}\cdot\tilde{\boldsymbol{\omega}}^{j}_{j} + \mathbf{f}^{j}_{i}\cdot\ddot{\mathbf{s}}^{j}_{0,j}\right]\cdot\mathbf{z}^{j}_{j} & j \le i \\ \mathbf{z}^{i}_{i}\cdot\mathbf{G}^{i}_{i,j} & j > i \end{cases} \tag{7.3.17}$$

In the following, for the implementation of equations (7.3.15)-(7.3.17), we shall assume that the nonlinear dynamic robot model which is described by Algorithm 5.7 is available. This assumption is justified since in control applications, the generalized force vector $\boldsymbol{\tau}$ is an integral part of most control laws. Thus, quantities such as: $\mathbf{u}^{i}_{o_i}$, $\tilde{\boldsymbol{\omega}}^{i}_{i}$, $\boldsymbol{\Omega}^{j}_{j}$, $\ddot{\mathbf{s}}^{j}_{0,j}$, $\mathbf{K}^{i}_{o_i}$ and $\hat{\mathbf{K}}^{i}_{o_i}$ are assumed to be available from Algorithm 5.7. Also, to simplify the structure of Algorithm

7.1, we shall assume that the simple equation $\dot{\mathbf{s}}^{i}_{i,i+1} = \tilde{\boldsymbol{\omega}}^{i}_{i}\mathbf{s}^{i}_{i,i+1}$ has been included in Algorithm 5.7, and thus the vectors $\dot{\mathbf{s}}^{i}_{i,i+1}$ are assumed to be available. From the foregoing, to compute the elements of the joint space coefficient sensitivity matrices $\mathbf{D}^{o}$, $\mathbf{V}^{o}$ and $\mathbf{P}^{o}$ of a linearized robot dynamic model, we can use the following algorithm.

ALGORITHM 7.1

Step 0 : Initialization

$$\mathbf{A}_{n+1} = \mathbf{I}\,, \quad \mathbf{U}^{n+1}_{\mathrm{o}_{n+1}} = \dot{\mathbf{U}}^{n+1}_{\mathrm{o}_{n+1}} = \mathbf{E}^{n+1}_{\mathrm{o}_{n+1}} = \mathbf{L}^{n+1}_{n+1} = 0$$

Step 1 : Backward Recursion :- $i = n, 1$

$$\mathbf{U}^{i}_{\mathrm{o}_i} = \mathbf{u}^{i}_{\mathrm{o}_i} + \mathbf{A}_{i+1}\mathbf{U}^{i+1}_{\mathrm{o}_{i+1}} \tag{7.3.18a}$$

$$\dot{\mathbf{U}}^{i}_{\mathrm{o}_i} = \tilde{\boldsymbol{\omega}}^{i}_{i}\mathbf{u}^{i}_{\mathrm{o}_i} + \mathbf{A}_{i+1}\dot{\mathbf{U}}^{i+1}_{\mathrm{o}_{i+1}} \tag{7.3.18b}$$

$$\mathbf{E}^{i}_{\mathrm{o}_i} = \mathbf{K}^{i}_{\mathrm{o}_i} - [\tilde{\mathbf{s}}^{i}_{i,i+1}\tilde{\mathbf{U}}^{i}_{\mathrm{o}_{i+1}} + (\tilde{\mathbf{s}}^{i}_{i,i+1}\tilde{\mathbf{U}}^{i}_{\mathrm{o}_{i+1}})^{T}] + \mathbf{A}_{i+1}\mathbf{E}^{i+1}_{\mathrm{o}_{i+1}}\mathbf{A}^{T}_{i+1} \tag{7.3.18c}$$

$$\mathbf{L}^{i}_{i} = \tilde{\boldsymbol{\omega}}^{i}_{i}\hat{\mathbf{K}}^{i}_{\mathrm{o}_i} + [\tilde{\mathbf{s}}^{i}_{i,i+1}\dot{\tilde{\mathbf{U}}}^{i}_{\mathrm{o}_{i+1}} + \tilde{\mathbf{U}}^{i}_{\mathrm{o}_{i+1}}\dot{\tilde{\mathbf{s}}}^{i}_{i,i+1}] + \mathbf{A}_{i+1}\mathbf{L}^{i+1}_{i+1}\mathbf{A}^{T}_{i+1} \tag{7.3.18d}$$

$$\mathbf{f}^{i}_{i} = \tilde{\mathbf{z}}^{i}_{i}\mathbf{U}^{i}_{\mathrm{o}_i} \tag{7.3.18e}$$

$$\hat{\mathbf{f}}^{i}_{i} = \left[\tilde{\mathbf{U}}^{i}_{\mathrm{o}_i}\tilde{\boldsymbol{\omega}}^{i}_{i} + \dot{\tilde{\mathbf{U}}}^{i}_{\mathrm{o}_i}\right]\cdot\mathbf{z}^{i}_{i} \tag{7.3.18f}$$

$$\overline{\mathbf{f}}^{i}_{i} = \left[\tilde{\mathbf{U}}^{i}_{\mathrm{o}_i}\boldsymbol{\Omega}^{i}_{i} + 2\dot{\tilde{\mathbf{U}}}^{i}_{\mathrm{o}_i}\tilde{\boldsymbol{\omega}}^{i}_{i} + \ddot{\tilde{\mathbf{U}}}^{i}_{\mathrm{o}_i}\right]\cdot\mathbf{z}^{i}_{i} \tag{7.3.18g}$$

$$\mathbf{B}^{i}_{i,i} = \mathbf{z}^{i}_{i}\cdot\mathbf{E}^{i}_{\mathrm{o}_i} \tag{7.3.18h}$$

$$\mathbf{C}^{i}_{i,i} = \mathbf{z}^{i}_{i}\cdot\mathbf{L}^{i}_{i} \tag{7.3.18i}$$

$$\mathbf{H}^{i}_{i,i} = \left[\mathbf{E}^{i}_{\mathrm{o}_i}\tilde{\boldsymbol{\omega}}^{i}_{i} - \mathbf{L}^{i}_{i}\right]\cdot\mathbf{z}^{i}_{i} \tag{7.3.18j}$$

$$\mathbf{G}^{i}_{i,i} = \left[\mathbf{E}^{i}_{\mathrm{o}_i}\boldsymbol{\Omega}^{i}_{i} - 2\mathbf{L}^{i}_{i}\tilde{\boldsymbol{\omega}}^{i}_{i} + \tilde{\mathbf{U}}^{i}_{\mathrm{o}_i}\ddot{\tilde{\mathbf{s}}}^{i}_{0,i} - \tilde{\boldsymbol{\eta}}^{i}_{i}\right]\cdot\mathbf{z}^{i}_{i} \tag{7.3.18k}$$

$$d_{ii} = \mathbf{B}^{i}_{i,i} \cdot \mathbf{z}^{i}_{i} \tag{7.3.18l}$$

$$v_{ii} = 2\left[\mathbf{B}^{i}_{i,i} \cdot \tilde{\omega}^{i}_{i} - \mathbf{C}^{i}_{i,i}\right] \cdot \mathbf{z}^{i}_{i} \tag{7.3.18m}$$

$$p_{ii} = \left[\mathbf{B}^{i}_{i,i} \cdot \Omega^{i}_{i} - 2\mathbf{C}^{i}_{i,i} \cdot \tilde{\omega}^{i}_{i} + \mathbf{f}^{i}_{i} \cdot \ddot{\tilde{\mathbf{s}}}^{i}_{0,i}\right] \cdot \mathbf{z}^{i}_{i} \tag{7.3.18n}$$

Step 2 : Backward Recursion :- $j = i\text{-}1, 1$

$$\mathbf{f}^{j}_{i} = \mathbf{A}_{j+1}\mathbf{f}^{j+1}_{i} \tag{7.3.19a}$$

$$\hat{\mathbf{f}}^{j}_{i} = \mathbf{A}_{j+1}\hat{\mathbf{f}}^{j+1}_{i} \tag{7.3.19b}$$

$$\overline{\mathbf{f}}^{j}_{i} = \mathbf{A}_{j+1}\overline{\mathbf{f}}^{j+1}_{i} \tag{7.3.19c}$$

$$\mathbf{B}^{j}_{j,i} = \mathbf{A}_{j+1}\mathbf{B}^{j+1}_{j+1,i} - \mathbf{f}^{j}_{i} \cdot \tilde{\mathbf{s}}^{j}_{j,j+1} \tag{7.3.19d}$$

$$\mathbf{C}^{j}_{j,i} = \mathbf{A}_{j+1}\mathbf{C}^{j+1}_{j+1,i} + \mathbf{f}^{j}_{i} \cdot \dot{\tilde{\mathbf{s}}}^{j}_{j,j+1} \tag{7.3.19e}$$

$$\mathbf{H}^{j}_{j,i} = \mathbf{A}_{j+1}\mathbf{H}^{j+1}_{j+1,i} - \tilde{\mathbf{s}}^{j}_{j,j+1}\hat{\mathbf{f}}^{j}_{i} \tag{7.3.19f}$$

$$\mathbf{G}^{j}_{j,i} = \mathbf{A}_{j+1}\mathbf{G}^{j+1}_{j+1,i} - \tilde{\mathbf{s}}^{j}_{j,j+1}\overline{\mathbf{f}}^{j}_{i} \tag{7.3.19g}$$

$$d_{ij} = \mathbf{B}^{j}_{j,i} \cdot \mathbf{z}^{j}_{j} \tag{7.3.19h}$$

$$d_{ji} = d_{ij} \tag{7.3.19i}$$

$$v_{ij} = 2\left[\mathbf{B}^{j}_{j,i} \cdot \tilde{\omega}^{j}_{j} - \mathbf{C}^{j}_{j,i}\right] \cdot \mathbf{z}^{j}_{j} \tag{7.3.19j}$$

$$v_{ji} = 2\mathbf{z}^{j}_{j} \cdot \mathbf{H}^{j}_{j,i} \tag{7.3.19k}$$

$$p_{ij} = \left[\mathbf{B}^{j}_{j,i} \cdot \Omega^{j}_{j} - 2\mathbf{C}^{j}_{j,i} \cdot \tilde{\omega}^{j}_{j} + \mathbf{f}^{j}_{i} \cdot \ddot{\tilde{\mathbf{s}}}^{j}_{0,j}\right] \cdot \mathbf{z}^{j}_{j} \tag{7.3.19l}$$

$$p_{ji} = \mathbf{z}^{j}_{j} \cdot \mathbf{G}^{j}_{j,i} \tag{7.3.19m}$$

end

7.3.2 Implementation and Computational Considerations

The computational complexity of Algorithm 7.1 is of course $o(n^2)$ since many of its equations need to be evaluated $n(n-1)/2$ times. However, the total number of scalar operations which are required for its numerical implementation can be reduced considerably with proper organization. As we can see, the structure of the equations in Algorithm 7.1 is the same as that of the equations in Algorithm 6.2. Therefore, observations made in Section 6.4 can also be used here. For example, since $\mathbf{z}_i^i$ is a unit coordinate vector, all the dot product operations with the vector $\mathbf{z}_i^i$ (or $\mathbf{z}_j^j$) can be implemented with no computational cost. Also, in Step 1, for $i = 1$ most equations either need not be computed or only certain of their entries are needed for evaluating the quantities d_{11}, v_{11} and p_{11}. Moreover, in Step 2, for $j = 1$ only the last entry of the vectors $\mathbf{C}_{j,i}^j$, $\mathbf{H}_{j,i}^j$ and $\mathbf{G}_{j,i}^j$ needs to be computed since, this is the only entry in these vectors which is actually used in the evaluation of the first column and the first row of the sensitivity matrices D, V and P. Therefore, these vectors need to be evaluated completely (i.e., all three components) $(n-1)(n-2)/2$ times. Obviously, all the other equations in Step 2 need to be evaluated $n(n-1)/2$ times.

Following the observations made above, a breakdown of the number of scalar multiplications and additions required for the implementation of each equation in Step 1 of Algorithm 7.1 is given in Table 7.1 and of each equation in Step 2 is given in Table 7.2. The total figure in each table represents the operations count for computing the corresponding equations when all the joints of a robot manipulator are of revolute type. Thus, based on these figures, to implement Algorithm 7.1 for a general 6 degrees-of-freedom revolute joint manipulator, one needs approximately 2056 scalar multiplications and 1762 scalar additions. When the *twist* angles of the manipulator are 0 or 90 degrees, the above numbers reduce to 1516 scalar multiplications and 1472 scalar additions, since in this case the coordinate transformations $\mathbf{A}_i$ are simpler. Note that these figures can be reduced further when the transformation matrix $\mathbf{A}_{n+1}$ is equal to $\mathbf{I}$ (as is usually the case in practice) or when the first revolute joint of the manipulator is assumed to be parallel to the gravity field.

Step 1	General manipulator with revolute joints		General manipulator with revolute joints & $\alpha=0^o$ or 90°	
Equation	Multipl.	Additions	Multipl.	Additions
7.3.18a	$8(n-1)$	$7(n-1)$	$4(n-1)$	$5(n-1)$
7.3.18b	$14(n-1)$	$10(n-1)$	$10(n-1)$	$8(n-1)$
7.3.18c	$30(n-1)+18$	$39(n-1)+8$	$20(n-1)+18$	$32(n-1)+8$
7.3.18d	$69(n-1)+18$	$74(n-1)+14$	$51(n-1)+18$	$63(n-1)+14$
7.3.18e	0	0	0	0
7.3.18f	$4(n-1)$	$n-1$	$4(n-1)$	$n-1$
7.3.18g	$9(n-1)$	$12(n-1)$	$9(n-1)$	$12(n-1)$
7.3.18h	0	0	0	0
7.3.18i	0	0	0	0
7.3.18j	$6(n-1)$	$6(n-1)$	$6(n-1)$	$6(n-1)$
7.3.18k	$18(n-1)$	$21(n-1)$	$18(n-1)$	$21(n-1)$
7.3.18l	0	0	0	0
7.3.18m	$2(n-1)$	$3(n-1)$	$2(n-1)$	$3(n-1)$
7.3.18n	$7(n-1)$	$7n-1)$	$7(n-1)$	$7(n-1)$
Total	$167n-131$	$180n-158$	$131n-95$	$158n-136$
$n=6$	871	922	691	812

Table 7.1: Operations counts for implementing Step 1 of Algorithm 7.1.

It is worth noting that the procedure which is proposed here for obtaining the coefficient sensitivity matrices of a joint space linearized dynamic robot model has significantly higher computational efficiency than most approaches available today.

Step 2	General manipulator with revolute joints		General manipulator with revolute joints & $\alpha=0^o$ or 90°	
Equation	Multipl.	Additions	Multipl.	Additions
7.3.19a	$4n(n-1)$	$2n(n-1)$	$2n(n-1)$	$n(n-1)$
7.3.19b	$4n(n-1)$	$2n(n-1)$	$2n(n-1)$	$n(n-1)$
7.3.19c	$4n(n-1)$	$2n(n-1)$	$2n(n-1)$	$n(n-1)$
7.3.19d	$7n(n-1)$	$5n(n-1)$	$5n(n-1)$	$4n(n-1)$
7.3.19e	$7n^2-16n+9$	$5n^2-11n+6$	$5n^2-10+5$	$4n^2-8n+4$
7.3.19f	$7n^2-16n+9$	$5n^2-11n+6$	$5n^2-10+5$	$4n^2-8n+4$
7.3.19g	$7n^2-16n+9$	$5n^2-11n+6$	$5n^2-10+5$	$4n^2-8n+4$
7.3.19h	0	0	0	0
7.3.19i	0	0	0	0
7.3.19j	$n(n-1)$	$1.5n(n-1)$	$n(n-1)$	$1.5n(n-1)$
7.3.19k	0	$0.5n(n-1)$	0	$0.5n(n-1)$
7.3.19l	$3n(n-1)$	$3n(n-1)$	$3n(n-1)$	$3n(n-1)$
7.3.19m	0	0	0	0
Total	$44n^2-71n+27$	$31n^2-49n+18$	$30n^2-45n+15$	$24n^2-36n+12$
$n=6$	1185	840	825	660

Table 7.2: Operations counts for implementing Step 2 of Algorithm 7.1.

7.4 Cartesian Space Robot Dynamic Models and their Linearization

As is well known [18], it is possible to describe the dynamics of a robot manipulator by using other sets of variables besides joint space variables. These variables are known as *operational space variables*, and among them, the *Cartesian configuration space variables* or simply *Cartesian space variables* are probably the most important. For example, in many cases, such as for end-effector motion and force control it may be desirable to express the dynamics of a manipulator in terms of "external" variables for direct, and thus better, measurements. In these cases, Cartesian space variables are obviously appropriate. Briefly, dynamic robot models described in terms of Cartesian space variables - henceforth called *Cartesian space dynamic robot models* can be introduced as follows.

7.4.1 Cartesian Space Dynamic Robot Models

As we mentioned in Chapter 2, the Cartesian space variables χ are defined to be the independent configuration parameters which specify the position and orientation of the end-effector relative to the inertia coordinate system. These Cartesian variables are functions of the joint space coordinates and this relationship is usually expressed by a "geometric" equation of the form

$$\chi = \mathbf{h}(\mathbf{q}). \tag{7.4.1}$$

In general, the vector function $\mathbf{h}$ is not one-to-one. However, for nonredundant manipulators in a restricted domain of the joint space, $\mathbf{h}$ can be assumed to be one-to-one. In this case the Cartesian space dynamic model of a manipulator can be defined [1,18] by the following equation

$$\mathbf{f} = \mathbf{D}_\chi(\mathbf{q})\ddot{\chi} + \mathbf{C}_\chi(\mathbf{q}, \dot{\mathbf{q}}) + \mathbf{G}_\chi(\mathbf{q}) \tag{7.4.2}$$

where $\mathbf{f}$ is an $n\times 1$ force-torque vector acting on the end-effector of the robot, and χ is a Cartesian space vector which describes the position and the orientation of the end-effector. The other terms in equation (7.4.2) are defined as follows: $\mathbf{D}_\chi(\mathbf{q})$ is the $n \times n$ Cartesian space generalized inertia tensor, $\mathbf{C}_\chi(\mathbf{q}, \dot{\mathbf{q}})$ is the $n \times 1$ Cartesian space vector of centrifugal and Coriolis terms, and $\mathbf{G}_\chi(\mathbf{q})$ is the $n \times 1$ Cartesian space vector of gravity terms. Obviously, all these Cartesian terms are implicit functions of the joint space coordinates. Actually, it can be shown [18] that if a closed-form joint space dynamic robot model is given by the equation

$$\boldsymbol{\tau} = \mathbf{D}(\mathbf{q})\ddot{\mathbf{q}} + \mathbf{C}(\mathbf{q}, \dot{\mathbf{q}}) + \mathbf{G}(\mathbf{q}) \tag{7.4.3}$$

where $\mathbf{D}(\mathbf{q})$ is the $n \times n$ joint space generalized inertia tensor of the manipulator, $\mathbf{C}(\mathbf{q}, \dot{\mathbf{q}})$ is the $n \times 1$ joint space vector of centrifugal and Coriolis terms, and $\mathbf{G}(\mathbf{q})$ is the $n \times 1$ joint space vector of gravity terms, then the aforementioned Cartesian space quantities are related to their joint space counterparts by the following equations:

$$\mathbf{D}_\chi(\mathbf{q}) = \mathbf{J}^{-T}(\mathbf{q})\mathbf{D}(\mathbf{q})\mathbf{J}^{-1}(\mathbf{q}) \tag{7.4.4a}$$

$$\mathbf{C}_\chi(\mathbf{q}, \dot{\mathbf{q}}) = \mathbf{J}^{-T}(\mathbf{q})\left[\mathbf{C}(\mathbf{q}, \dot{\mathbf{q}}) - \mathbf{D}(\mathbf{q})\mathbf{J}^{-1}(\mathbf{q})\dot{\mathbf{J}}(\mathbf{q})\dot{\mathbf{q}}\right] \tag{7.4.4b}$$

$$\mathbf{G}_\chi(\mathbf{q}) = \mathbf{J}^{-T}(\mathbf{q})\mathbf{G}(\mathbf{q}) \tag{7.4.4c}$$

$$\mathbf{f} = \mathbf{J}^{-T}(\mathbf{q})\boldsymbol{\tau} \tag{7.4.4d}$$

where $\mathbf{J}(\mathbf{q})$ is the manipulator Jacobian which has been assumed here to be nonsingular. When the Jacobian is locally singular, it is still possible [18] to define equation (7.4.2) by considering the manipulator to be a redundant manipulator locally. However, since in this monograph we are dealing with nonredundant manipulators, we shall assume that the manipulator Jacobian is nonsingular.

Now, since in practical applications we cannot actually cause a Cartesian force to be applied to the end-effector of a manipulator, we use equation (7.4.4d) to transfer the Cartesian force vector $\mathbf{f}$ to an equivalent joint torque vector $\boldsymbol{\tau}$ which effectively will cause the end-effector to follow the required motion. Therefore, instead of first computing the force vector $\mathbf{f}$ and then transferring it to $\boldsymbol{\tau}$, we can compute directly the joint torque vector $\boldsymbol{\tau}$. To achieve this, we combine equations (7.4.2) and (7.4.4d) and write the following *Cartesian configuration space torque equation.*

$$\boldsymbol{\tau} = \hat{\mathbf{D}}_{\chi}(\mathbf{q})\ddot{\boldsymbol{\chi}} + \hat{\mathbf{C}}_{\chi}(\mathbf{q}, \dot{\mathbf{q}}) + \hat{\mathbf{G}}_{\chi}(\mathbf{q}) \tag{7.4.5}$$

which defines directly the vector of the joint torques $\boldsymbol{\tau}$ when the dynamics of a robot are expressed in terms of the Cartesian space variables $\boldsymbol{\chi}$.

From the foregoing, it is possible to define the generalized force vector $\boldsymbol{\tau}$ in terms of either joint or Cartesian space variables. Therefore, it may be required in practice (e.g., for Cartesian based control applications) to define the perturbations $\delta\boldsymbol{\tau}$ of $\boldsymbol{\tau}$ in terms of perturbations of the Cartesian space variables, i.e., to define *Cartesian (configuration) space linearized dynamic robot models.*

7.4.2 Cartesian Space Linearized Dynamic Robot Models

Direct application of the Taylor series expansion to the nonlinear equations of a robot manipulator written in terms of the Cartesian space variables, as in equations (7.4.2) or (7.4.5), is rather difficult because it involves implicit differentiation in terms of the joint space variables. To avoid this complex differentiation, we can follow a similar approach to that used to derive Cartesian space nonlinear dynamic models from the joint space ones. In particular, in this approach, to define Cartesian space linearized dynamic robot models we first define a joint space linearized dynamic robot model and then, by expressing the joint space perturbations in terms of the Cartesian space perturbations, we algebraically manipulate the joint space linearized dynamic robot model to a Cartesian space one. Therefore, in order to derive a Cartesian space linearized dynamic robot model, we shall assume that the joint space linearized dynamic robot model of equation (7.2.1) is available.

As we mentioned in Section 7.4.1, the end-effector Cartesian space coordinates χ of a nonredundant manipulator can be considered as (Cartesian space) generalized coordinates which are related to the joint space generalized coordinates $\mathbf{q}$ by the nonlinear equation (7.4.1). As is well known, the time derivatives of these two sets of generalized coordinates are related by the equation

$$\dot{\chi} = \mathrm{J}(\mathbf{q})\dot{\mathbf{q}} \tag{7.4.6}$$

where $\mathrm{J}(\mathbf{q})$ is the manipulator Jacobian. For general operational spaces, the manipulator Jacobian is defined by the equation

$$\mathrm{J}(\mathbf{q}) = \frac{\partial \mathbf{h}(\mathbf{q})}{\partial \mathbf{q}} \tag{7.4.7}$$

where $\mathbf{h}$ is defined by equation (7.4.2). Unfortunately, this is not true for the Cartesian space variables (i.e., the linear and angular velocities) since there is no 3×1 orientation vector whose derivative is the vector of the angular velocity. However, in the case of Cartesian space variables, the manipulator Jacobian can be easily extracted from the equations which define the linear and angular velocity of the end-effector. Based on these equations, several methods for defining the manipulator Jacobian have been proposed in [19].

Equation (7.4.6) implies that infinitesimal Cartesian displacements or small perturbations of the end-effector Cartesian vector χ are related to the joint space perturbations $\delta\mathbf{q}$ by the equation

$$\delta\chi = \mathrm{J}(\mathbf{q})\delta\mathbf{q}. \tag{7.4.8}$$

Furthermore, by differentiating equation (7.4.8), we get

$$\delta\dot{\chi} = \dot{\mathrm{J}}(\mathbf{q})\delta\mathbf{q} + \mathrm{J}(\mathbf{q})\delta\dot{\mathbf{q}} \tag{7.4.9}$$

and

$$\delta\ddot{\chi} = \ddot{\mathrm{J}}(\mathbf{q})\delta\mathbf{q} + 2\dot{\mathrm{J}}(\mathbf{q})\delta\dot{\mathbf{q}} + \mathrm{J}(\mathbf{q})\delta\ddot{\mathbf{q}}. \tag{7.4.10}$$

Now, since in the definition of the Cartesian space nonlinear dynamic equations, we have assumed $\mathbf{J(q)}$ to be nonsingular, $\mathbf{J}^{-1}$ exists and therefore we can solve equations (7.4.8)-(7.4.10) for $\delta\mathbf{q}$, $\delta\dot{\mathbf{q}}$ and $\delta\ddot{\mathbf{q}}$ to get

$$\delta\mathbf{q} = \mathbf{J}^{-1}\delta\boldsymbol{\chi} \tag{7.4.11}$$

$$\delta\dot{\mathbf{q}} = \mathbf{J}^{-1}\delta\dot{\boldsymbol{\chi}} - \mathbf{J}^{-1}\dot{\mathbf{J}}\mathbf{J}^{-1}\delta\boldsymbol{\chi} \tag{7.4.12}$$

$$\delta\ddot{\mathbf{q}} = \mathbf{J}^{-1}\delta\ddot{\boldsymbol{\chi}} - 2\mathbf{J}^{-1}\dot{\mathbf{J}}\mathbf{J}^{-1}\delta\dot{\boldsymbol{\chi}} - \left[\mathbf{J}^{-1}\ddot{\mathbf{J}}\mathbf{J}^{-1} - 2\mathbf{J}^{-1}\dot{\mathbf{J}}\mathbf{J}^{-1}\dot{\mathbf{J}}\mathbf{J}^{-1}\right]\delta\boldsymbol{\chi}. \tag{7.4.13}$$

Expressions (7.4.11)-(7.4.13) can be used in (7.2.1) to yield

$$\begin{aligned}\delta\boldsymbol{\tau} = &\left[\mathbf{D}^o\mathbf{J}^{-1}\right]\delta\ddot{\boldsymbol{\chi}} + \left[\mathbf{V}^o\mathbf{J}^{-1} - 2\mathbf{D}^o\mathbf{J}^{-1}\dot{\mathbf{J}}\mathbf{J}^{-1}\right]\delta\dot{\boldsymbol{\chi}} \\ &+ \left[\mathbf{P}^o\mathbf{J}^{-1} - \mathbf{V}^o\mathbf{J}^{-1}\dot{\mathbf{J}}\mathbf{J}^{-1} - \mathbf{D}^o\left(\mathbf{J}^{-1}\ddot{\mathbf{J}}\mathbf{J}^{-1} - 2\mathbf{J}^{-1}\dot{\mathbf{J}}\mathbf{J}^{-1}\dot{\mathbf{J}}\mathbf{J}^{-1}\right)\right]\delta\boldsymbol{\chi}.\end{aligned} \tag{7.4.14}$$

or, if we define

$$\hat{\mathbf{D}}^o_\chi = \mathbf{D}^o\mathbf{J}^{-1} \tag{7.4.15}$$

$$\hat{\mathbf{V}}^o_\chi = \left(\mathbf{V}^o - 2\hat{\mathbf{D}}^o_\chi\dot{\mathbf{J}}\right)\mathbf{J}^{-1} \tag{7.4.16}$$

and

$$\hat{\mathbf{P}}^o_\chi = \left(\mathbf{P}^o - \hat{\mathbf{V}}^o_\chi\dot{\mathbf{J}} - \hat{\mathbf{D}}^o_\chi\ddot{\mathbf{J}}\right)\mathbf{J}^{-1}, \tag{7.4.17}$$

equation (7.4.14) can be written in a compact form as

$$\delta\boldsymbol{\tau} = \hat{\mathbf{D}}^o_\chi\delta\ddot{\boldsymbol{\chi}} + \hat{\mathbf{V}}^o_\chi\delta\dot{\boldsymbol{\chi}} + \hat{\mathbf{P}}^o_\chi\delta\boldsymbol{\chi}. \tag{7.4.18}$$

Equation (7.4.18) defines the perturbation in the vector of joint torques $\boldsymbol{\tau}$ as a result of perturbations in the vectors of Cartesian space positions, velocities and accelerations, i.e., it defines a *Cartesian space linearized dynamic robot model.* Moreover, by analogy with joint space linearized dynamic robot models, we may refer to the matrix coefficients of equations (7.4.18) as the *Cartesian space coefficient sensitivity matrices.*

Now, as we can see from equations (7.4.15)-(7.4.17), most of the quantities which are involved in the definitions of the Cartesian space coefficient sensitivity matrices $\hat{\mathbf{D}}_{\chi}^{o}$, $\hat{\mathbf{V}}_{\chi}^{o}$ and $\hat{\mathbf{P}}_{\chi}^{o}$ may be considered to be known, since they are available either from the joint space linearized dynamic robot model or from the Cartesian space nonlinear dynamic equations (e.g., see equation (7.4.4)). Therefore, the implementation of equation (7.4.18) is similar to that of equation (7.4.2).

7.5 Concluding Remarks

In this chapter, the linearization of the dynamic equations of rigid-link serial-type robot manipulators has been considered. Based on the Taylor series expansion and using Cartesian tensor analysis, we have proposed a procedure for obtaining the elements of the joint space coefficient sensitivity matrices. Also, we have shown that this procedure leads to a recursive algorithm which can be implemented numerically more efficiently than other similar algorithms existing in the literature.

The problem of obtaining Cartesian space linearized dynamic robot models has also been addressed in this chapter and, to the best of our knowledge, this is the first time where Cartesian space linearized dynamic robot models have been considered. To simplify our analysis, we have assumed that the manipulator is operating in a region of the work space where the Jacobian is nonsingular. We have shown that in these singularity free Cartesian configuration space, linearized dynamic robot models can be readily obtained using the joint space linearized models and the manipulator Jacobian.

7.6 References

[1] J. J. Craig, *Introduction to Robotics : Mechanics & Control*, Addison-Wesley, Reading. MA 1986.

[2] P. Misra, R. V. Patel, and C. A. Balafoutis, "Robust Control of Linearized Dynamic Robot Models", in *'Robot Manipulators: Modeling, Control and Education'*, M. Jamshidi, J.Y.S. Luh and M. Shahinpur, *Eds.*, 1986.

[3] P. Misra, R. V. Patel and C. A. Balafoutis, "Robust Control of Robot Manipulators in Cartesian Space", in *Proc. American Control Conf.*, pp. 1351-1356, Atlanta, Georgia, June 15-17, 1988.

[4] C. S. G. Lee and M. J. Chung, "An Adaptive Control Strategy for Mechanical Manipulators", in *Tutorial on Robotics*, C. S. G. Lee, R. C. Gonzales and K. S. Fu, *Eds.*, IEEE Computer Society, 1983.

[5] J. Y. S. Luh, M. W. Walker and R. P. Paul, "Resolved-Acceleration Control for Mechanical Manipulators", *J. Dyn. Sys., Meas., & Cont.*, Vol. 102, pp. 69-76, 1980.

[6] E. Freund, "Fast Nonlinear Control with Arbitrary Pole-Placement for Industrial Robots and Manipulators", *Int. J. Robotics Research*, pp. 65-78, Vol. 1, 1982.

[7] P. M. Frank, *Introduction to System Sensitivity Theory*, Academic Press, New York, 1978.

[8] C. P. Neuman and J. J. Murray, "Linearization and Sensitivity Functions of Dynamic Robot Models", *IEEE Trans. Syst. Man and Cyber.*, SMC-14, pp. 805-818, 1984.

[9] J. J. Murray and C. P. Neuman, "Linearization and Sensitivity Models of the Newton-Euler Dynamic Robot Models", *J. Dyn. Sys. Meas. & Contr.*, Vol. 108, pp. 272-276, 1986.

[10] C. P. Neuman, and P. K. Khosla, "Identification of Robot Dynamics : An Application of Recursive Estimation," in *Adaptive and Learning Systems : Theory and Applications*, K. S. Narendra, Eds., Plenum Publishing Corporation, New York, 1986.

[11] R. Su, "On the Linear Equivalents of Nonlinear Systems" *Systems and Control Letters*, Vol. 2, No. 1, pp. 48-52, 1982.

[12] L. R. Hunt, R. Su, and G. Meyer, "Global Transformations of Nonlinear Systems", *IEEE Trans. on Automatic Control,* Vol. AC-28, No. 1, pp.24-31, 1983.

[13] Y. Chen, *Nonlinear Feedback and Computer Control of Robot Arms*, Ph. D. Thesis, Washington University, St. Louis, MO, 1984.

[14] M. Vukobratovic, and N. Kircanski, *Scientific Fundamentals of Robotics 4 : Real-Time Dynamics of Manipulation Robots*, Springer-Verlag, Berlin, 1985.

[15] C. A. Balafoutis, P. Misra, and R. V. Patel, "Recursive Evaluation of Linearized Dynamic Robot Models", *IEEE J. Robotics and Automation,* RA-2, pp. 146-155, 1986.

[16] C. A. Balafoutis and R. V. Patel, "Linearized Robot Models in Joint and Cartesian Spaces", *Trans. of the Canadian Society of Mechanical Engineering* , Vol. 13, No. 4, pp. 103-112, 1989; also in *Proc. 9th Symposium on Engineering Applications of Mechanics : Current and Emerging Technologies*, London, Ontario, May 29-31, pp. 587-594, 1988.

[17] J. J. Murray and C. P. Neuman, "ARM : An Algebraic Robot Dynamic Modeling Program", in *Proc. IEEE Conf. on Robotics,* pp. 103-114, Atlanta, CA, Mar. 13-15, 1984.

[18] O. Khatib, "A Unified Approach for Motion and Force Control of Robot Manipulators : The Operational Space Formulation", *IEEE Journal of Robotics and Automation,* Vol. 3, No. 1, pp. 43-53, 1987.

[19] D. E. Orin and W. W. Schrader, "Efficient Computation of the Jacobian for Robot Manipulators", *Int. J. Robotics Research,* Vol. 3, pp. 66-75, 1984.

Appendix A

Recursive Lagrangian Formulation

In a dynamical system, the Lagrangian $\mathbf{L}$ is defined as the difference between the total kinetic energy of the system, $\boldsymbol{\Phi}$, and the total potential energy of the system, $\mathbf{P}$, assuming no dissipation of energy, i.e.,

$$\mathbf{L} = \boldsymbol{\Phi} - \mathbf{P} \tag{A.1}$$

The corresponding Euler-Lagrange equations are written as

$$\tau_i = \frac{d}{dt}\left[\frac{\partial \mathbf{L}}{\partial \dot{q}_i}\right] - \frac{\partial \mathbf{L}}{\partial q_i}, \qquad i = 1, \cdots, n, \tag{A.2}$$

where τ_i are the generalized forces, q_i are the generalized coordinates, and $\dot{q}_i$ is the generalized velocity.

Now, following an analysis similar to that used by Hollerbach in [14] (see references in Chapter 5), we can show that the total kinetic energy of the manipulator is given by the equation

$$\boldsymbol{\Phi} = \frac{1}{2}\sum_{j=1}^{n} tr\left\{ m_j \dot{\mathbf{s}}_{0j} \dot{\mathbf{s}}_{0j}^T + 2(\dot{\mathbf{W}}_j \mathbf{n}_j^j)\dot{\mathbf{s}}_{0j}^T + \dot{\mathbf{W}}_j \mathbf{J}_{o_j}^j \dot{\mathbf{W}}_j^T \right\} \tag{A.3}$$

where m_j is the mass of the j-th link, $\dot{\mathbf{s}}_{0j}$ is the absolute velocity of the origin o_j of the j-th link coordinate system, $\mathbf{n}_j^j = m_j \mathbf{r}_{jj}^j$ is the first moment of the j-th link about o_j expressed in the link coordinate system orientation, $\dot{\mathbf{W}}_j$ is the absolute derivative of the orientation tensor of the j-th link coordinate system and $\mathbf{J}_{\mathrm{o}_j}^j$ is the Euler tensor of the j-th link about o_j, expressed in the link coordinate system orientation.

The total potential energy $\mathbf{P}$ is equal to the sum of the work required to transport the mass center of each link from a reference plane, i.e.,

$$\begin{aligned}\mathbf{P} &= \text{constant} - \sum_{j=1}^{n} m_j \mathbf{g}^T \mathbf{r}_{0j} \\ &= \text{constant} - \sum_{j=1}^{n} m_j \mathbf{g}^T \left[\mathbf{s}_{0j} + \mathbf{W}_j \mathbf{r}_{jj}^j \right], \end{aligned} \tag{A.4}$$

where $\mathbf{g}$ is the acceleration due to gravity with reference to the base coordinate system. This form for potential energy is different from equation (A.6) in [14] and leads to the modified analysis which is presented in this appendix. Since potential energy is only position dependent, equation (A.2) can be written as

$$\tau_i = \frac{d}{dt}\left[\frac{\partial \mathbf{\Phi}}{\partial \dot{q}_i}\right] - \frac{\partial \mathbf{\Phi}}{\partial q_i} + \frac{\partial \mathbf{P}}{\partial q_i}, \qquad i = 1, \cdots, n. \tag{A.5}$$

Moreover, as in [14], we can write

$$\begin{aligned}\frac{d}{dt}\left[\frac{\partial \mathbf{\Phi}}{\partial \dot{q}_i}\right] - \frac{\partial \mathbf{\Phi}}{\partial q_i} = \sum_{j=1}^{n} tr \Bigg[& m_j \frac{\partial \mathbf{s}_{0j}}{\partial q_i} \ddot{\mathbf{s}}_{0j}^T + \frac{\partial \mathbf{s}_{0j}}{\partial q_i} (\mathbf{n}_j^j)^T \ddot{\mathbf{W}}_j^T \\ & + \frac{\partial \mathbf{W}_j}{\partial q_i} \mathbf{n}_j^j \ddot{\mathbf{s}}_{0j}^T + \frac{\partial \mathbf{W}_j}{\partial q_i} \mathbf{J}_{\mathrm{o}_j}^j \ddot{\mathbf{W}}_j^T \Bigg]. \end{aligned} \tag{A.6}$$

For the partial derivative of the potential energy we have from equation (A.4)

$$\frac{\partial P}{\partial q_i} = -\sum_{j=1}^{n} m_j \mathbf{g}^T \frac{\partial \mathbf{r}_{0,j}}{\partial q_i}$$

$$= -\sum_{j=1}^{n} m_j \mathbf{g}^T \left[\frac{\partial \mathbf{s}_{0,j}}{\partial q_i} + \frac{\partial \mathbf{W}_j}{\partial q_i} {}^j\mathbf{r}_{j,j} \right]. \tag{A.7}$$

Moreover, since for $j \geq i$ we have

$$\frac{\partial \mathbf{s}_{0,j}}{\partial q_i} = \frac{\partial \mathbf{W}_i}{\partial q_i} {}^i\mathbf{s}_{i,j} \tag{A.8}$$

and

$$\frac{\partial \mathbf{W}_j}{\partial q_i} = \frac{\partial \mathbf{W}_i}{\partial q_i} {}^i\mathbf{W}_j, \tag{A.9}$$

equation (A.7) can be written as

$$\frac{\partial P}{\partial q_i} = -\mathbf{g}^T \frac{\partial \mathbf{W}_i}{\partial q_i} \sum_{j=1}^{n} m_j ({}^i\mathbf{s}_{i,j} + {}^i\mathbf{W}_j {}^j\mathbf{r}_{j,j})$$

$$= -\mathbf{g}^T \frac{\partial \mathbf{W}_i}{\partial q_i} \sum_{j=1}^{n} m_j {}^i\mathbf{r}_{i,j}. \tag{A.10}$$

Now, using (A.6) and (A.10) we can write equation (A.5) as follows :

$$\tau_i = \sum_{j=1}^{n} \left\{ tr \left[m_j \frac{\partial \mathbf{s}_{0,j}}{\partial q_i} \ddot{\mathbf{s}}_{0,j}^T + \frac{\partial \mathbf{s}_{0,j}}{\partial q_i} ({}^j\mathbf{n}_j)^T \ddot{\mathbf{W}}_j^T + \frac{\partial \mathbf{W}_j}{\partial q_i} {}^j\mathbf{n}_j \ddot{\mathbf{s}}_{0,j}^T \right.\right.$$

$$\left.\left. + \frac{\partial \mathbf{W}_j}{\partial q_i} {}^j\mathbf{J}_{o_j} \ddot{\mathbf{W}}_j^T \right] - m_j \mathbf{g}^T \frac{\partial \mathbf{W}_i}{\partial q_i} {}^i\mathbf{r}_{i,j} \right\}, \quad i = 1, \cdots, n. \tag{A.11}$$

Moreover, if we use equations (A.8) and (A.9), we can simplify equation (A.11) as follows

$$\tau_i = tr\left\{\frac{\partial \mathbf{W}_i}{\partial q_i}\sum_{j=1}^{n}\left[m_j \mathbf{s}^i_{i,j}\ddot{\mathbf{s}}^T_{0,j} + \mathbf{s}^i_{i,j}(\mathbf{n}^j_j)^T\ddot{\mathbf{W}}^T_j + {}^i\mathbf{W}_j\mathbf{n}^j_j\ddot{\mathbf{s}}^T_{0,j}\right.\right.$$

$$\left.\left. + {}^i\mathbf{W}_j\mathbf{J}^j_{o_j}\ddot{\mathbf{W}}^T_j\right]\right\} - \mathbf{g}^T\frac{\partial \mathbf{W}_i}{\partial q_i}\sum_{j=1}^{n} m_j \mathbf{r}^i_{i,j}, \qquad i = 1, \cdots, n. \tag{A.12}$$

Now, following Hollerbach's approach [14], the first summation term on the right-hand side of (A.12) can be computed by the recurrence relations

$$\mathbf{D}_i = \sum_{j=1}^{n}\left[m_j \mathbf{s}^i_{i,j}\ddot{\mathbf{s}}^T_{0,j} + \mathbf{s}^i_{i,j}(\mathbf{n}^j_j)^T\ddot{\mathbf{W}}^T_j + {}^i\mathbf{W}_j\mathbf{n}^j_j\ddot{\mathbf{s}}^T_{0,j} + {}^i\mathbf{W}_j\mathbf{J}^j_{o_j}\ddot{\mathbf{W}}^T_j\right]$$

$$= \mathbf{A}_{i+1}\mathbf{D}_{i+1} + \mathbf{s}^i_{i,i+1}\mathbf{e}_{i+1} + \mathbf{n}^i_i\ddot{\mathbf{s}}^T_{0,i} + \mathbf{J}^i_{o_i}\ddot{\mathbf{W}}^T_i \tag{A.13}$$

where

$$\mathbf{e}_i = \sum_{j=1}^{n}\left[m_j\ddot{\mathbf{s}}^T_{0,j} + (\mathbf{n}^j_j)^T\ddot{\mathbf{W}}^T_j\right]$$

$$= m_i\ddot{\mathbf{s}}^T_{0,i} + (\mathbf{n}^i_i)^T\ddot{\mathbf{W}}^T_i + \mathbf{e}_{i+1}. \tag{A.14}$$

Similarly, the second summation term can also be computed by a recurrence relation. However, the recurrence relation presented here is different from equation (13) derived by Hollerbach in [14]. We proceed as follows :

$$\mathbf{c}^i_i = \sum_{j=1}^{n} m_j \mathbf{r}^i_{i,j}$$

$$= m_i\mathbf{r}^i_{i,i} + \sum_{j=1+1}^{n} m_j\left[\mathbf{s}^i_{i,i+1} + \mathbf{A}_{i+1}\mathbf{r}^{i+1}_{i+1,j}\right]$$

$$= m_i\mathbf{r}^i_{i,i} + \mathbf{s}^i_{i,i+1}\sum_{j=1+1}^{n} m_j + \mathbf{A}_{i+1}\sum_{j=1+1}^{n} m_j\mathbf{r}^{i+1}_{i+1,j}$$

$$= m_i \mathbf{r}_{i,i}^{i} + \bar{m}_{i+1} \mathbf{s}_{i,i+1}^{i} + \mathbf{A}_{i+1} \mathbf{c}_{i+1}^{i+1} \tag{A.15}$$

Substituting equations (A.13) and (A.15) into equation (A.12), we finally get Hollerbach's recurrence equation

$$\tau_i = tr \left[\frac{\partial \mathbf{W}_i}{\partial q_i} \mathbf{D}_i \right] - \mathbf{g}^T \frac{\partial \mathbf{W}_i}{\partial q_i} \mathbf{c}_i^i, \qquad i = 1, \cdots, n . \tag{A.16}$$

Appendix B

On Moment Vectors and Generalized Forces

As is well known, due to the rotational motion of a manipulator link, say the j-th one, forces and moments will be developed at all joints i for $i \leq j$. In this appendix we analyze the contribution of these moment vectors on the generalized force vector τ.

From the Newton-Euler formulation of the equations of motion of a robot manipulator, it is known that when the j-th joint is of revolute type, then as a result of the rotation of the j-th link, a moment vector $\mathbf{M}_{o_j}$ is developed with respect to the center of rotation. This vector is defined by the equation

$$\mathbf{M}_{o_j} = \mathbf{I}_{o_j} \cdot \dot{\boldsymbol{\omega}}_j + \tilde{\boldsymbol{\omega}}_j \mathbf{I}_{o_j} \cdot \boldsymbol{\omega}_j \tag{B.1}$$

where $\mathbf{I}_{o_j}$ is the inertia tensor of the j-th link with respect to the origin o_j, expressed in base frame orientation, and $\boldsymbol{\omega}_j$ ($\dot{\boldsymbol{\omega}}_j$) is the angular velocity (acceleration) of the j-th link. The contribution of this vector on the j-th component of the generalized force vector is defined by the equation

$$\tau_j = \mathbf{z}_j \cdot \mathbf{M}_{o_j} \tag{B.2}$$

where $\mathbf{z}_j$ is the unit vector which is parallel to the j-th axis of rotation. Moreover, from the structure of the recurrence equation (5.3.8b) of Algorithm 5.4 (or equation (5.3.46b) of Algorithm 5.5), the moment vector $\mathbf{M}_{o_j}$ will also contribute to all components τ_i, for $i \leq j$, of the generalized force vector τ and this contribution is given by

$$\tau_i = \mathbf{z}_i \cdot \mathbf{M}_{o_j} \tag{B.3}$$

From the foregoing, in the Newton-Euler formulation, the contribution of the moment vector $\mathbf{M}_{o_j}$ on the generalized force vector τ is explicitly defined by equations (B.2) and (B.3). However, this is not explicit or obvious, in the Lagrangian formulation of the dynamic equations of motion of a robot manipulator. In the following we shall show that

$$\mathbf{M}_{o_j} \cdot \mathbf{z}_i = tr\left[\frac{\partial \mathbf{W}_j}{\partial q_i} \mathbf{J}^j_{o_j} \ddot{\mathbf{W}}_j^T\right] \tag{B.4}$$

where $\mathbf{W}_j$ is the rotation tensor which specifies the orientation of the j-th link with reference to the base frame and $\mathbf{J}^j_{o_j}$ is the pseudo-inertia tensor of the j-th link with reference to o_j. Equation (B.4) is important since, by using the physical interpretation of the left-hand side of equation (B.4), we can gain valuable insight into the structure of the Lagrangian formulation. To derive equation (B.4) we proceed as follows.

First we notice from equation (4.2.11) that the rotation tensor $\mathbf{W}_j$ satisfies the equation

$$\begin{aligned} \ddot{\mathbf{W}}_j &= \mathbf{\Omega}_j \mathbf{W}_j \\ &= \left[\dot{\tilde{\omega}}_j + \tilde{\omega}_j \tilde{\omega}_j\right] \mathbf{W}_j . \end{aligned} \tag{B.5}$$

Moreover, for the partial derivative of the rotation tensor $\mathbf{W}_j$, we can write

$$\frac{\partial \mathbf{W}_j}{\partial q_i} = \tilde{\mathbf{z}}_i \mathbf{W}_j \tag{B.6}$$

Now, using equations (B.5) and (B.6) we can manipulate the right-hand side of equation (B.4) as follows

$$tr\left[\frac{\partial \mathbf{W}_j}{\partial q_i}\mathbf{J}^j_{\mathbf{o}_j}\ddot{\mathbf{W}}_j^T\right] = tr\left[\ddot{\mathbf{W}}_j\mathbf{J}^j_{\mathbf{o}_j}\frac{\partial \mathbf{W}_j^T}{\partial q_i}\right]$$

$$= tr\left[\left(\dot{\tilde{\boldsymbol{\omega}}}_j + \tilde{\boldsymbol{\omega}}_j\tilde{\boldsymbol{\omega}}_j\right)\mathbf{W}_j\mathbf{J}^j_{\mathbf{o}_j}\mathbf{W}_j^T\tilde{\mathbf{z}}_i^T\right]$$

$$= tr\left[\dot{\tilde{\boldsymbol{\omega}}}_j\mathbf{J}_{\mathbf{o}_j}\tilde{\mathbf{z}}_i^T\right] + tr\left[\tilde{\boldsymbol{\omega}}_j\tilde{\boldsymbol{\omega}}_j\mathbf{J}_{\mathbf{o}_j}\tilde{\mathbf{z}}_i^T\right]. \tag{B.7}$$

Furthermore, since the inertia and the pseudo-inertia tensors of a link satisfy equation (4.3.6), i.e., they satisfy the equation

$$\mathbf{J}_{\mathbf{o}_j} = -\mathbf{I}_{\mathbf{o}_j} + \frac{1}{2}tr[\mathbf{I}_{\mathbf{o}_j}]\mathbf{1},$$

we can use Proposition 3.12 of Chapter 3 to write

$$tr\left[\dot{\tilde{\boldsymbol{\omega}}}_j\mathbf{J}_{\mathbf{o}_j}\tilde{\mathbf{z}}_i^T\right] = \dot{\boldsymbol{\omega}}_j \cdot \mathbf{I}_{\mathbf{o}_j} \cdot \mathbf{z}_i \tag{B.8}$$

Also, since from equation (4.3.19) we can write

$$\tilde{\boldsymbol{\omega}}_j\mathbf{J}_{\mathbf{o}_j} = \dot{\mathbf{J}}_{\mathbf{o}_j} + \mathbf{J}_{\mathbf{o}_j}\tilde{\boldsymbol{\omega}}_j,$$

the second term on the right-hand side of (B.7) becomes

$$tr\left[\tilde{\boldsymbol{\omega}}_j\tilde{\boldsymbol{\omega}}_j\mathbf{J}_{\mathbf{o}_j}\tilde{\mathbf{z}}_i^T\right] = tr\left[\tilde{\boldsymbol{\omega}}_j\dot{\mathbf{J}}_{\mathbf{o}_j}\tilde{\mathbf{z}}_i^T\right] + tr\left[\tilde{\boldsymbol{\omega}}_j\mathbf{J}_{\mathbf{o}_j}\tilde{\boldsymbol{\omega}}_j\tilde{\mathbf{z}}_i^T\right].$$

However, the tensor $\tilde{\boldsymbol{\omega}}_j\mathbf{J}_{\mathbf{o}_j}\tilde{\boldsymbol{\omega}}_j$ is a symmetric tensor and therefore, by Proposition 3.17 we have

$$tr\left[\tilde{\boldsymbol{\omega}}_j\mathbf{J}_{\mathbf{o}_j}\tilde{\boldsymbol{\omega}}_j\tilde{\mathbf{z}}_i^T\right] = 0.$$

Thus, we can write

$$tr\left[\tilde{\omega}_j\tilde{\omega}_j\mathbf{J}_{o_j}\tilde{\mathbf{z}}_i^T\right] = tr\left[\tilde{\omega}_j\dot{\mathbf{J}}_{o_j}\tilde{\mathbf{z}}_i^T\right]$$

which, on using Proposition 3.12, may be simplified to yield

$$tr\left[\tilde{\omega}_j\tilde{\omega}_j\mathbf{J}_{o_j}\tilde{\mathbf{z}}_i^T\right] = \omega_j \cdot \dot{\mathbf{I}}_{o_j} \cdot \mathbf{z}_i. \tag{B.9}$$

Therefore, by substituting equations (B.8) and (B.9) into (B.7), we get

$$tr\left[\frac{\partial \mathbf{W}_j}{\partial q_i}\mathbf{J}^j_{o_j}\ddot{\mathbf{W}}_j^T\right] = \left[\dot{\omega}_j \cdot \mathbf{I}_{o_j} + \omega_j \cdot \dot{\mathbf{I}}_{o_j}\right] \cdot \mathbf{z}_i \tag{B.10}$$

Moreover, since $\mathbf{I}_{o_j}$ and $\dot{\mathbf{I}}_{o_j}$ are symmetric, by using equations (3.2.24) and (4.3.17) we can write equation (B.10) as follows

$$\begin{aligned} tr\left[\frac{\partial \mathbf{W}_j}{\partial q_i}\mathbf{J}^j_{o_j}\ddot{\mathbf{W}}_j^T\right] &= \left[\mathbf{I}_{o_j} \cdot \dot{\omega}_j + \dot{\mathbf{I}}_{o_j} \cdot \omega_j\right] \cdot \mathbf{z}_i \\ &= \left[\mathbf{I}_{o_j} \cdot \dot{\omega}_j + \left(\tilde{\omega}_j\mathbf{I}_{o_j} - \mathbf{I}_{o_j}\tilde{\omega}_j\right) \cdot \omega_j\right] \cdot \mathbf{z}_i \\ &= \left[\mathbf{I}_{o_j} \cdot \dot{\omega}_j + \tilde{\omega}_j\mathbf{I}_{o_j} \cdot \omega_j\right] \cdot \mathbf{z}_i \\ &= \mathbf{M}_{o_j} \cdot \mathbf{z}_i. \end{aligned}$$

which is equation (B.4).

Appendix C

On Partial Differentiation

To derive the Jacobians which define the joint space coefficient sensitivity matrices $\mathbf{D}^o$, $\mathbf{V}^o$ and $\mathbf{P}^o$ of a linearized robot model, we need to compute the partial derivatives of a number of tensor and vector functions involved in the definition of the nonlinear robot model. In this appendix, the partial derivatives of the tensor and vector functions appearing in Algorithm 5.5 are defined and some important lemmas are proved.

First to compute the partial derivatives of the angular velocity and angular acceleration tensors $\tilde{\boldsymbol{\omega}}_i^i$ and $\boldsymbol{\Omega}_i^i$, we can use either the recursive equations (5.3.45a)-(5.3.45c) or the following equations

$$\tilde{\boldsymbol{\omega}}_i^i = \mathbf{W}_i^T \dot{\mathbf{W}}_i \tag{C.1}$$

and

$$\boldsymbol{\Omega}_i^i = \mathbf{W}_i^T \ddot{\mathbf{W}}_i \tag{C.2}$$

which define these tensors (see Chapter IV) in terms of the orientation (transformation) tensor $\mathbf{W}_i = \mathbf{A}_1 \mathbf{A}_2 \cdots \mathbf{A}_i$. Here we shall use equations (C.1) and (C.2). Before we proceed to define these partial derivatives we need to state the following simple facts. First, for $j \leq i$, the following equations are obviously true:

$$\mathbf{W}_i = \mathbf{W}_j\, {}^j\mathbf{W}_i \tag{C.3}$$

$$\dot{\mathbf{W}}_i = \dot{\mathbf{W}}_j\, {}^j\mathbf{W}_i + \mathbf{W}_j\, {}^j\dot{\mathbf{W}}_i \tag{C.4}$$

$$\ddot{\mathbf{W}}_i = \ddot{\mathbf{W}}_j\, {}^j\mathbf{W}_i + 2\dot{\mathbf{W}}_j\, {}^j\dot{\mathbf{W}}_i + \mathbf{W}_j\, {}^j\ddot{\mathbf{W}}_i \tag{C.5}$$

where ${}^j\mathbf{W}_i = \mathbf{A}_{j+1}\mathbf{A}_{j+2} \cdots \mathbf{A}_i$. Furthermore, from (C.1), (C.3) and (C.4) we have

$$\begin{aligned}
\tilde{\omega}_i^i &= \mathbf{W}_i^T(\dot{\mathbf{W}}_j\, {}^j\mathbf{W}_i + \mathbf{W}_j\, {}^j\dot{\mathbf{W}}_i) \\
&= {}^j\mathbf{W}_i^T(\mathbf{W}_j^T\, \dot{\mathbf{W}}_j)\, {}^j\mathbf{W}_i + {}^j\mathbf{W}_i^T\, {}^j\dot{\mathbf{W}}_i \\
&= {}^j\mathbf{W}_i^T \tilde{\omega}_j^j\, {}^j\mathbf{W}_i + {}^j\mathbf{W}_i^T\, {}^j\dot{\mathbf{W}}_i \\
&= \tilde{\omega}_j^i + {}^j\mathbf{W}_i^T\, {}^j\dot{\mathbf{W}}_i
\end{aligned}$$

which implies that for $j \leq i$ we can write

$${}^j\mathbf{W}_i^T\, {}^j\dot{\mathbf{W}}_i = \tilde{\omega}_i^i - \tilde{\omega}_j^i. \tag{C.6}$$

Similarly, from (C.2), (C.3) and (C.5), for $j \leq i$, we have

$$\begin{aligned}
\mathbf{\Omega}_i^i &= \mathbf{W}_i^T(\ddot{\mathbf{W}}_j\, {}^j\mathbf{W}_i + 2\dot{\mathbf{W}}_j\, {}^j\dot{\mathbf{W}}_i + \mathbf{W}_j\, {}^j\ddot{\mathbf{W}}_i) \\
&= {}^j\mathbf{W}_i^T(\mathbf{W}_j^T\, \ddot{\mathbf{W}}_j)\, {}^j\mathbf{W}_i + 2\, {}^j\mathbf{W}_i^T(\mathbf{W}_j^T\, \dot{\mathbf{W}}_j)\, {}^j\mathbf{W}_i({}^j\mathbf{W}_i^T\, {}^j\dot{\mathbf{W}}_i) + {}^j\mathbf{W}_i^T\, {}^j\ddot{\mathbf{W}}_i \\
&= \dot{\tilde{\omega}}_j^i + \tilde{\omega}_j^i\tilde{\omega}_j^i + 2\tilde{\omega}_j^i(\tilde{\omega}_i^i - \tilde{\omega}_j^i) + {}^j\mathbf{W}_i^T\, {}^j\ddot{\mathbf{W}}_i \\
&= \dot{\tilde{\omega}}_j^i - \tilde{\omega}_j^i\tilde{\omega}_j^i + 2\tilde{\omega}_j^i\tilde{\omega}_i^i + {}^j\mathbf{W}_i^T\, {}^j\ddot{\mathbf{W}}_i \\
&= -\,\mathbf{\Omega}_j^{i^T} + 2\tilde{\omega}_j^i\tilde{\omega}_i^i + {}^j\mathbf{W}_i^T\, {}^j\ddot{\mathbf{W}}_i,
\end{aligned}$$

which finally gives,

$${}^j\mathbf{W}_i^T\, {}^j\ddot{\mathbf{W}}_i = \mathbf{\Omega}_i^i + \mathbf{\Omega}_j^{i^T} - 2\tilde{\omega}_j^i\tilde{\omega}_i^i. \tag{C.7}$$

Furthermore, since $\mathbf{z}_j^j$ is constant relative to the j-th frame, its absolute derivative satisfies the equation

$$\dot{\mathbf{z}}_j^j = \tilde{\omega}_j^j \mathbf{z}_j^j,$$

which in the i-th frame orientation is simply written as

$$\dot{\mathbf{z}}_j^i = \tilde{\omega}_j^i \mathbf{z}_j^i. \tag{C.8}$$

Similarly, the absolute acceleration $\ddot{\mathbf{z}}_j^j$ of $\mathbf{z}_j^j$, written in the i-th frame orientation, satisfies the following equation

$$\ddot{\mathbf{z}}_j^i = \Omega_j^i \mathbf{z}_j^i$$
$$= \dot{\tilde{\omega}}_j^i \mathbf{z}_j^i + \tilde{\omega}_j^i \tilde{\omega}_j^i \mathbf{z}_j^i.$$

Moreover, the dual tensors $\dot{\tilde{\mathbf{z}}}_j^i$ and $\ddot{\tilde{\mathbf{z}}}_j^i$ can be computed from

$$\dot{\tilde{\mathbf{z}}}_j^i = dual(\tilde{\omega}_j^i \mathbf{z}_j^i)$$
$$= \tilde{\omega}_j^i \tilde{\mathbf{z}}_j^i - \tilde{\mathbf{z}}_j^i \tilde{\omega}_j^i \tag{C.9}$$

and

$$\ddot{\tilde{\mathbf{z}}}_j^i = dual(\dot{\tilde{\omega}}_j^i \mathbf{z}_j^i) + dual(\tilde{\omega}_j^i \tilde{\omega}_j^i \mathbf{z}_j^i). \tag{C.10}$$

Now, using equation (3.4.15) and (3.4.25) and some algebraic manipulations, equation (C.10) can be simplified as follows:

$$\ddot{\tilde{\mathbf{z}}}_j^i = [\dot{\tilde{\omega}}_j^i - \tilde{\omega}_j^i \tilde{\omega}_j^i]\tilde{\mathbf{z}}_j^i - \tilde{\mathbf{z}}_j^i[\dot{\tilde{\omega}}_j^i + \tilde{\omega}_j^i \tilde{\omega}_j^i] + tr[\tilde{\omega}_j^i \tilde{\omega}_j^i]\tilde{\mathbf{z}}_j^i$$
$$= -\Omega_j^{i^T} \tilde{\mathbf{z}}_j^i - \tilde{\mathbf{z}}_j^i \Omega_j^i + tr[\Omega_j^i]\tilde{\mathbf{z}}_j^i \tag{C.11}$$

since $tr[\Omega_j^i] = tr[\tilde{\omega}_j^i \tilde{\omega}_j^i]$.

After these preliminaries, we are in a position to prove the following lemmas. First, due to the nature of the transformation tensors $\mathbf{A}_i$ and $\mathbf{W}_i$, the following two results can be easily shown:

Lemma C-1: The partial derivatives of $\mathbf{A}_i$, $\dot{\mathbf{A}}_i$ and $\ddot{\mathbf{A}}_i$, with respect to the generalized coordinate q_i and its time derivatives $\dot{q}_i$, $\ddot{q}_i$ satisfy the following equations:

$$i)\ \frac{\partial \mathbf{A}_i}{\partial q_i} = \frac{\partial \dot{\mathbf{A}}_i}{\partial \dot{q}_i} = \frac{\partial \ddot{\mathbf{A}}_i}{\partial \ddot{q}_i} = \mathbf{A}_i \bar{\mathbf{z}}_i^i$$

$$ii)\ \frac{\partial \dot{\mathbf{A}}_i}{\partial q_i} = \dot{\mathbf{A}}_i \bar{\mathbf{z}}_i^i, \quad \frac{\partial \ddot{\mathbf{A}}_i}{\partial q_i} = \ddot{\mathbf{A}}_i \bar{\mathbf{z}}_i^i$$

$$iii)\ \frac{\partial \ddot{\mathbf{A}}_i}{\partial \dot{q}_i} = 2\frac{\partial \dot{\mathbf{A}}_i}{\partial q_i}.$$

Lemma C-2: The partial derivatives of $\mathbf{W}_i$ and its time derivatives $\dot{\mathbf{W}}_i$ and $\ddot{\mathbf{W}}_i$, with respect to the generalized coordinate q_j and its time derivatives $\dot{q}_j, \ddot{q}_j$ satisfy the following equations:

$$i)\ \frac{\partial \mathbf{W}_i}{\partial \dot{q}_j} = \frac{\partial \mathbf{W}_i}{\partial \ddot{q}_j} = \frac{\partial \dot{\mathbf{W}}_i}{\partial \ddot{q}_j} = 0$$

$$ii)\ \frac{\partial \mathbf{W}_i}{\partial q_j} = \frac{\partial \dot{\mathbf{W}}_i}{\partial \dot{q}_j} = \frac{\partial \ddot{\mathbf{W}}_i}{\partial \ddot{q}_j}$$

$$iii)\ \frac{\partial \ddot{\mathbf{W}}_i}{\partial \dot{q}_j} = 2\frac{\partial \dot{\mathbf{W}}_i}{\partial q_j}.$$

The proofs of these two lemmas are straightforward and are therefore omitted.

Lemma C-3 The partial derivatives of the transformation matrix $\mathbf{W}_i$ and its time derivatives $\dot{\mathbf{W}}_i$ and $\ddot{\mathbf{W}}_i$, with respect to the generalized coordinate q_j, satisfy the following tensor equations:

$$i)\ \frac{\partial \mathbf{W}_i}{\partial q_j} = \begin{cases} \mathbf{W}_i \tilde{\mathbf{z}}_j^i & j \le i \\ 0 & j > i \end{cases}$$

$$ii)\ \frac{\partial \dot{\mathbf{W}}_i}{\partial q_j} = \begin{cases} \mathbf{W}_i \left[\dot{\tilde{\mathbf{z}}}_j^i + \tilde{\mathbf{z}}_j^i \tilde{\omega}_i^i \right] & j \le i \\ 0 & j > i \end{cases}$$

$$iii)\ \frac{\partial \ddot{\mathbf{W}}_i}{\partial q_j} = \begin{cases} \mathbf{W}_i \left[\ddot{\tilde{\mathbf{z}}}_j^i + 2\dot{\tilde{\mathbf{z}}}_j^i \tilde{\omega}_i^i + \tilde{\mathbf{z}}_j^i \Omega_i^i \right] & j \le i \\ 0 & j > i \end{cases}$$

Proof:

$i)$ For $j \le i$, since $\mathbf{W}_i = \mathbf{W}_j {}^j\mathbf{W}_i$ and ${}^j\mathbf{W}_i$ is independent of q_j, we have

$$\frac{\partial \mathbf{W}_i}{\partial q_j} = \frac{\partial \mathbf{W}_j}{\partial q_j} {}^j\mathbf{W}_i$$

$$= \mathbf{W}_{j-1} \frac{\partial \mathbf{A}_j}{\partial q_j} {}^j\mathbf{W}_i$$

$$= \mathbf{W}_j \tilde{\mathbf{z}}_j^j {}^j\mathbf{W}_i$$

$$= \mathbf{W}_i \tilde{\mathbf{z}}_j^i.$$

$ii)$ For $j \le i$, from $\dot{\mathbf{W}}_i = \dot{\mathbf{W}}_j {}^j\mathbf{W}_i + \mathbf{W}_j {}^j\dot{\mathbf{W}}_i$, and since

$$\frac{\partial \dot{\mathbf{W}}_j}{\partial q_j} = \dot{\mathbf{W}}_{j-1}\frac{\partial \mathbf{A}_j}{\partial q_j} + \mathbf{W}_{j-1}\frac{\partial \dot{\mathbf{A}}_j}{\partial q_j}$$

$$= \dot{\mathbf{W}}_j \tilde{\mathbf{z}}_j^j,$$

we have

$$\frac{\partial \dot{\mathbf{W}}_i}{\partial q_j} = \dot{\mathbf{W}}_j \tilde{\mathbf{z}}_j^{j\,j}\mathbf{W}_i + \mathbf{W}_j \tilde{\mathbf{z}}_j^{j\,j}\dot{\mathbf{W}}_i$$

$$= \dot{\mathbf{W}}_j\, {}^j\mathbf{W}_i \tilde{\mathbf{z}}_j^i + \mathbf{W}_i \tilde{\mathbf{z}}_j^{i\,j}\mathbf{W}_i^{Tj}\dot{\mathbf{W}}_i .$$

Now since $\dot{\mathbf{W}}_j = \mathbf{W}_j \tilde{\boldsymbol{\omega}}_j^j$, using (C.6) we can write

$$\frac{\partial \dot{\mathbf{W}}_i}{\partial q_j} = \mathbf{W}_j \tilde{\boldsymbol{\omega}}_j^{j\,j}\mathbf{W}_i \tilde{\mathbf{z}}_j^i + \mathbf{W}_i \tilde{\mathbf{z}}_j^i \left[\tilde{\boldsymbol{\omega}}_i^i - \tilde{\boldsymbol{\omega}}_j^i\right]$$

$$= \mathbf{W}_i \left[\tilde{\boldsymbol{\omega}}_j^i \tilde{\mathbf{z}}_j^i - \tilde{\mathbf{z}}_j^i \tilde{\boldsymbol{\omega}}_j^i + \tilde{\mathbf{z}}_j^i \tilde{\boldsymbol{\omega}}_i^i\right]$$

$$= \mathbf{W}_i \left[dual(\tilde{\boldsymbol{\omega}}_j^i \mathbf{z}_j^i) + \tilde{\mathbf{z}}_j^i \tilde{\boldsymbol{\omega}}_i^i\right]$$

$$= \mathbf{W}_i \left[\tilde{\dot{\mathbf{z}}}_j^i + \tilde{\mathbf{z}}_j^i \tilde{\boldsymbol{\omega}}_i^i\right].$$

iii) For $j \le i$, from equation $\ddot{\mathbf{W}}_i = \ddot{\mathbf{W}}_j\, {}^j\mathbf{W}_i + 2\dot{\mathbf{W}}_j\, {}^j\dot{\mathbf{W}}_i + \mathbf{W}_j\, {}^j\ddot{\mathbf{W}}_i$, since

$$\frac{\partial \ddot{\mathbf{W}}_j}{\partial q_j} = \ddot{\mathbf{W}}_j \tilde{\mathbf{z}}_j^j,$$

we have

$$\frac{\partial \ddot{\mathbf{W}}_i}{\partial q_j} = \ddot{\mathbf{W}}_j \tilde{\mathbf{z}}_j^j \, {}^j\mathbf{W}_i + 2\dot{\mathbf{W}}_j \tilde{\mathbf{z}}_j^j \, {}^j\dot{\mathbf{W}}_i + \mathbf{W}_j \tilde{\mathbf{z}}_j^j \, {}^j\ddot{\mathbf{W}}_i .$$

Also, since $\ddot{\mathbf{W}}_j = \mathbf{W}_j \boldsymbol{\Omega}_j^j$ and $\dot{\mathbf{W}}_j = \mathbf{W}_j \tilde{\boldsymbol{\omega}}_j^j$, the above expression becomes

$$\frac{\partial \ddot{\mathbf{W}}_i}{\partial q_j} = \mathbf{W}_j [\boldsymbol{\Omega}_j^j \tilde{\mathbf{z}}_j^j \, {}^j\mathbf{W}_i + 2\tilde{\boldsymbol{\omega}}_j^j \tilde{\mathbf{z}}_j^j \, {}^j\dot{\mathbf{W}}_i + \tilde{\mathbf{z}}_j^j \, {}^j\ddot{\mathbf{W}}_i]$$

$$= \mathbf{W}_i \left[\boldsymbol{\Omega}_j^i \tilde{\mathbf{z}}_j^i + 2\tilde{\boldsymbol{\omega}}_j^i \tilde{\mathbf{z}}_j^i \, {}^j\mathbf{W}_i^T \, {}^j\dot{\mathbf{W}}_i + \tilde{\mathbf{z}}_j^i \, {}^j\mathbf{W}_i^T \, {}^j\ddot{\mathbf{W}}_i \right].$$

Moreover, using (C.6) , (C.7), we have

$$\frac{\partial \ddot{\mathbf{W}}_i}{\partial q_j} = \mathbf{W}_i \left[\boldsymbol{\Omega}_j^i \tilde{\mathbf{z}}_j^i + 2\tilde{\boldsymbol{\omega}}_j^i \tilde{\mathbf{z}}_j^i \left(\tilde{\boldsymbol{\omega}}_i^i - \tilde{\boldsymbol{\omega}}_j^i \right) + \tilde{\mathbf{z}}_j^i [\boldsymbol{\Omega}_i^i + \boldsymbol{\Omega}_j^{i^T} - 2\tilde{\boldsymbol{\omega}}_j^i \tilde{\boldsymbol{\omega}}_i^i] \right]$$

$$= \mathbf{W}_i \left[\boldsymbol{\Omega}_j^i \tilde{\mathbf{z}}_j^i + \tilde{\mathbf{z}}_j^i \boldsymbol{\Omega}_j^{i^T} - 2\tilde{\boldsymbol{\omega}}_j^i \tilde{\mathbf{z}}_j^i \tilde{\boldsymbol{\omega}}_j^i + 2[\tilde{\boldsymbol{\omega}}_j^i \tilde{\mathbf{z}}_j^i - \tilde{\mathbf{z}}_j^i \tilde{\boldsymbol{\omega}}_j^i] \tilde{\boldsymbol{\omega}}_i^i + \tilde{\mathbf{z}}_j^i \boldsymbol{\Omega}_i^i \right]. \quad \text{(C.12)}$$

Now, the term $\boldsymbol{\Omega}_j^i \tilde{\mathbf{z}}_j^i + \tilde{\mathbf{z}}_j^i \boldsymbol{\Omega}_j^{i^T} - 2\tilde{\boldsymbol{\omega}}_j^i \tilde{\mathbf{z}}_j^i \tilde{\boldsymbol{\omega}}_j^i$ may be simplified as follows: Using the relation $\boldsymbol{\Omega}_j^i = \dot{\tilde{\boldsymbol{\omega}}}_j^i + \tilde{\boldsymbol{\omega}}_j^i \tilde{\boldsymbol{\omega}}_j^i$ and equation (3.4.19b), we have

$$\boldsymbol{\Omega}_j^i \tilde{\mathbf{z}}_j^i + \tilde{\mathbf{z}}_j^i \boldsymbol{\Omega}_j^{i^T} - 2\tilde{\boldsymbol{\omega}}_j^i \tilde{\mathbf{z}}_j^i \tilde{\boldsymbol{\omega}}_j^i = \dot{\tilde{\boldsymbol{\omega}}}_j^i \tilde{\mathbf{z}}_j^i - \tilde{\boldsymbol{\omega}}_j^i \tilde{\boldsymbol{\omega}}_j^i \tilde{\mathbf{z}}_j^i - \tilde{\mathbf{z}}_j^i \dot{\tilde{\boldsymbol{\omega}}}_j^i - \tilde{\mathbf{z}}_j^i \tilde{\boldsymbol{\omega}}_j^i \tilde{\boldsymbol{\omega}}_i^i + tr[\tilde{\boldsymbol{\omega}}_j^i \tilde{\boldsymbol{\omega}}_j^i] \tilde{\mathbf{z}}_j^i$$

$$= - \boldsymbol{\Omega}_j^{i^T} \tilde{\mathbf{z}}_j^i - \tilde{\mathbf{z}}_j^i \boldsymbol{\Omega}_j^i + tr[\boldsymbol{\Omega}_j^i] \tilde{\mathbf{z}}_j^i$$

$$= \ddot{\tilde{\mathbf{z}}}_j^i \quad \text{(C.13)}$$

where the last step follows from (C.11). Therefore, using (C.9) and (C.13), equation (C.12) can be written as

$$\frac{\partial \ddot{\mathbf{W}}_i}{\partial q_j} = \mathbf{W}_i [\ddot{\tilde{\mathbf{z}}}_j^i + 2\dot{\tilde{\mathbf{z}}}_j^i \tilde{\boldsymbol{\omega}}_i^i + \tilde{\mathbf{z}}_j^i \boldsymbol{\Omega}_i^i]$$

and this completes the proof. □

Lemma C-4

$$i)\ \mathbf{W}_i^T \frac{\partial \mathbf{W}_i}{\partial q_j} = \begin{cases} \tilde{\mathbf{z}}_j^i & j \leq i \\ 0 & j > i \end{cases}$$

$$ii)\ \mathbf{W}_i^T \frac{\partial \dot{\mathbf{W}}_i}{\partial q_j} = \begin{cases} \dot{\tilde{\mathbf{z}}}_j^i + \tilde{\mathbf{z}}_j^i \tilde{\boldsymbol{\omega}}_i^i & j \leq i \\ 0 & j > i \end{cases}$$

$$iii)\ \mathbf{W}_i^T \frac{\partial \ddot{\mathbf{W}}_i}{\partial q_j} = \begin{cases} \ddot{\tilde{\mathbf{z}}}_j^i + 2\dot{\tilde{\mathbf{z}}}_j^i \tilde{\boldsymbol{\omega}}_i^i + \tilde{\mathbf{z}}_j^i \boldsymbol{\Omega}_i^i & j \leq i \\ 0 & j > i \end{cases}$$

Proof: Follows from Lemma C-3. □

Now, we can evaluate the partial derivatives of the angular velocity and angular acceleration tensors $\tilde{\boldsymbol{\omega}}_i^i$ and $\boldsymbol{\Omega}_i^i$ with respect to the generalized coordinates.

Lemma C-5 The partial derivatives of the angular velocity tensor $\tilde{\boldsymbol{\omega}}_i^i$, with respect to the generalized coordinates q_j, $\dot{q}_j$ and $\ddot{q}_j$, are given by

$$i)\ \frac{\partial \tilde{\boldsymbol{\omega}}_i^i}{\partial q_j} = \begin{cases} \dot{\tilde{\mathbf{z}}}_j^i & j \leq i \\ 0 & j > i \end{cases}$$

$$ii)\quad \frac{\partial \tilde{\omega}_i^i}{\partial \dot{q}_j} = \begin{cases} \tilde{\mathbf{z}}_j^i & j \le i \\ 0 & j > i \end{cases}$$

$$iii)\quad \frac{\partial \tilde{\omega}_i^i}{\partial \ddot{q}_j} = 0 \qquad \text{for } all\ j$$

Proof:

i) For $j \le i$, since $\tilde{\omega}_i^i = \mathbf{W}_i^T \dot{\mathbf{W}}_i$, we have

$$\frac{\partial \tilde{\omega}_i^i}{\partial q_j} = \left(\frac{\partial \mathbf{W}_i}{\partial q_j}\right)^T \dot{\mathbf{W}}_i + \mathbf{W}_i^T \frac{\partial \dot{\mathbf{W}}_i}{\partial q_j}$$

$$= \left(\mathbf{W}_i^T \frac{\partial \mathbf{W}_i}{\partial q_j}\right)^T \mathbf{W}_i^T \dot{\mathbf{W}}_i + \mathbf{W}_i^T \frac{\partial \dot{\mathbf{W}}_i}{\partial q_j}$$

$$= -\tilde{\mathbf{z}}_j^i \tilde{\omega}_i^i + \dot{\tilde{\mathbf{z}}}_j^i + \tilde{\mathbf{z}}_j^i \tilde{\omega}_i^i$$

$$= \dot{\tilde{\mathbf{z}}}_j^i$$

ii) For $j \le i$, we have

$$\frac{\partial \tilde{\omega}_i^i}{\partial \dot{q}_j} = \mathbf{W}_i^T \frac{\partial \dot{\mathbf{W}}_i}{\partial \dot{q}_j}$$

$$= \mathbf{W}_i^T \frac{\partial \mathbf{W}_i}{\partial q_j}$$

$$= \tilde{\mathbf{z}}_j^i$$

□

Lemma C-6: The partial derivatives of the angular acceleration tensor $\mathbf{\Omega}_i^i$ with respect to the generalized coordinates $q_j, \dot{q}_j$ and $\ddot{q}_j$ are given by

$$i)\ \frac{\partial \mathbf{\Omega}_i^i}{\partial q_j} = \begin{cases} \tilde{\ddot{\mathbf{z}}}_j^i + 2\tilde{\dot{\mathbf{z}}}_j^i \tilde{\boldsymbol{\omega}}_i^i & j \le i \\ 0 & j > i \end{cases}$$

$$ii)\ \frac{\partial \mathbf{\Omega}_i^i}{\partial \dot{q}_j} = \begin{cases} 2[\tilde{\dot{\mathbf{z}}}_j^i + \tilde{\mathbf{z}}_j^i \tilde{\boldsymbol{\omega}}_i^i] & j \le i \\ 0 & j > i \end{cases}$$

$$iii)\ \frac{\partial \mathbf{\Omega}_i^i}{\partial \ddot{q}_j} = \begin{cases} \tilde{\mathbf{z}}_j^i & j \le i \\ 0 & j > i \end{cases}$$

Proof:

i) for $j \le i$, since $\mathbf{\Omega}_i^i = \mathbf{W}_i^T \ddot{\mathbf{W}}_i$, we have

$$\frac{\partial \mathbf{\Omega}_i^i}{\partial q_j} = \left[\frac{\partial \mathbf{W}_i}{\partial q_j}\right]^T \ddot{\mathbf{W}}_i + \mathbf{W}_i^T \frac{\partial \ddot{\mathbf{W}}_i}{\partial q_j}$$

$$= \left[\mathbf{W}_i^T \frac{\partial \mathbf{W}_i}{\partial q_j}\right]^T \mathbf{W}_i^T \ddot{\mathbf{W}}_i + \mathbf{W}_i^T \frac{\partial \ddot{\mathbf{W}}_i}{\partial q_j}$$

$$= - \tilde{\mathbf{z}}_j^i \mathbf{\Omega}_i^i + \tilde{\ddot{\mathbf{z}}}_j^i + 2\tilde{\dot{\mathbf{z}}}_j^i \tilde{\boldsymbol{\omega}}_i^i + \tilde{\mathbf{z}}_j^i \mathbf{\Omega}_i^i$$

$$= \tilde{\ddot{\mathbf{z}}}_j^i + 2\tilde{\dot{\mathbf{z}}}_j^i \tilde{\boldsymbol{\omega}}_i^i .$$

ii) for $j \le i$, we have

$$\frac{\partial \Omega_i^i}{\partial \dot{q}_j} = \mathbf{W}_i^T \frac{\partial \ddot{\mathbf{W}}_i}{\partial \dot{q}_j}$$

$$= 2\mathbf{W}_i^T \frac{\partial \dot{\mathbf{W}}_i}{\partial q_j}$$

$$= 2[\dot{\tilde{\mathbf{z}}}_j^i + \tilde{\mathbf{z}}_j^i \tilde{\omega}_i^i].$$

iii) for $j \leq i$, we have

$$\frac{\partial \Omega_i^i}{\partial \ddot{q}_j} = \mathbf{W}_i^T \frac{\partial \ddot{\mathbf{W}}_i}{\partial \ddot{q}_j}$$

$$= \mathbf{W}_i^T \frac{\partial \mathbf{W}_i}{\partial q_j}$$

$$= \tilde{\mathbf{z}}_j^i \qquad \square$$

Furthermore, by using Lemma C-6, we can derive formulas for computing the partial derivatives of the dual tensor of the angular acceleration vector $\dot{\omega}_i^i$ with respect to the generalized coordinates.

Lemma C-7

$$i) \quad \frac{\partial \dot{\tilde{\omega}}_i^i}{\partial q_j} = \begin{cases} \ddot{\tilde{\mathbf{z}}}_j^i + dual(\dot{\tilde{\mathbf{z}}}_j^i \omega_i^i) & j \leq i \\ 0 & j > i \end{cases}$$

$$ii) \quad \frac{\partial \dot{\tilde{\omega}}_i^i}{\partial \dot{q}_j} = \begin{cases} 2\dot{\tilde{\mathbf{z}}}_j^i + dual(\tilde{\mathbf{z}}_j^i \omega_i^i) & j \leq i \\ 0 & j > i \end{cases}$$

$$iii)\ \frac{\partial \dot{\tilde{\omega}}_i^i}{\partial \ddot{q}_j} = \begin{cases} \tilde{\mathbf{z}}_j^i & j \le i \\ 0 & j > i \end{cases}$$

Proof: Follows from the equation

$$\dot{\tilde{\omega}}_i^i = \frac{\Omega_i^i - \Omega_i^{i^T}}{2}$$

and Lemma C-6. □

Note that the partial derivatives with respect to the generalized coordinates of the angular velocity and angular acceleration vectors are readily available from Lemmas C-5 and C-7.

Now, from Algorithm 5.5, it is obvious that to compute the Jacobians which define the coefficient sensitivity matrices $\mathbf{D}^o$, $\mathbf{V}^o$ and $\mathbf{P}^o$, we need to compute the partial derivatives (with respect to the generalized coordinates $\ddot{q}_j$, $\dot{q}_j$ and q_j) of the vectors η_i^i (for revolute joints) which are defined by (5.3.46b). Equation (5.3.46b), for $i \le j \le n$, can also be written as

$$\eta_i^i = \sum_{k=i}^{j-1} {}^i\mathbf{W}_k[\mu_k^k + \tilde{\mathbf{u}}_{\mathrm{o}_k}^k \ddot{\mathbf{s}}_{0,k}^k + \tilde{\mathbf{s}}_{k,k+1}^k \ddot{\mathbf{U}}_{\mathrm{o}_{k+1}}^k] + {}^i\mathbf{W}_j \eta_j^j. \qquad \text{(C.14)}$$

Therefore, to compute the partial derivatives of the vectors η_i^i, we need to compute the partial derivatives of various vector functions which appear in equation (C.14). These partial derivatives are obtained as follows.

Lemma C-8: The partial derivatives, with respect to the generalized coordinates, of the vector function μ_i^i are given by:

$$i)\quad \frac{\partial \mu_i^i}{\partial \ddot{q}_j} = \begin{cases} \mathbf{K}_{o_i}^i \mathbf{z}_j^i & j \le i \\ 0 & j > i \end{cases}$$

$$ii)\quad \frac{\partial \mu_i^i}{\partial \dot{q}_j} = \begin{cases} 2[\mathbf{K}_{o_i}^i \dot{\mathbf{z}}_j^i - \hat{\mathbf{L}}_{o_i}^i \mathbf{z}_j^i] & j \le i \\ 0 & j > i \end{cases}$$

$$iii)\quad \frac{\partial \mu_i^i}{\partial q_j} = \begin{cases} \mathbf{K}_{o_i}^i \ddot{\mathbf{z}}_j^i - 2\hat{\mathbf{L}}_{o_i}^i \dot{\mathbf{z}}_j^i & j \le i \\ 0 & j > i \end{cases}$$

where $\mathbf{K}_{o_i}^i$ is the inertia tensor of the i-th augmented link, $\hat{\mathbf{L}}_{o_i}^i = \tilde{\omega}_i^i \hat{\mathbf{K}}_{o_i}^i$ and $\hat{\mathbf{K}}_{o_i}^i = \frac{1}{2} tr[\mathbf{K}_{o_i}^i]\mathbf{1} - \mathbf{K}_{o_i}^i$.

Proof: As we have shown in Chapter 4, the vector function μ_i^i may be defined by the equation

$$\mu_i^i = \mathbf{K}_{o_i}^i \dot{\omega}_i^i + \tilde{\omega}_i^i \mathbf{K}_{o_i}^i \omega_i^i$$

Now, since the inertia tensor of the i-th augmented link $\mathbf{K}_{o_i}^i$ is independent of the generalized coordinates, for the case of revolute joints we have that

i) follows from Lemmas (C-5) and (C-7).

ii) is obvious for $j > i$. For $j \le i$, by using Lemmas (C-5), (C-7) and the tensor equations (3.4.4) and (3.4.25), we get

$$\frac{\partial \mu_i^i}{\partial \dot{q}_j} = \mathbf{K}_{o_i}^i [2\dot{\mathbf{z}}_j^i + (\tilde{\mathbf{z}}_j^i \omega_i^i)] + \tilde{\mathbf{z}}_j^i (\mathbf{K}_{o_i}^i \omega_i^i) + \tilde{\omega}_i^i \mathbf{K}_{o_i}^i \mathbf{z}_j^i$$

$$= 2\mathbf{K}^i_{o_i}\dot{\mathbf{z}}^i_j - \mathbf{K}^i_{o_i}\tilde{\omega}^i_i\mathbf{z}^i_j - dual(\mathbf{K}^i_{o_i}\omega^i_i)\mathbf{z}^i_j + \tilde{\omega}^i_i\mathbf{K}^i_{o_i}\mathbf{z}^i_j \qquad \text{by (3.4.4)}$$

$$= 2\mathbf{K}^i_{o_i}\dot{\mathbf{z}}^i_j + [-\mathbf{K}^i_{o_i}\tilde{\omega}^i_i - dual(\mathbf{K}^i_{o_i}\omega^i_i) + \tilde{\omega}^i_i\mathbf{K}^i_{o_i}]\mathbf{z}^i_j$$

$$= 2\mathbf{K}^i_{o_i}\dot{\mathbf{z}}^i_j + [2\tilde{\omega}^i_i\mathbf{K}^i_{o_i} - tr(\mathbf{K}^i_{o_i})\tilde{\omega}^i_i]\mathbf{z}^i_j \qquad \text{by (3.4.25)}$$

$$= 2\mathbf{K}^i_{o_i}\dot{\mathbf{z}}^i_j - 2\tilde{\omega}^i_i\hat{\mathbf{K}}^i_{o_i}\mathbf{z}^i_j$$

$$= 2[\mathbf{K}^i_{o_i}\dot{\mathbf{z}}^i_j - \hat{\mathbf{L}}^i_{o_i}\mathbf{z}^i_j]$$

where, $\hat{\mathbf{L}}^i_{o_i} = \tilde{\omega}^i_i\hat{\mathbf{K}}^i_{o_i}$ and $\hat{\mathbf{K}}^i_{o_i} = \frac{1}{2}tr(\mathbf{K}^i_{o_i})\mathbf{1} - \mathbf{K}^i_{o_i}$.

iii) the proof is similar to that of *ii*). □

Lemma C-9: The partial derivatives with respect to the generalized coordinates of the vector function $\ddot{\mathbf{U}}^i_{o_i}$ are given by:

$$i)\quad \frac{\partial\ddot{\mathbf{U}}^i_{o_i}}{\partial q_j} = \begin{cases} \ddot{\tilde{\mathbf{z}}}^i_j\mathbf{U}^i_{o_i} + 2\dot{\tilde{\mathbf{z}}}^i_j\dot{\mathbf{U}}^i_{o_i} & j \le i \\ \ddot{\tilde{\mathbf{z}}}^i_j\mathbf{U}^i_{o_j} + 2\dot{\tilde{\mathbf{z}}}^i_j\dot{\mathbf{U}}^i_{o_j} + \tilde{\mathbf{z}}^i_j\ddot{\mathbf{U}}^i_{o_j} & j > i \end{cases}$$

$$ii)\quad \frac{\partial\ddot{\mathbf{U}}^i_{o_i}}{\partial \dot{q}_j} = \begin{cases} 2[\dot{\tilde{\mathbf{z}}}^i_j\mathbf{U}^i_{o_i} + \tilde{\mathbf{z}}^i_j\dot{\mathbf{U}}^i_{o_i}] & j \le i \\ 2[\dot{\tilde{\mathbf{z}}}^i_j\mathbf{U}^i_{o_j} + \tilde{\mathbf{z}}^i_j\dot{\mathbf{U}}^i_{o_j}] & j > i \end{cases}$$

$$iii)\quad \frac{\partial\ddot{\mathbf{U}}^i_{o_i}}{\partial \ddot{q}_j} = \begin{cases} \tilde{\mathbf{z}}^i_j\mathbf{U}^i_{o_i} & j \le i \\ \tilde{\mathbf{z}}^i_j\mathbf{U}^i_{o_j} & j > i \end{cases}$$

Proof: From equation (5.3.46a), we have $\ddot{\mathbf{U}}^i_{o_i} = \sum_{k=i}^{n} {}^i\mathbf{W}_k\Omega^k_k\mathbf{u}^k_{o_k}$, which, for

$i < j$, can be written as $\ddot{\mathbf{U}}_{o_i}^i = \sum_{k=i}^{j-1} {}^i\mathbf{W}_k \boldsymbol{\Omega}_k^k \mathbf{u}_{o_k}^k + {}^i\mathbf{W}_j \ddot{\mathbf{U}}_{o_j}^j$. Then, we proceed as follows:

i) From Lemma C-6 and the fact that $\mathbf{u}_{o_i}^i$ is independent of the generalized coordinates (for revolute joints), we have

$$\frac{\partial \ddot{\mathbf{U}}_{o_i}^i}{\partial q_j} = \begin{cases} \sum_{k=i}^{n} {}^i\mathbf{W}_k \dfrac{\partial \boldsymbol{\Omega}_k^k}{\partial q_j} \mathbf{u}_{o_k}^k & j \le i \\ \dfrac{\partial\, {}^i\mathbf{W}_j}{\partial q_j} \ddot{\mathbf{U}}_{o_j}^j + {}^i\mathbf{W}_j \dfrac{\partial \ddot{\mathbf{U}}_{o_j}^j}{\partial q_j} & j > i \end{cases}$$

$$\frac{\partial \ddot{\mathbf{U}}_{o_i}^i}{\partial q_j} = \begin{cases} \sum_{k=i}^{n} {}^i\mathbf{W}_k [\,\ddot{\tilde{\mathbf{z}}}_j^k + 2\dot{\tilde{\mathbf{z}}}_j^k \tilde{\boldsymbol{\omega}}_k^k\,]\mathbf{u}_{o_k}^k = \ddot{\tilde{\mathbf{z}}}_j^i \mathbf{U}_{o_i}^i + 2\dot{\tilde{\mathbf{z}}}_j^i \dot{\mathbf{U}}_{o_i}^i & j \le i \\ {}^i\mathbf{W}_j \tilde{\mathbf{z}}_j^j \ddot{\mathbf{U}}_{o_j}^j + {}^i\mathbf{W}_j [\ddot{\tilde{\mathbf{z}}}_j^j \mathbf{U}_{o_j}^j + 2\dot{\tilde{\mathbf{z}}}_j^j \dot{\mathbf{U}}_{o_j}^j] = \ddot{\tilde{\mathbf{z}}}_j^i \mathbf{U}_{o_j}^i + 2\dot{\tilde{\mathbf{z}}}_j^i \dot{\mathbf{U}}_{o_j}^i + \tilde{\mathbf{z}}_j^i \ddot{\mathbf{U}}_{o_j}^i & j > i \end{cases}$$

ii) and *iii*) can be proved in a similar manner. □

To compute the partial derivatives of $\ddot{\mathbf{s}}_{0,i}^i$, we need the following results.

Lemma C-10:

$$i)\quad \frac{\partial \ddot{\mathbf{s}}_{i,i+1}^i}{\partial q_j} = \begin{cases} \ddot{\tilde{\mathbf{z}}}_j^i \mathbf{s}_{i,i+1}^i + 2\dot{\tilde{\mathbf{z}}}_j^i \dot{\mathbf{s}}_{i,i+1}^i & j \le i \\ 0 & j > i \end{cases}$$

$$ii)\quad \frac{\partial \ddot{\mathbf{s}}^{\,i}_{i,i+1}}{\partial \dot{q}_j} = \begin{cases} 2[\dot{\tilde{\mathbf{z}}}^{i}_{j}\,\mathbf{s}^{i}_{i,i+1} + \tilde{\mathbf{z}}^{i}_{j}\,\dot{\mathbf{s}}^{i}_{i,i+1}] & j \le i \\ 0 & j > i \end{cases}$$

$$iii)\quad \frac{\partial \ddot{\mathbf{s}}^{\,i}_{i,i+1}}{\partial \ddot{q}_j} = \begin{cases} \tilde{\mathbf{z}}^{i}_{j}\,\mathbf{s}^{i}_{i,i+1} & j \le i \\ 0 & j > i \end{cases}$$

Proof: The results follow from the relation $\ddot{\mathbf{s}}^{\,i}_{i,i+1} = \mathbf{\Omega}^{i}_{i}\mathbf{s}^{i}_{i,i+1}$ and the partial derivatives of $\mathbf{\Omega}^{i}_{i}$. □

Lemma C-11: The partial derivatives with respect to the generalized coordinates of the vector function $\ddot{\mathbf{s}}^{\,i}_{0,i}$ are given by:

$$i)\quad \frac{\partial \ddot{\mathbf{s}}^{\,i}_{0,i}}{\partial q_j} = \begin{cases} \ddot{\tilde{\mathbf{z}}}^{i}_{j}\,\mathbf{s}^{i}_{j,i} + 2\dot{\tilde{\mathbf{z}}}^{i}_{j}\,\dot{\mathbf{s}}^{i}_{j,i} - \tilde{\mathbf{z}}^{i}_{j}\,\ddot{\mathbf{s}}^{i}_{0,j} & j \le i \\ 0 & j > i \end{cases}$$

$$ii)\quad \frac{\partial \ddot{\mathbf{s}}^{\,i}_{0,i}}{\partial \dot{q}_j} = \begin{cases} 2[\dot{\tilde{\mathbf{z}}}^{i}_{j}\,\mathbf{s}^{i}_{j,i} + \tilde{\mathbf{z}}^{i}_{j}\,\dot{\mathbf{s}}^{i}_{j,i}] & j \le i \\ 0 & j > i \end{cases}$$

$$iii)\quad \frac{\partial \ddot{\mathbf{s}}^{\,i}_{0,i}}{\partial \ddot{q}_j} = \begin{cases} \tilde{\mathbf{z}}^{i}_{j}\mathbf{s}^{i}_{j,i} & j \le i \\ 0 & j > i \end{cases}$$

Proof: For $j \le i$ we can write

$$\ddot{\mathbf{s}}^{i}_{0,i} = \ddot{\mathbf{s}}^{i}_{0,j} + \ddot{\mathbf{s}}^{i}_{j,i} = {}^{(j-1)}\mathbf{W}_i^T\,\ddot{\mathbf{s}}^{j-1}_{0,j} + \sum_{k=j}^{i-1} {}^{k}\mathbf{W}_i^T\,\ddot{\mathbf{s}}^{k}_{k,k+1}$$

i) From Lemmas C-1 and C-8, we get

$$\frac{\partial \ddot{\mathbf{s}}^{i}_{0,i}}{\partial q_j} = {}^{j}\mathbf{W}_i^T\,\tilde{\mathbf{z}}_j^{j^T}\mathbf{A}_j^T\,\ddot{\mathbf{s}}^{j-1}_{0,j} + \sum_{k=j}^{i-1} {}^{k}\mathbf{W}_i^T\,[\ddot{\tilde{\mathbf{z}}}_j^{k}\,\mathbf{s}^{k}_{k,k+1} + 2\dot{\tilde{\mathbf{z}}}_j^{k}\,\dot{\mathbf{s}}^{k}_{k,k+1}]$$

$$= -\,\tilde{\mathbf{z}}_j^{i}\,\ddot{\mathbf{s}}^{i}_{0,j} + 2\dot{\tilde{\mathbf{z}}}_j^{i}\,\dot{\mathbf{s}}^{i}_{j,i} + \ddot{\tilde{\mathbf{z}}}_j^{i}\,\mathbf{s}^{i}_{j,i}$$

ii) and *iii*) can be proved similarly. □

Now, we are in a position to derive the partial derivatives with respect to the generalized coordinates of the vector η_i^i defined by equation (C.14).

Lemma C-12: The partial derivatives with respect to the generalized coordinates of the vector function η_i^i are given by:

$$i)\ \frac{\partial \eta_i^i}{\partial \ddot{q}_j} = \begin{cases} [\mathbf{E}^{i}_{o_i} - \tilde{\mathbf{U}}^{i}_{o_i}\,\tilde{\mathbf{s}}^{i}_{j,i}]\mathbf{z}_j^{i} & j \le i \\ {}^{i}\mathbf{W}_j[\mathbf{E}^{j}_{o_j} - \tilde{\mathbf{s}}^{j}_{i,j}\tilde{\mathbf{U}}^{j}_{o_j}]\mathbf{z}_j^{j} & j > i \end{cases}$$

$$ii)\ \frac{\partial \eta_i^i}{\partial \dot{q}_j} = 2\begin{cases} [\mathbf{E}^{i}_{o_i} - \tilde{\mathbf{U}}^{i}_{o_i}\,\tilde{\mathbf{s}}^{i}_{j,i}]\dot{\mathbf{z}}_j^{i} - [\mathbf{L}_i^{i} + \tilde{\mathbf{U}}^{i}_{o_i}\,\dot{\tilde{\mathbf{s}}}^{i}_{j,i}]\mathbf{z}_j^{i} & j \le i \\ {}^{i}\mathbf{W}_j\left([\mathbf{E}^{j}_{o_j} - \tilde{\mathbf{s}}^{j}_{i,j}\tilde{\mathbf{U}}^{j}_{o_j}]\dot{\mathbf{z}}_j^{j} - [\mathbf{L}_j^{j} + \tilde{\mathbf{s}}^{j}_{i,j}\dot{\tilde{\mathbf{U}}}^{j}_{o_j}]\mathbf{z}_j^{j}\right) & j > i \end{cases}$$

and iii)

$$\frac{\partial \eta_i^i}{\partial q_j} = \begin{cases} [\mathbf{E}^{i}_{o_i} - \tilde{\mathbf{U}}^{i}_{o_i}\,\tilde{\mathbf{s}}^{i}_{j,i}]\ddot{\mathbf{z}}_j^{i} - 2[\mathbf{L}_i^{i} + \tilde{\mathbf{U}}^{i}_{o_i}\,\dot{\tilde{\mathbf{s}}}^{i}_{j,i}]\dot{\mathbf{z}}_j^{i} + \tilde{\mathbf{U}}^{i}_{o_i}\,\ddot{\tilde{\mathbf{s}}}^{i}_{0,j}\mathbf{z}_j^{i} & j \le i \\ {}^{i}\mathbf{W}_j([\mathbf{E}^{j}_{o_j} - \tilde{\mathbf{s}}^{j}_{i,j}\tilde{\mathbf{U}}^{j}_{o_j}]\ddot{\mathbf{z}}_j^{j} - 2[\mathbf{L}_j^{j} + \tilde{\mathbf{s}}^{j}_{i,j}\dot{\tilde{\mathbf{U}}}^{j}_{o_j}]\dot{\mathbf{z}}_j^{j} + [\tilde{\mathbf{U}}^{j}_{o_j}\ddot{\tilde{\mathbf{s}}}^{j}_{0,j} - \tilde{\mathbf{s}}^{j}_{i,j}\ddot{\tilde{\mathbf{U}}}^{j}_{o_j} - \tilde{\eta}_j^{j}]\mathbf{z}_j^{j}) & j > i \end{cases}$$

where $\mathbf{E}^i_{o_i}$ ($\mathbf{U}^i_{o_i}$) is the second (first) moment of the i-th generalized link and the tensor $\mathbf{L}^i_i$ satisfies the equation

$$\mathbf{L}^i_i = \hat{\mathbf{L}}^i_{o_i} + [\tilde{\mathbf{s}}^i_{i,i+1}\tilde{\mathbf{U}}^i_{o_{i+1}} + \tilde{\mathbf{U}}^i_{o_{i+1}}\tilde{\mathbf{s}}^i_{i,i+1}] + \mathbf{A}_{i+1}\mathbf{L}^{i+1}_{i+1}\mathbf{A}^T_{i+1} \tag{C.15}$$

Proof: Using Lemmas C-8, C-9 and C-11, the partial derivatives of $\boldsymbol{\eta}^i_i$ can be derived as follows:

i) There are two cases:

a) For $j \le i$, we have

$$\frac{\partial \boldsymbol{\eta}^i_i}{\partial \ddot{q}_j} = \sum_{k=i}^{n} {}^i\mathbf{W}_k \left(\frac{\partial \boldsymbol{\mu}^k_k}{\partial \ddot{q}_j} + \tilde{\mathbf{u}}^k_{o_k} \frac{\partial \ddot{\mathbf{s}}^k_{0,k}}{\partial \ddot{q}_j} + \tilde{\mathbf{s}}^k_{k,k+1} \frac{\partial \ddot{\mathbf{U}}^k_{o_{k+1}}}{\partial \ddot{q}_j} \right)$$

$$= \sum_{k=i}^{n} {}^i\mathbf{W}_k \left(\mathbf{K}^k_{o_k}\mathbf{z}^k_j + \tilde{\mathbf{u}}^k_{o_k}\tilde{\mathbf{z}}^k_j \mathbf{s}^k_{j,k} + \tilde{\mathbf{s}}^k_{k,k+1}\tilde{\mathbf{z}}^k_j\mathbf{U}^k_{o_{k+1}} \right)$$

$$= \sum_{k=i}^{n} {}^i\mathbf{W}_k \left(\mathbf{K}^k_{o_k} - \tilde{\mathbf{u}}^k_{o_k}\tilde{\mathbf{s}}^k_{j,k} - \tilde{\mathbf{s}}^k_{k,k+1}\tilde{\mathbf{U}}^k_{o_{k+1}} \right) {}^i\mathbf{W}^T_k\,\mathbf{z}^i_j \qquad \text{by (C-3).}$$

Now, for $j \le i \le k$, $\tilde{\mathbf{s}}^k_{j,k} = \tilde{\mathbf{s}}^k_{j,i} + \tilde{\mathbf{s}}^k_{i,k}$. Therefore, we have

$$\frac{\partial \boldsymbol{\eta}^i_i}{\partial \ddot{q}_j} = \sum_{k=i}^{n} {}^i\mathbf{W}_k \left(\mathbf{K}^k_{o_k} - \tilde{\mathbf{u}}^k_{o_k}\tilde{\mathbf{s}}^k_{i,k} - \tilde{\mathbf{s}}^k_{k,k+1}\tilde{\mathbf{U}}^k_{o_{k+1}} \right) {}^i\mathbf{W}^T_k\mathbf{z}^i_j - \sum_{k=i}^{n} {}^i\mathbf{W}_k \left(\tilde{\mathbf{u}}^k_{o_k}\tilde{\mathbf{s}}^k_{j,i} \right) {}^i\mathbf{W}^T_k\mathbf{z}^i_j$$

$$= [\mathbf{E}^i_{o_i} - \tilde{\mathbf{U}}^i_{o_i}\tilde{\mathbf{s}}^i_{j,i}]\mathbf{z}^i_j$$

where

$$\mathbf{E}^i_{o_i} = \sum_{k=i}^{n} {}^i\mathbf{W}_k[\mathbf{K}^k_{o_k} - \tilde{\mathbf{u}}^k_{o_k}\tilde{\mathbf{s}}^k_{i,k} - \tilde{\mathbf{s}}^k_{k,k+1}\tilde{\mathbf{U}}^k_{o_{k+1}}]^i\mathbf{W}^T_k \tag{C.16}$$

is the inertia tensor of the i-th generalized link and

$$\tilde{\mathbf{U}}^i_{o_i} = \sum_{k=i}^{n} {}^i\mathbf{W}_k\tilde{\mathbf{u}}^k_{o_k}\,{}^i\mathbf{W}^T_k \tag{C.17}$$

is the dual tensor of the first moment of the i-th generalized link. To see that $\mathbf{E}^i_{o_i}$ is the inertia tensor of the i-th generalized link, we shall show that equation (C.16) is equivalent to equation (6.3.6). Since $\tilde{\mathbf{s}}^i_{i,i} = 0$, we have

$$\mathbf{E}^i_{o_i} = \mathbf{K}^i_{o_i} - \tilde{\mathbf{s}}^i_{i,i+1}\tilde{\mathbf{U}}^i_{o_{i+1}}$$

$$+ \sum_{k=i+1}^{n} {}^i\mathbf{W}_k\left[\mathbf{K}^k_{o_k} - \tilde{\mathbf{u}}^k_{o_k}(\tilde{\mathbf{s}}^k_{i,i+1} + \tilde{\mathbf{s}}^k_{i+1,k}) - \tilde{\mathbf{s}}^k_{k,k+1}\tilde{\mathbf{U}}^k_{o_{k+1}}\right]{}^i\mathbf{W}^T_k$$

$$= \mathbf{K}^i_{o_i} - \tilde{\mathbf{s}}^i_{i,i+1}\tilde{\mathbf{U}}^i_{o_{i+1}} - \sum_{k=i+1}^{n} {}^i\mathbf{W}_k\tilde{\mathbf{u}}^k_{o_k}\tilde{\mathbf{s}}^k_{i,i+1}\,{}^i\mathbf{W}^T_k + \mathbf{A}_{i+1}\mathbf{E}^{i+1}_{o_{i+1}}\mathbf{A}^T_{i+1}$$

$$= \mathbf{K}^i_{o_i} - \tilde{\mathbf{s}}^i_{i,i+1}\tilde{\mathbf{U}}^i_{o_{i+1}} - \tilde{\mathbf{U}}^i_{o_{i+1}}\tilde{\mathbf{s}}^i_{i,i+1} + \mathbf{A}_{i+1}\mathbf{E}^{i+1}_{o_{i+1}}\mathbf{A}^T_{i+1}.$$

which implies that indeed equation (C.16) is equivalent to equation (6.3.6). Thus case a) of part i) has been proved.

b) For $j > i$, we have

$$\frac{\partial\boldsymbol{\eta}^i_i}{\partial\ddot{q}_j} = \sum_{k=i}^{j-1} {}^i\mathbf{W}_k\left[\frac{\partial\boldsymbol{\mu}^k_k}{\partial\ddot{q}_j} + \tilde{\mathbf{u}}^k_{o_k}\frac{\partial\ddot{\mathbf{s}}^k_{0,k}}{\partial\ddot{q}_j} + \tilde{\mathbf{s}}^k_{k,k+1}\frac{\partial\ddot{\mathbf{U}}^k_{o_{k+1}}}{\partial\ddot{q}_j}\right]{}^i\mathbf{W}_j\frac{\partial\boldsymbol{\eta}^j_j}{\partial\ddot{q}_j}$$

$$= \sum_{k=i}^{j-1} {}^i\mathbf{W}_k[\tilde{\mathbf{s}}^k_{k,k+1}\tilde{\mathbf{z}}^k_j\mathbf{U}^k_{o_j}] + {}^i\mathbf{W}_j\frac{\partial\boldsymbol{\eta}^j_j}{\partial\ddot{q}_j}$$

$$= {}^i\mathbf{W}_j[-\tilde{\mathbf{s}}^{\,j}_{i,j}\tilde{\mathbf{U}}^j_{\mathrm{o}_j}\mathbf{z}^j_j] + {}^i\mathbf{W}_j[\mathbf{E}^j_{\mathrm{o}_j} - \tilde{\mathbf{U}}^j_{\mathrm{o}_j}\tilde{\mathbf{s}}^{\,j}_{j,j}]\mathbf{z}^j_j \qquad \text{(by part a)}$$

$$= {}^i\mathbf{W}_j[\mathbf{E}^j_{\mathrm{o}_j} - \tilde{\mathbf{s}}^{\,j}_{i,j}\tilde{\mathbf{U}}^j_{\mathrm{o}_j}]\mathbf{z}^j_j, \qquad \text{since } \tilde{\mathbf{s}}^{\,j}_{j,j} = 0$$

ii) As in part *i*), we have two cases:

a) For $j \leq i$, from Lemmas C-8, C-9 and C-11 we get

$$\frac{\partial \eta^i_i}{\partial \dot{q}_j} = \sum_{k=i}^{n} {}^i\mathbf{W}_k \Big\{ 2[\mathbf{K}^k_{\mathrm{o}_k}\dot{\mathbf{z}}^k_j - \hat{\mathbf{L}}^{i+1}_{\mathrm{o}_{i+1}}\mathbf{z}^k_j] + 2\tilde{\mathbf{u}}^k_{\mathrm{o}_k}[\dot{\tilde{\mathbf{z}}}^k_j\mathbf{s}^k_{j,k} + \tilde{\mathbf{z}}^k_j\dot{\mathbf{s}}^k_{j,k}]$$

$$+ 2\tilde{\mathbf{s}}^k_{k,k+1}[\dot{\tilde{\mathbf{z}}}^k_j\mathbf{U}^k_{\mathrm{o}_{k+1}} + \tilde{\mathbf{z}}^k_j\dot{\mathbf{U}}^k_{\mathrm{o}_{k+1}}] \Big\}$$

$$= 2\sum_{k=i}^{n} {}^i\mathbf{W}_k[\mathbf{K}^k_{\mathrm{o}_k} - \tilde{\mathbf{u}}^k_{\mathrm{o}_k}\tilde{\mathbf{s}}^k_{j,k} - \tilde{\mathbf{s}}^k_{k,k+1}\tilde{\mathbf{U}}^k_{\mathrm{o}_{k+1}}]\,{}^i\mathbf{W}^T_k\dot{\mathbf{z}}^i_j$$

$$- 2\sum_{k=i}^{n} {}^i\mathbf{W}_k[\hat{\mathbf{L}}^{i+1}_{\mathrm{o}_{i+1}} + \tilde{\mathbf{u}}^k_{\mathrm{o}_k}\dot{\tilde{\mathbf{s}}}^k_{i,k} + \tilde{\mathbf{s}}^k_{k,k+1}\dot{\tilde{\mathbf{U}}}^k_{\mathrm{o}_{k+1}}]\,{}^i\mathbf{W}^T_k\mathbf{z}^i_j - 2\sum_{k=i}^{n} {}^i\mathbf{W}_k[\tilde{\mathbf{u}}^k_{\mathrm{o}_k}\dot{\tilde{\mathbf{s}}}^k_{j,i}]\,{}^i\mathbf{W}^T_k\mathbf{z}^i_j$$

$$= 2\left([\mathbf{E}^i_{\mathrm{o}_i} - \tilde{\mathbf{U}}^i_{\mathrm{o}_i}\tilde{\mathbf{s}}^i_{j,i}]\dot{\mathbf{z}}^i_j - [\mathbf{L}^i_i + \tilde{\mathbf{U}}^i_{\mathrm{o}_i}\dot{\tilde{\mathbf{s}}}^i_{j,i}]\mathbf{z}^i_j\right)$$

where

$$\mathbf{L}^i_i = \sum_{k=i}^{n} {}^i\mathbf{W}_k[\hat{\mathbf{L}}^{i+1}_{\mathrm{o}_{i+1}} + \tilde{\mathbf{u}}^k_{\mathrm{o}_k}\dot{\tilde{\mathbf{s}}}^k_{i,k} + \tilde{\mathbf{s}}^k_{k,k+1}\dot{\tilde{\mathbf{U}}}^k_{\mathrm{o}_{k+1}}]\,{}^i\mathbf{W}^T_k$$

To see that $\mathbf{L}_i^i$ satisfies equation (C.15) we proceed as follows:

Since $\dot{\tilde{\mathbf{s}}}_{i,i}^{i} = 0$, we have

$$\mathbf{L}_i^i = \hat{\mathbf{L}}_{o_i}^i + \tilde{\mathbf{s}}_{i,i+1}^{i}\dot{\tilde{\mathbf{U}}}_{o_{i+1}}^i$$

$$+ \sum_{k=i+1}^{n} {}^i\mathbf{W}_k \left[\hat{\mathbf{L}}_{o_{i+1}}^{i+1} + \tilde{\mathbf{u}}_{o_k}^{k}(\dot{\tilde{\mathbf{s}}}_{i,i+1}^{k} + \dot{\tilde{\mathbf{s}}}_{i+1,k}^{k}) + \tilde{\mathbf{s}}_{k,k+1}^{k}\dot{\tilde{\mathbf{U}}}_{o_{k+1}}^k\right] {}^i\mathbf{W}_k^T$$

$$= \hat{\mathbf{L}}_{o_i}^i + \tilde{\mathbf{s}}_{i,i+1}^{i}\dot{\tilde{\mathbf{U}}}_{o_{i+1}}^i + \sum_{k=i+1}^{n} {}^i\mathbf{W}_k \tilde{\mathbf{u}}_{o_k}^{k}\dot{\tilde{\mathbf{s}}}_{i,i+1}^{k}\, {}^i\mathbf{W}_k^T + \mathbf{A}_{i+1}\mathbf{L}_{i+1}^{i+1}\mathbf{A}_{i+1}^T$$

$$= \hat{\mathbf{L}}_{o_i}^i + [\tilde{\mathbf{s}}_{i,i+1}^{i}\dot{\tilde{\mathbf{U}}}_{o_{i+1}}^i + \tilde{\mathbf{U}}_{o_{i+1}}^i\dot{\tilde{\mathbf{s}}}_{i,i+1}^{i}] + \mathbf{A}_{i+1}\mathbf{L}_{i+1}^{i+1}\mathbf{A}_{i+1}^T$$

Therefore, case a) of part *ii*) has been proved.

b) For $j > i$, using (C.14) we get

$$\frac{\partial \eta_i^i}{\partial \dot{q}_j} = 2\sum_{k=i}^{j-1} {}^i\mathbf{W}_k \tilde{\mathbf{s}}_{k,k+1}^{k}(\dot{\tilde{\mathbf{z}}}_j^k \mathbf{U}_{o_j}^k + \tilde{\mathbf{z}}_j^k \dot{\mathbf{U}}_{o_j}^k) + {}^i\mathbf{W}_j \frac{\partial \eta_j^j}{\partial \dot{q}_j}$$

$$= 2\,{}^i\mathbf{W}_j[-\tilde{\mathbf{s}}_{i,j}^{j}\tilde{\mathbf{U}}_{o_j}^j\dot{\mathbf{z}}_j^j - \tilde{\mathbf{s}}_{i,j}^{j}\dot{\tilde{\mathbf{U}}}_{o_j}^j\mathbf{z}_j^j] + 2\,{}^i\mathbf{W}_j[\mathbf{E}_{o_j}^j\dot{\mathbf{z}}_j^j - \mathbf{L}_j^j\mathbf{z}_j^j]$$

$$= 2\,{}^i\mathbf{W}_j \left([\mathbf{E}_{o_j}^j - \tilde{\mathbf{s}}_{i,j}^{j}\tilde{\mathbf{U}}_{o_j}^j]\dot{\mathbf{z}}_j^j - [\mathbf{L}_j^j + \tilde{\mathbf{s}}_{i,j}^{j}\dot{\tilde{\mathbf{U}}}_{o_j}^j]\mathbf{z}_j^j\right)$$

which completes the proof of part *ii*).

iii) As before using Lemmas C-8, C-9 and C-11, we have

a) for $j \leq i$,

$$\frac{\partial \eta_i^i}{\partial q_j} = \sum_{k=i}^{n} {}^i\mathbf{W}_k \left[\frac{\partial \mu_k^k}{\partial q_j} + \tilde{\mathbf{u}}_{o_k}^k \frac{\partial \ddot{\mathbf{s}}_{0,k}^k}{\partial q_j} + \tilde{\mathbf{s}}_{k,k+1}^k \frac{\partial \ddot{\mathbf{U}}_{o_{k+1}}^k}{\partial q_j} \right]$$

$$= \sum_{k=i}^{n} {}^i\mathbf{W}_k [\mathbf{K}_{o_k}^k - \tilde{\mathbf{u}}_{o_k}^k \tilde{\mathbf{s}}_{j,k}^k - \tilde{\mathbf{s}}_{k,k+1}^k \tilde{\mathbf{U}}_{o_{k+1}}^k] {}^i\mathbf{W}_k^T \ddot{\mathbf{z}}_j^i$$

$$- 2\sum_{k=i}^{n} {}^i\mathbf{W}_k [\hat{\mathbf{L}}_{o_{i+1}}^{i+1} + \tilde{\mathbf{u}}_{o_k}^k \bar{\tilde{\mathbf{s}}}_{j,k}^k + \tilde{\mathbf{s}}_{k,k+1}^k \bar{\tilde{\mathbf{U}}}_{o_{k+1}}^k] {}^i\mathbf{W}_k^T \dot{\mathbf{z}}_j^i + \sum_{k=i}^{n} {}^i\mathbf{W}_k [\tilde{\mathbf{u}}_{o_k}^k \ddot{\tilde{\mathbf{s}}}_{0,j}^k] {}^i\mathbf{W}_k^T \mathbf{z}_j^i$$

$$= [\mathbf{E}_{o_i}^i - \tilde{\mathbf{U}}_{o_i}^i \tilde{\mathbf{s}}_{j,i}^i] \ddot{\mathbf{z}}_j^i - 2[\mathbf{L}_i^i + \tilde{\mathbf{U}}_{o_i}^i \bar{\tilde{\mathbf{s}}}_{j,i}^i] \dot{\mathbf{z}}_j^i + \tilde{\mathbf{U}}_{o_i}^i \ddot{\tilde{\mathbf{s}}}_{0,j}^i \mathbf{z}_j^i$$

b) for $j > i$,

$$\frac{\partial \eta_i^i}{\partial q_j} = \sum_{k=i}^{j-1} {}^i\mathbf{W}_k [\frac{\partial \mu_k^k}{\partial q_j} + \tilde{\mathbf{u}}_{o_k}^k \frac{\partial \ddot{\mathbf{s}}_{0,k}^k}{\partial q_j} + \tilde{\mathbf{s}}_{k,k+1}^k \frac{\partial \ddot{\mathbf{U}}_{o_{k+1}}^k}{\partial q_j}] + \frac{\partial {}^i\mathbf{W}_j}{\partial q_j} \eta_j^j + {}^i\mathbf{W}_j \frac{\partial \eta_j^j}{\partial q_j}$$

$$= {}^i\mathbf{W}_j [- \tilde{\mathbf{s}}_{i,j}^j \tilde{\mathbf{U}}_{o_j}^j \ddot{\mathbf{z}}_j^j - 2\tilde{\mathbf{s}}_{i,j}^j \bar{\tilde{\mathbf{U}}}_{o_j}^j \dot{\mathbf{z}}_j^j - \tilde{\mathbf{s}}_{i,j}^j \ddot{\tilde{\mathbf{U}}}_{o_j}^j \mathbf{z}_j^i]$$

$$+ {}^i\mathbf{W}_j \tilde{\mathbf{z}}_j^j \eta_j^j + {}^i\mathbf{W}_j [\mathbf{E}_{o_j}^j \ddot{\mathbf{z}}_j^j - 2\mathbf{L}_j^j \dot{\mathbf{z}}_j^j + \tilde{\mathbf{U}}_{o_j}^j \ddot{\tilde{\mathbf{s}}}_{0,j}^j \mathbf{z}_j^j]$$

$$= {}^i\mathbf{W}_j \left([\mathbf{E}_{o_j}^j - \tilde{\mathbf{s}}_{i,j}^j \tilde{\mathbf{U}}_{o_j}^j] \ddot{\mathbf{z}}_j^j - 2[\mathbf{L}_j^j + \tilde{\mathbf{s}}_{i,j}^j \bar{\tilde{\mathbf{U}}}_{o_j}^j] \dot{\mathbf{z}}_j^j + [\tilde{\mathbf{U}}_{o_j}^j \ddot{\tilde{\mathbf{s}}}_{0,j}^j - \tilde{\mathbf{s}}_{i,j}^j \ddot{\tilde{\mathbf{U}}}_{o_j}^j - \tilde{\eta}_j^j] \mathbf{z}_j^j \right)$$

and this complete the proof of Lemma C-12. □

Appendix D

List of Symbols and Abbreviations

n	Number of degrees-of-freedom of a manipulator
m_i	Mass of the i-th link
$\overline{m}_i$	Composite mass of links i to n
o_i	Point denoting the origin of the i-th link coordinate system
c_i	Point denoting the center of mass (c.m.) of the i-th link
1	The unity (identity) tensor
$\mathbf{I}^k_{c_i}$	The inertia tensor of the i-th link about c_i expressed in the k-th coordinate system orientation
$\mathbf{J}^k_{c_i}$	Euler's inertia tensor of the i-th link about c_i expressed in the k-th coordinate system orientation
$\mathbf{K}^k_{o_i}$	The inertia tensor of the i-th augmented link about o_i expressed in the k-th coordinate system orientation
$\hat{\mathbf{K}}^k_{o_i}$	Euler's inertia tensor of the i-th augmented link about o_i expressed in the k-th coordinate system orientation

$\mathbf{E}^{k}_{o_i}$	The inertia tensor of the i-th generalized link about o_i expressed in the k-th coordinate system orientation
$\mathbf{D}$	The joint space generalized inertia tensor of a robot manipulator
$\mathbf{u}^{i}_{o_i}$	The first moment of the i-th augmented link about o_i, expressed in the i-th coordinate system orientation
$\mathbf{U}^{i}_{o_i}$	The first moment of the i-th generalized link about o_i, expressed in the i-th coordinate system orientation
$\mathbf{F}^{i}_{c_i}$	Force vector acting on c_i expressed in the i-th coordinate system orientation
$\mathbf{M}^{i}_{c_i}$	Moment vector about c_i expressed in the i-th coordinate system orientation
$\mathbf{r}^{i}_{i\,j}$	Position vector from o_i to c_j expressed in the i-th coordinate system orientation
$\mathbf{s}^{i}_{i\,j}$	Position vector from o_i to o_j expressed in the i-th coordinate system orientation
$\dot{\mathbf{r}}$ $(\ddot{\mathbf{r}})$	The absolute velocity (acceleration) of vector $\mathbf{r}$
$\boldsymbol{\omega}^{i}_{i}$ $(\dot{\boldsymbol{\omega}}^{i}_{i})$	Absolute angular velocity (acceleration) of the i-th coordinate system expressed in the i-th coordinate system orientation
$\tilde{\boldsymbol{\omega}}^{i}_{i}$	The angular velocity tensor of the i-th link, expressed in the i-th coordinate system orientation
$\boldsymbol{\Omega}^{i}_{i}$	The angular acceleration tensor of the i-th link, expressed in the i-th coordinate system orientation
$\mathbf{q}$ $(\dot{\mathbf{q}},\ddot{\mathbf{q}})$	Joint space position (velocity, acceleration) vector
$\boldsymbol{\chi}$ $(\dot{\boldsymbol{\chi}},\ddot{\boldsymbol{\chi}})$	Cartesian space position (velocity, acceleration) vector
$\boldsymbol{\tau}$	Joint space generalized force vector
A_i	The 3×3 coordinate (or the 4×4 homogeneous) transformation matrix relating the i-th frame to the $(i-1)$-th frame

W_i — The 3×3 coordinate (or the 4×4 homogeneous) transformation matrix relating the i-th frame to the base frame

dual $(\cdot)$ — A tensor-valued vector operator (or a vector-valued tensor operator)

$\tilde{\mathbf{v}}$ — Skew-symmetric tensor which denotes the action of the *dual* operator on a vector **v**

$\mathbf{D}^o$ — The inertia force-acceleration sensitivity tensor of a linearized robot dynamic model

$\mathbf{V}^o$ — The centrifugal and Coriolis force-velocity sensitivity tensor of a linearized robot dynamic model

$\mathbf{P}^o$ — The force-position sensitivity tensor of a linearized robot dynamic model

□ — Denotes the end of a proof

IDP — The Inverse Dynamics Problem

FDP — The Forward Dynamics Problem

Index